MIXED-PHASE CLOUDS

MIXED-PHASE CLOUDS

Observations and Modeling

Edited by

CONSTANTIN ANDRONACHE

Elsevier
Radarweg 29, PO Box 211, 1000 AE Amsterdam, Netherlands
The Boulevard, Langford Lane, Kidlington, Oxford OX5 1GB, United Kingdom
50 Hampshire Street, 5th Floor, Cambridge, MA 02139, United States

Notices
Knowledge and best practice in this field are constantly changing. As new research and experience broaden
our understanding, changes in research methods, professional practices, or medical treatment may
become necessary.

Practitioners and researchers must always rely on their own experience and knowledge in evaluating
and using any information, methods, compounds, or experiments described herein. In using such
information or methods they should be mindful of their own safety and the safety of others, including
parties for whom they have a professional responsibility.

To the fullest extent of the law, neither the Publisher nor the authors, contributors, or editors, assume
any liability for any injury and/or damage to persons or property as a matter of products liability,
negligence or otherwise, or from any use or operation of any methods, products, instructions, or ideas
contained in the material herein.

Library of Congress Cataloging-in-Publication Data
A catalog record for this book is available from the Library of Congress

British Library Cataloguing-in-Publication Data
A catalogue record for this book is available from the British Library

ISBN: 978-0-12-810549-8

For information on all Elsevier publications
visit our website at https://www.elsevier.com/books-and-journals

Working together
to grow libraries in
developing countries

www.elsevier.com • www.bookaid.org

Publisher: Candice Janco
Acquisition Editor: Laura S. Kelleher
Editorial Project Manager: Tasha Frank
Production Project Manager: Anitha Sivaraj
Cover Designer: Christian J. Bilbow

Typeset by SPi Global, India

CONTENTS

CONTRIBUTORS

Andrew S. Ackerman
National Aeronautics and Space Administration, Goddard Institute for Space Studies, New York, NY, United States

Constantin Andronache
Boston College, Chestnut Hill, MA, United States

Joseph Finlon
University of Illinois at Urbana–Champaign, Urbana, IL, United States

Jeffrey French
University of Wyoming, Laramie, WY, United States

Ann M. Fridlind
National Aeronautics and Space Administration, Goddard Institute for Space Studies, New York, NY, United States

Kalli Furtado
Met Office, Exeter, United Kingdom

Dennis L. Hartmann
University of Washington, Seattle, WA, United States

Robert Jackson
Argonne National Laboratory, Environmental Sciences Division, Lemont, IL, United States

Olivier Jourdan
Université Clermont Auvergne, Clermont-Ferrand; CNRS, Aubière, France

Daniel T. McCoy
University of Leeds, Leeds, United Kingdom

Steven D. Miller
Colorado State University, Fort Collins, CO, United States

Guillaume Mioche
Université Clermont Auvergne, Clermont-Ferrand; CNRS, Aubière, France

Yoo-Jeong Noh
Colorado State University, Fort Collins, CO, United States

Trude Storelvmo
Yale University, New Haven, CT, United States

Ivy Tan
Yale University, New Haven, CT, United States

Thomas F. Whale
University of Leeds, Leeds, United Kingdom

Mark D. Zelinka
Lawrence Livermore National Laboratory, Livermore, CA, United States

PREFACE

The objective of this book is to present a series of advanced research topics on mixed-phase clouds. The motivation of this project is the recognized important role clouds play in weather and climate. Clouds influence the atmospheric radiative balance and hydrological cycle of the Earth. Reducing uncertainties in weather forecasting and climate projections requires accurate cloud observations and improved representation in numerical cloud models. In this effort to better understand the role of cloud systems, the mixed-phase clouds present particular challenges, which are illustrated in this book.

The book has two parts, covering a wide range of topics. The first part, "Observations," contains articles on cloud microphysics, in situ and ground-based observations, passive and active satellite measurements, and synergistic use of aircraft data with space-borne measurements. The second part, "Modeling," covers numerical modeling using large eddy simulations to analyze Arctic mixed-phase clouds, and global climate models to address cloud feedbacks and climate sensitivity to mixed-phase cloud characteristics. It is my hope that this book will give some indication of the enormous power and future potential of increasing refined observation techniques and numerical modeling at multiple scales to solve the complex problems of the role of cloud systems in Earth Sciences.

The publication of this book would not have been possible without the help, interest, and enthusiasm of the contributing authors. I would like to thank all of the authors and their supporting institutions for making this project possible. I am particularly grateful to Ann Fridlind, Michael Folmer, Daniel McCoy, Ivy Tan, and Michael Tjernström who offered many useful suggestions during the review process. Finally, it is a great pleasure to acknowledge Candice Janco, Laura Kelleher, Louisa Hutchins, Tasha Frank, Anitha Sivaraj, and Anita Mercy Vethakkan from Elsevier for their willing, dedicated, and continuous help during the project.

Constantin Andronache
Boston Massachusetts

CHAPTER 1

Introduction

Constantin Andronache
Boston College, Chestnut Hill, MA, United States

Contents

Clouds have a significant influence on the atmospheric radiation balance and hydrological cycle. By interacting with incoming shortwave radiation and outgoing longwave radiation, clouds impact the energy budget of the Earth. They also have an important role in the Earth's hydrological cycle by affecting water transport and precipitation (Gettelman and Sherwood, 2016). The interaction of clouds with atmospheric radiation depends on hydrometeor phase, size, and shape. Under favorable humidity conditions, the cloud phase is determined largely by the temperature, condensation nuclei, and ice nuclei in the atmosphere. When the cloud temperature is above 0°C, clouds are formed of liquid water droplets. Ice clouds consist of ice crystals and can be found at temperatures well below 0°C. In some clouds, supercooled liquid droplets coexist with ice crystals, most frequently at temperatures from −35°C to 0°C. These are mixed-phase clouds, which are particularly difficult to observe and describe in numerical weather prediction (NWP) and climate models.

Mixed-phase clouds cover a large area of the Earth's surface, and are often persistent, with a liquid layer on top of ice clouds. Many observations have documented the presence of these clouds in all regions of the world and in all seasons (Shupe et al., 2008). They tend to be more frequent at mid- and high-latitudes, where temperatures are favorable to the formation and persistence of supercooled liquid clouds. Global climatology is available from CloudSat and Cloud-Aerosol Lidar and Infrared Pathfinder Satellite Observations (CALIPSO) data, accumulated in recent years (Stephens et al., 2002; Winker et al., 2009; Zhang et al., 2010). Earlier observations, based on aircraft in situ measurements, detected single layer mixed-phase clouds characterized by a layer of supercooled liquid droplets at the top of an ice cloud (Rauber and Tokay, 1991). Over the last decades, in situ observations using instrumented aircraft (Baumgardner et al., 2011), have provided very detailed insights in cloud microphysics and dynamical conditions that form and maintain these clouds. Such data are essential for calibration of ground-based and spaceborne remote sensing instruments, as well as for the validation of numerical models.

Given the importance of mixed-phase clouds in a number of applications, such as the prediction and prevention of aircraft icing, weather modification, and improvement of NWP and climate projections, a series of research programs have contributed to rapid progress in these areas. Selected results are illustrated in this volume, accompanied by references to the most recent studies. The chapters of this book present research on various aspects of mixed-phase clouds, from cloud microphysics to GCM simulations. Chapters 2–6 focus mainly on observational aspects, while Chapters 7–10 illustrate modeling work from small scales using LES to a global scale using GCMs. The next sections give a short description of each chapter.

1. OBSERVATIONS

Chapter 2 discusses the relevance of ice nucleation to mixed-phase clouds, and current research on ice nuclei particles (INPs) in the atmosphere. The existence of mixed-phase clouds is possible because liquid water droplets can exist in a supercooled state at temperatures as low as $-38°C$. For lower temperatures, in the absence of INPs, the process of homogeneous ice nucleation can start. The coexistence of liquid water droplets and ice particles in mixed-phase clouds requires specific microphysical and dynamical conditions. When a cloud consisting of supercooled liquid water droplets evolves to a state containing some ice crystals, the process of ice nucleation is involved. Despite decades of research, the process of heterogeneous ice nucleation is not sufficiently known (Phillips et al., 2008, 2013; DeMott et al., 2011; Atkinson et al., 2013). A better characterization of the heterogeneous ice nucleation process is needed for the understanding of mixed-phase clouds. This chapter reviews a series of topics relevant for the study of mixed-phase clouds. First, the modes of heterogeneous ice nucleation are described, with a focus on deposition ice nucleation and freezing ice nucleation. Second, the ice nucleation in the atmosphere—particularly in mixed-phase clouds—is summarized and discussed. Third, the experimental methods for examining ice nucleation are presented with a focus on wet and dry dispersion methods. Fourth, the nucleation theory is concisely explained in both homogeneous and heterogeneous cases. Fifth, the properties of good heterogeneous ice nucleators are discussed, including the direct measurement of INP concentration in the atmosphere. This information on direct measurements is particularly important for (a) providing atmospheric model input data, and (b) allowing comparisons between models and observations, thus contributing to the understanding of the ice nucleation processes in the atmosphere.

Chapter 3 introduces a method for the detection of liquid-top mixed-phase (LTMP) clouds from satellite passive radiometer observations. While in situ measurements of mixed-phase clouds provide detailed information for these clouds, such observations are limited and insufficient for many applications. Satellite remote-sensing techniques are efficient for the continuous monitoring and characterization of mixed-phase clouds. Active satellite sensor measurements, such as CloudSat and CALIPSO have the capability to observe detailed vertical structures of mixed-phase clouds. Nevertheless, they are limited to a spatial domain along the satellite path (Stephens et al., 2002; Winker et al., 2009) and have limited applicability for some short-term purposes. Thus, there is great interest in developing methods for mixed-phase clouds detection using passive radiometry. If adequate methods are developed, satellite remote sensing will provide an ideal venue for observing the global distribution of mixed-phase clouds and the detailed structures such as LTMP clouds. This chapter introduces a method of daytime detection of LTMP clouds from passive radiometer observations, which utilizes reflected sunlight in narrow bands at 1.6 and 2.25 μm to probe below liquid-topped clouds. The basis of the algorithm is established on differential absorption properties of liquid and ice particles and accounts for varying sun/sensor geometry and cloud optical properties (Miller et al., 2014). The algorithm has been applied to the Visible/Infrared Imaging Radiometer Suite (VIIRS) on the Suomi National Polar-orbiting Partnership VIIRS/S-NPP and Himawari-8 Advanced Himawari Imager (Himawari-8 AHI). The measurements with the active sensors from CloudSat and CALIPSO were used for evaluation. The results showed that the algorithm has potential to distinguish LTMP clouds under a wide range of conditions, with possible practical applications for the aviation community.

Chapter 4 illustrates some of the problems associated with the microphysical properties of convectively forced mixed-phase clouds. Field experiments are conducted using aircraft with particle measurement probes to obtain direct observations of the microphysical properties of clouds. Such experiments have been carried out to study various types of cloud systems, including supercooled clouds and mixed-phase clouds. One particular subset of these clouds is the convectively forced mixed-phase clouds. Analysis of observations based on retrievals from CloudSat, CALIPSO, and Moderate Resolution Imaging Spectroradiometer (MODIS) show that about 30%–60% of precipitating clouds in the mid- and high-latitudes contain mixed-phase (Mülmenstädt et al., 2015). In this chapter, authors describe in detail the methodology used in aircraft campaigns, what quantities are typically measured, the importance of particle size distribution (PSD) of hydrometeors, and its moments. The primary in situ measurement methods reviewed include bulk measurements, single particle probes, and imaging probes, with references to recent field campaigns (Jackson et al., 2012, 2014; Jackson and McFarquhar, 2014). Examples of observations made during the COnvective Precipitation Experiment (COPE) in southwest England during summer 2013 are presented, with a detailed analysis of liquid water content (LWC), ice water content (IWC), and PSD characterization. In general, the microphysical properties of convective clouds can be widely variable due to numerous factors that include temperature, position in the cloud, vertical velocity, strength of entrainment, and the amount of cloud condensation nuclei loaded into the cloud. The study illustrates that determining IWC from the airborne measurement is much more challenging than determining LWC. Therefore, reducing the uncertainty in IWC from airborne cloud microphysical measurements remains an important research priority.

Chapter 5 provides an overview of the characterization of mixed-phase clouds from field campaigns and ground-based networks. Earlier field campaigns focused on measurements of the microphysical and dynamical conditions of mixed-phase cloud formation and evolution (Rauber and Tokay, 1991; Heymsfield et al., 1991; Heymsfield and Miloshevich, 1993). These studies contributed to solving problems such as aircraft icing and cloud seeding for weather modification. In situ aircraft measurements documented the presence of mixed-phase clouds with a layer of supercooled liquid water on the top of an ice cloud. The US Department of Energy (DOE) Atmospheric Radiation Measurement (ARM) program and its focus on the role of clouds in the climate system facilitated many field missions. Some were directed to observations in Arctic regions, aiming to establish a permanent observational station in Barrow, Alaska (Verlinde et al., 2016). Advances in ground-based remote sensing capabilities developed by the ARM program, aided by field campaigns, produced accurate methods to observe atmospheric processes related to water vapor, aerosol, clouds, and radiation. The ability to detect and characterize mixed-phase clouds at ARM sites provided the basis for developing additional observation stations in other parts of the world. One significant development in Europe was the Cloudnet program, which established a standard set of ground-based remote sensing instruments capable of providing cloud parameters that can be compared with current operational NWP models (Illingworth et al., 2007). Developments following the Cloudnet program and the expansion of ARM capabilities and collaborations have resulted in a more comprehensive approach for monitoring cloud systems— including mixed-phase clouds—at a variety of sites, enabling the evaluation and improvement of high-resolution numerical models (Haeffelin et al., 2016).

Chapter 6 focuses on the characterization of mixed-phase clouds in the Arctic region, using aircraft in situ measurements and satellite observations. Data from the CALIPSO and CloudSat satellites are used to determine the frequency of mixed-phase clouds. Results show that mixed-phase clouds exhibit a frequent and nearly constant presence in the Atlantic side of the Arctic region. In contrast, the Pacific side of the Arctic region has a distinct seasonal variability, with mixed-phase clouds less frequent in winter and spring and more frequent in summer and fall. The vertical distribution of mixed-phase clouds showed that generally, they are present below 3 km, except in summer when these clouds are frequently observed at mid-altitudes (3–6 km). Results indicate that the North Atlantic Ocean and the melting of sea ice influence the spatial, vertical, and seasonal variability of mixed-phase clouds (Mioche et al., 2015, 2017). The microphysical and optical properties of the ice crystals and liquid droplets within mixed-phase clouds and the associated formation and growth processes responsible for the cloud life cycle are evaluated based on in situ airborne observations. Lastly, the authors show that the coupling of in situ mixed-phase clouds airborne measurements with the collocated satellite active remote sensing from CloudSat radar and CALIOP lidar measurements are useful in validating remote sensing observations.

2. MODELING

Chapter 7 provides an overview of numerical simulations of mixed-phase boundary layer clouds using large eddy simulation (LES) modeling. Atmospheric turbulent mixing characterizes boundary layer clouds, and the LES modeling has been extensively used to represent the coupling between dynamical and mixed-phase microphysical processes. Many detailed LES and intercomparison studies have been based on specific cloud systems observed during field campaigns (McFarquhar et al., 2007; Fridlind et al., 2007, 2012; Morrison et al., 2011). The focus of this chapter is mainly on modeling results from the three major field campaigns on which intercomparison studies have been based: the First International Satellite Cloud Climatology Project (ISCCP) Regional Experiment-Arctic Cloud Experiment (FIRE-ACE)/Surface Heat Budget in the Arctic (SHEBA) campaign (Curry et al., 2000), the Mixed-Phase Arctic Cloud Experiment (M-PACE) (Verlinde et al., 2007), and the Indirect and Semi-Direct Aerosol Campaign (ISDAC) (McFarquhar et al., 2011). The chapter presents detailed results from each case study and discusses outstanding questions about fundamental microphysical processes of Arctic mixed-phase clouds.

Chapter 8 presents efforts toward a parametrization of mixed-phase clouds in general circulation models. Observations show that mid- and high-latitude mixed-phase clouds have a prolonged existence, considerably longer than most models predict. A series of simplified physical models and LES simulations have been applied to data from aircraft observations to understand the factors that lead to the longevity of mixed-phase clouds. The results from many case studies indicate that the persistence of mixed-phase conditions is the result of the competition between small-scale turbulent air motions and ice microphysical processes (Korolev and Field, 2008; Hill et al., 2014; Field et al., 2014; Furtado et al., 2016). Under certain situations, this competition can sustain a steady state in which water saturated conditions are maintained for an extended period of time in a constant fraction of the cloud volume. This chapter examines previous work on understanding this mechanism and explains how it can be elaborated into a parametrization of mixed-phase clouds. The parametrization is constructed on exact, steady state solutions for the statistics of supersaturation variations in a turbulent cloud layer, from which expressions for the liquid-cloud properties can be obtained. The chapter reviews the implementation of the parametrization in a general circulation model. It has been shown to correct the representation of Arctic stratus, compared to in situ observations, and improve the distribution of liquid water at high latitudes. Some important consequences of these enhancements are the reduction in the recognized radiative biases over the Southern Ocean and improvement of the sea surface temperatures in fully coupled climate simulations.

Chapter 9 introduces and examines cloud feedback in the climate system. The reflected shortwave (SW) radiation by the oceanic boundary layer (BL) clouds leads to a negative cloud radiative effect (CRE) that strongly affects the Earth's radiative balance. The response of the BL clouds to climate warming represents a cloud feedback that is highly uncertain in current global climate models. This situation impacts the uncertainty in the estimation of equilibrium climate sensitivity (ECS), defined as the change in the equilibrated surface temperature response to a doubling of atmospheric CO_2 concentrations. This chapter considers cloud feedback, with a focus on the mid- and high-latitudes where cloud albedo increases with warming, as simulated by global climate models. In these regions, the increase in cloud albedo appears to be caused by mixed-phase clouds transitioning from a more ice-dominated to a more liquid-dominated state (McCoy et al., 2014, 2015, 2016). The chapter discusses problems in constraining mixed-phase clouds in global climate models due to: (a) uncertainties in ice nucleation—a fundamental microphysical process in mixed-phase clouds formation, and (b) current difficulties in measuring the cloud ice mass. Another feature of global climate models is that they use a parameterization of mixed-phase clouds. A frequent approach is to use a phase partition with temperature based on aircraft measurements. One serious limitation of this method is that it cannot account for the regional variability of ice nuclei (IN) (DeMott et al., 2011). Comparisons with satellite data suggest that this behavior appears to be, at least to some extent, due to an inability to maintain supercooled liquid water at sufficiently low temperatures in current global climate models.

Chapter 10 addresses the impact of mixed-phase clouds' supercooled liquid fraction (SLF) on ECS. The ECS is a measure of the ultimate response of the climate system to doubled atmospheric CO_2 concentrations. Recent work involving GCM simulations aimed to determine ECS due to changes in the cloud system in a warming climate. This chapter examines the impact of mixed-phase clouds SLF on ECS using a series of coupled climate simulations constrained by satellite observations. It follows a series of recent studies on mixed-phase cloud feedback as determined by GCM simulations (Storelvmo et al., 2015; Tan and Storelvmo, 2016; Tan et al., 2016; Zelinka et al., 2012a,b). This study presents non–cloud feedbacks (Planck, water vapor, lapse rate, and albedo) and cloud feedbacks (cloud optical depth, height, and amount). The cloud phase feedback is a subcategory within the cloud optical depth feedback. It relates to how the repartitioning of cloud liquid droplets and ice crystals affects the reflectivity of mixed-phase clouds. Results suggest that cloud thermodynamic phase plays a significant role in the SW optical depth feedback in the extratropical regions, and ultimately influences climate change.

3. CONCLUDING REMARKS

The recent research on mixed-phase clouds presented in this volume, as well as the selected references for each chapter, provide an overview of current efforts to appreciate cloud systems and their role in weather and climate. Understanding the role of clouds in the atmosphere is increasingly imperative for applications such as short-term weather forecast, prediction and prevention of aircraft icing, weather modification, assessment of the effects of cloud phase partition on climate models, and accurate climate projections. In response to these challenges, there is a constant need to refine atmospheric observation techniques and numerical models. These efforts are sustained by many evolving research programs and by a vibrant community of scientists. The book "Mixed-phase Clouds: Observations and Modeling" provides the essential information to help readers understand the current status of observations, simulations, and applications of mixed-phase clouds, and their implications for weather and climate.

ACKNOWLEDGMENTS

I want to express my sincere gratitude to all of the authors and reviewers who contributed to this volume.

REFERENCES

Atkinson, J.D., Murray, B.J., Woodhouse, M.T., Whale, T.F., Baustian, K.J., Carslaw, K.S., Dobbie, S., O'Sullivan, D., Malkin, T.L., 2013. The importance of feldspar for ice nucleation by mineral dust in mixed-phase clouds. Nature 498, 355–358.

Baumgardner, D., Brenguier, J.-L., Bucholtz, A., Coe, H., DeMott, P., Garrett, T.J., Gayet, J.F., Hermann, M., Heymsfield, A., Korolev, A., Kramer, M., Petzold, A., Strapp, W., Pilewskie, P., Taylor, J., Twohy, C., Wendisch, M., Bachalo, W., Chuang, P., 2011. Airborne instruments to measure atmospheric aerosol particles, clouds and radiation: a cook's tour of mature and emerging technology. Atmos. Res. 102 (1-2), 10–29. https://doi.org/10.1016/j.atmosres.2011.06.021.

Curry, J.A., Hobbs, P.V., King, M.D., Randall, D., Minnis, P., Isaac, G.A., Pinto, J.O., Uttal, T., Bucholtz, A., Cripe, D., Gerber, H., Fairall, C.W., Garrett, T.J., Hudson, J., Intrieri, J., Jakob, C., Jensen, T., Lawson, P., Marcotte, D., Nguyen, L., Pilewskie, P., Rangno, A., Rogers, D.C., Strawbridge, K.B., Valero, F.P.J., Williams, A.G., Wylie, D., 2000. FIRE arctic clouds experiment. Bull. Am. Meteorol. Soc. 81 (1), 5–29.

DeMott, P.J., Mohler, O., Stetzer, O., Vali, G., Levin, Z., Petters, M.D., Murakami, M., Leisner, T., Bundke, U., Klein, H., Kanji, Z.A., Cotton, R., Jones, H., Benz, S., Brinkmann, M., Rzesanke, D., Saatho, H., Nicolet, M., Saito, A., Nillius, B., Bingemer, H., Abbatt, J., Ardon, K., Ganor, E., Georgakopoulos, D.G., Saunders, C., 2011. Resurgence in ice nuclei measurement research. Bull. Am. Meteorol. Soc. 92 (12), 1623–1635. https://doi.org/10.1175/2011BAMS3119.1.

Field, P.R., Hill, A., Furtado, K., Korolev, A., 2014. Mixed phase clouds in a turbulent environment. Part 2: analytic treatment. Q. J. Roy. Meteor. Soc. 21, 2651–2663. https://doi.org/10.1002/qj.2175.

Fridlind, A.M., Ackerman, A.S., McFarquhar, G.M., Zhang, G., Poellot, M.R., DeMott, P.J., Prenni, A.J., Heymsfield, A.J., 2007. Ice properties of single-layer stratocumulus during the Mixed-Phase Arctic Cloud Experiment: 2. Model results. J. Geophys. Res. 112 (D24), D24202. https://doi.org/10.1029/2007JD008646.

Fridlind, A.M., van Diedenhoven, B., Ackerman, A.S., Avramov, A., Mrowiec, A., Morrison, H., Zuidema, P., Shupe, M.D., 2012. A FIRE-ACE/SHEBA case study of mixed-phase Arctic boundary layer clouds: Entrainment rate limitations on rapid primary ice nucleation processes. J. Atmos. Sci. 69 (1), 365–389. https://doi.org/10.1175/JAS-D-11-052.1.

Furtado, K., Field, P.R., Boutle, I.A., Morcrette, C.R., Wilkinson, J., 2016. A physically-based, subgrid parametrization for the production and maintenance of mixed-phase clouds in a general circulation model. J. Atmos. Sci. 73, 279–291. https://doi.org/10.1175/JAS-D-15-0021.

Gettelman, A., Sherwood, S.C., 2016. Processes responsible for cloud feedback. Curr. Clim. Change Rep. 2, 179–189. https://doi.org/10.1007/s40641-016-0052-8.

Haeffelin, M., et al., 2016. Parallel developments and formal collaboration between European atmospheric profiling observatories and the U.S. ARM research program. The Atmospheric Radiation Measurement (ARM) program: the first 20 years. In: Meteorological Monographs. vol. 57. American Meteorological Society. https://doi.org/10.1175/AMSMONOGRAPHS-D-15-0045.1.

Heymsfield, A.J., Miloshevich, L.M., 1993. Homogeneous ice nucleation and supercooled liquid water in orographic wave clouds. J. Atmos. Sci. 50, 2235–2353.

Heymsfield, A.J., Miloshevich, L.M., Slingo, A., Sassen, K., Starr, D.O'.C., 1991. An observational and theoretical study of highly supercooled altocumulus. J. Atmos. Sci. 48, 923–945.

Hill, A.A., Field, P.R., Furtado, K., Korolev, A., Shipway, B.J., 2014. Mixed-phase clouds in a turbulent environment. Part 1: large-eddy simulation experiments. Q. J. R. Meteorol. Soc. 140, 855–869. https://doi.org/10.1002/qj.2177.

Illingworth, A.J., et al., 2007. CloudNet: continuous evaluations of cloud profiles in seven operational models using ground-based observations. Bull. Am. Meteorol. Soc. 88, 883–898.

Jackson, R.C., McFarquhar, G.M., 2014. An assessment of the impact of antishattering tips and artifact removal techniques on bulk cloud ice microphysical and optical properties measured by the 2D cloud probe. J. Atmos. Ocean. Technol. 31, 2131–2144. https://doi.org/10.1175/JTECH-D-14-00018.1.

Jackson, R.C., McFarquhar, G.M., Korolev, A.V., Earle, M.E., Liu, P.S.K., Lawson, R.P., Brooks, S., Wolde, M., Laskin, A., Freer, M., 2012. The dependence of ice microphysics on aerosol concentration in arctic mixed-phase stratus clouds during ISDAC and M-PACE. J. Geophys. Res. 117, D15207. https://doi.org/10.1029/2012JD017668.

Jackson, R.C., McFarquhar, G.M., Stith, J., Beals, M., Shaw, R.A., Jensen, J., Fugal, J., Korolev, A., 2014. An assessment of the impact of antishattering tips and artifact removal techniques on cloud ice size distributions measured by the 2D cloud probe. J. Atmos. Ocean. Technol. 31, 2567–2590. https://doi.org/10.1175/JTECH-D-13-00239.1.

Korolev, A., Field, P.R., 2008. The effect of dynamics on mixed-phase clouds: theoretical considerations. J. Atmos. Sci. 65, 66–86.

McCoy, D.T., Hartmann, D.L., Grosvenor, D.P., 2014. Observed southern ocean cloud properties and shortwave reflection. part ii: phase changes and low cloud feedback. J. Clim. 27 (23), 8858–8868.

McCoy, D.T., Hartmann, D.L., Zelinka, M.D., Ceppi, P., Grosvenor, D.P., 2015. Mixed phase cloud physics and southern ocean cloud feedback in climate models. J. Geophys. Res. Atmos. 120 (18), 9539–9554.

McCoy, D., Tan, I., Hartmann, D., Zelinka, M., Storelvmo, T., 2016. On the relationships among cloud cover, mixed-phase partitioning, and planetary albedo in GCMs. J. Adv. Model. Earth Syst. 8, 650–668. https://doi.org/10.1002/2015MS000589.

McFarquhar, G., Zhang, G., Poellot, M., Kok, G., McCoy, R., Tooman, T., Fridlind, A., Heymsfield, A., 2007. Ice properties of single-layer stratocumulus during the Mixed-Phase Arctic Cloud Experiment: 1. Observations. J. Geophys. Res. 112.

McFarquhar, G.M., Ghan, S., Verlinde, J., Korolev, A., Strapp, J.W., Schmid, B., Tomlinson, J.M., Wolde, M., Brooks, S.D., Cziczo, D., Dubey, M.K., Fan, J., Flynn, C., Gultepe, I., Hubbe, J., Gilles, M.K., Laskin, A., Lawson, P., Leaitch, W.R., Liu, P., Liu, X., Lubin, D., Mazzoleni, C., Macdonald, A.-M., Moffet, R.C., Morrison, H., Ovchinnikov, M., Shupe, M.D., Turner, D.D., Xie, S., Zelenyuk, A., Bae, K., Freer, M., Glen, A., 2011. Indirect and semi-direct aerosol campaign: the impact of arctic aerosols on clouds. Bull. Am. Meteorol. Soc. 92, 183–201. https://doi.org/10.1175/2010BAMS2935.1.

Miller, S.D., Noh, Y.J., Heidinger, A.K., 2014. Liquid-top mixed-phase cloud detection from shortwave-infrared satellite radiometer observations: a physical basis. J. Geophys. Res. 119. https://doi.org/10.1002/2013JD021262.

Mioche, G., Jourdan, O., Ceccaldi, M., Delanoë, J., 2015. Variability of mixed-phase clouds in the Arctic with a focus on the Svalbard region: a study based on spaceborne active remote sensing. Atmos. Chem. Phys. 15, 2445–2461. https://doi.org/10.5194/acp-15-2445-2015.

Mioche, G., Jourdan, O., Delanoë, J., Gourbeyre, C., Febvre, G., Dupuy, R., Szczap, F., Schwarzenboeck, A., Gayet, J.-F., 2017. Characterization of Arctic mixed-phase cloud properties at small scale and coupling with satellite remote sensing. Atmos. Chem. Phys. Discuss, 1–52. https://doi.org/10.5194/acp-2017-93.

Morrison, H., Zuidema, P., Ackerman, A.S., Avramov, A., De Boer, G., Fan, J., Fridlind, A.M., Hashino, T., Harrington, J.Y., Luo, Y., Ovchinnikov, M., Shipway, B., 2011. Intercomparison of cloud model simulations of Arctic mixed-phase boundary layer clouds observed during SHEBA/FIRE-ACE. J. Adv. Model. Earth Syst. 3, 1–23. https://doi.org/10.1029/2011MS000066.

Mülmenstädt, J., Sourdeval, O., Delanoë, J., Quaas, J., 2015. Frequency of occurrence of rain from liquid-, mixed-, and ice-phase clouds derived from A-Train satellite retrievals. Geophys. Res. Lett. 42, 6502–6509. https://doi.org/10.1002/2015GL064604.

Phillips, V.T.J., DeMott, P.J., Andronache, C., 2008. An empirical parameterization of heterogeneous ice nucleation for multiple chemical species of aerosol. J. Atmos. Sci. 65 (9), 2757–2783.

Phillips, V.T.J., DeMott, P.J., Andronache, C., Pratt, K., Prather, K.A., Subramanian, R., Twohy, C., 2013. Improvements to an empirical parameterization of heterogeneous ice nucleation and its comparison with observations. J. Atmos. Sci. 70, 378–409.

Rauber, R.M., Tokay, A., 1991. An explanation for the existence of supercooled water at the tops of cold clouds. J. Atmos. Sci. 48, 1005–1023.

Shupe, M., et al., 2008. A focus on mixed-phase clouds: the status of ground-based observational methods. Bull. Am. Meteorol. Soc. 87, 1549–1562.

Stephens, G.K., et al., 2002. The CLOUDSAT Mission and the A-Train—a new dimension of space-based observations of clouds and precipitation. Bull. Am. Meteorol. Soc. 83, 1771–1790.

Storelvmo, T., Tan, I., Korolev, A.V., 2015. Cloud phase changes induced by co2 warming—a powerful yet poorly constrained cloud-climate feedback. Curr. Clim. Change Rep. 1 (4), 288–296.

Tan, I., Storelvmo, T., 2016. Sensitivity study on the influence of cloud microphysical parameters on mixed-phase cloud thermodynamic phase partitioning in cam5. J. Atmos. Sci. 73 (2), 709–728.

Tan, I., Storelvmo, T., Zelinka, M., 2016. Observational constraints on mixed-phase clouds imply higher climate sensitivity. Science 352. https://doi.org/10.1126/science/aad530.

Verlinde, J., et al., 2007. The mixed-phase arctic cloud experiment. Bull. Am. Meteorol. Soc. 88, 205–221.

Verlinde, J., Zak, B., Shupe, M.D., Ivey, M., Stamnes, K., 2016. The ARM North Slope of Alaska (NSA) sites. The Atmospheric Radiation Measurement (ARM) program: the first 20 years. In: Meteorological Monographs. 57. American Meteorological Society. https://doi.org/10.1175/AMSMONOGRAPHS-D-15-0023.1.

Winker, D.M., Vaughan, M.A., Omar, A.H., Hu, Y., Powell, K.A., Liu, Z., Hunt, W.H., Young, S.A., 2009. Overview of the CALIPSO Mission and CALIOP data processing algorithms. J. Atmos. Ocean. Technol. 26, 2310–2323. https://doi.org/10.1175/2009JTECHA1281.1.

Zelinka, M.D., Klein, S.A., Hartmann, D.L., 2012a. Computing and partitioning cloud feedbacks using cloud property histograms. Part i: cloud radiative kernels. J. Clim. 25 (11), 3715–3735.

Zelinka, M.D., Klein, S.A., Hartmann, D.L., 2012b. Computing and partitioning cloud feedbacks using cloud property histograms. Part ii: attribution to changes in cloud amount, altitude, and optical depth. J. Clim. 25 (11), 3736–3754.

Zhang, D., Wang, Z., Liu, D., 2010. A global view of mid-level liquid-layer topped stratiform cloud distribution and phase partition from CALIPSO and CloudSat measurements. J. Geophys. Res. 115, D00H13. https://doi.org/10.1029/2009JD012143.

PART 1

Observations

CHAPTER 2

Ice Nucleation in Mixed-Phase Clouds

Thomas F. Whale
University of Leeds, Leeds, United Kingdom

Contents

1. THE RELEVANCE OF ICE NUCLEATION TO MIXED-PHASE CLOUDS

At atmospheric pressure ice I_h is the thermodynamically stable form of water below $0°C$. Pure water does not freeze at $0°C$ because the stable phase must nucleate before crystal growth can occur. Liquid water can supercool to temperatures below $-35°C$ (Herbert et al., 2015; Koop and Murray, 2016; Riechers et al., 2013) before ice nucleation occurs homogenously. On some level, this fact must be responsible for the existence of mixed-phase clouds. If liquid water supercooled only slightly, much of the variability and interest caused by the coexistence liquid water droplets and ice particles would not occur. The progress of a cloud from consisting entirely of supercooled liquid water to a state also containing ice must at some point involve an ice nucleation process. Despite decades of research, heterogeneous ice nucleation remains poorly understood. Improved insight into the process is of great importance for understanding of mixed-phase clouds.

Mixed-Phase Clouds
https://doi.org/10.1016/B978-0-12-810549-8.00002-7

13

1.1 Modes of Heterogeneous Ice Nucleation

There are several pathways by which ice can form on a heterogeneous ice nucleating particle (INP). These are known as modes. Historically, several different sets of definitions have been used for these modes. Notably, the definitions of Vali (1985) and Pruppacher and Klett (1997) are a little different. Recently, Vali et al. (2014) led an online discussion by the ice nucleation community on terminology and published a document outlining new definitions (Vali et al., 2015). These definitions that are described here are used throughout this chapter.

The two principle modes of ice nucleation are deposition and freezing. Deposition ice nucleation is defined as ice nucleation from supersaturated vapor on an INP or equivalent without prior formation of liquid (a phase transition from gas to solid). Freezing ice nucleation is defined as ice nucleation within a body of supercooled liquid ascribed to the presence of an INP, or equivalent (a phase transition from liquid to solid). Freezing nucleation is subdivided into immersion freezing, where the entire INP is covered in liquid water, contact freezing, where freezing is initiated at the air-water interface as the INP comes into contact with supercooled liquid water and condensation freezing, where freezing occurs concurrently with formation of liquid water. It is challenging to differentiate condensation freezing from both deposition nucleation and immersion freezing in a strict physical sense, as the microscopic mechanism of ice formation is not known in most cases. It is entirely plausible that many, most, or all cases of deposition nucleation are preceded by formation of microscopic quantities of water which then freezes, followed by depositional growth (Christenson, 2013; Marcolli, 2014). Mechanisms of this sort are known to occur for organic vapors (e.g., Campbell ct al., 2013; Kovács et al., 2012). Similarly, it is not clear how condensation freezing differs from immersion freezing in cases where liquid water does form prior to freezing (which may be most or all cases). Happily, it is thought that immersion mode freezing is likely to be the dominant freezing mode in most mixed-phase clouds (Cui et al., 2006; de Boer et al., 2011) so we need not concern ourselves with nucleation of ice below water saturation. The remainder of this chapter is therefore solely concerned with immersion mode ice nucleation, where particles are clearly immersed in water. The following section briefly describes the relevance of immersion mode ice nucleation to the atmosphere in general, to determine the role of ice nucleation in mixed-phase clouds within the broader field of ice nucleation studies.

1.2 Ice Nucleation in the Atmosphere

Clouds are made up of water droplets or ice crystals, or a mixture of thereof, suspended in the atmosphere. By interacting with incoming shortwave radiation and outgoing longwave radiation, they can impact the energy budget of the earth and thereby play a key role in the earth's climate. They also strongly influence the earth's hydrological cycle by controlling water transport and precipitation (Hartmann et al., 1992). The magnitude of the impact of clouds on the global energy budget remains highly uncertain despite decades of research (Lohmann and Feichter, 2005). The latest Intergovernmental Panel on Climate Change (IPCC) report suggests a net cooling effect from clouds of $-20 \ \mathrm{Wm}^{-2}$ (Boucher et al., 2013).

Much of this uncertainty stems from the poorly understood nature of interactions between atmospheric aerosol and clouds (Field et al., 2014). Atmospheric aerosol consists of solid or liquid particles suspended in the air. There are many different types of aerosol in the atmosphere. Primary aerosol is emitted directly from both natural and anthropogenic sources as particles, and includes mineral dust, sea salt, black carbon, and primary biological particles. Secondary aerosol forms from gaseous precursors that are often emitted by plants and oceanic processes. Clouds form when moist air rises through the atmosphere and cools down. Typically, water droplets form on aerosol particles called cloud condensation nuclei (CCN)(Pruppacher and Klett, 1997).

As the majority of clouds are formed via processes involving aerosol particles, cloud properties such as lifetime, composition, and size are highly dependent on the properties of the aerosol particles with which the cloud interacts. These effects are known as aerosol indirect effects (Denman et al., 2007). Cloud glaciation, which is dependent on the ice nucleation properties of the aerosol in clouds, (Denman et al., 2007) is one of these effects. In the latest IPCC report, these effects have been grouped together, and confidence in the assessment of the impact of aerosol-cloud interactions is rated as low. The potential scale of the impact ranges from a very slight warming effect to a relatively substantial cooling of $2 \ \mathrm{Wm}^{-2}$ (Field et al., 2014).

There are two overarching categories of tropospheric clouds in which ice nucleation is most relevant. These are cirrus clouds and mixed-phase clouds. Cirrus clouds form in the upper troposphere at temperatures below $-38°C$, and consist of concentrated solution droplets, which can be frozen via immersion mode ice nucleation, or ice formed by deposition nucleation. Mixed-phase clouds form lower down in the troposphere between $0°C$ and about $-38°C$ (the approximate temperature of homogeneous ice nucleation). Ice formation in these clouds is generally thought to be controlled by immersion mode ice nucleation (Cui et al., 2006; de Boer et al., 2011) although the contact mode may also play a role (Ansmann et al., 2005).

1.3 Ice Nucleation in Mixed-Phase Clouds

Ice nucleation processes have the potential to alter mixed-phase cloud properties in several ways. Liquid water clouds may occasionally supercool to temperatures where homogenous freezing is important before any ice is formed, below about $-35°C$ (Herbert et al., 2015), but generally glaciate at warmer temperatures (Ansmann et al., 2009; Kanitz et al., 2011). This indicates heterogeneous ice nucleation controls mixed-phase cloud glaciation in many cases. Satellite observations have indicated that at −20°C about half of mixed-phase clouds globally are glaciated (Choi et al., 2010).

The presence of ice crystals in a cloud can change its radiative properties significantly compared to a liquid cloud and the size and concentration of ice crystals are also important (Lohmann and Feichter, 2005). Cloud thickness, spatial extent, and lifetime can also alter radiative forcing and can potentially depend on INP concentration. Precipitation processes are closely linked to ice formation as ice I is more stable than liquid water below 0°C. As such, ice particles in mixed-phase clouds tend to grow at the expense of super-cooled liquid water droplets. This process is known as the Wegener-Bergeron-Findeisen process and is thought to be the most important route for precipitation from mixed-phase clouds as larger particles will fall faster than smaller ones (Pruppacher and Klett, 1997). Clouds which contain relatively small ice crystal concentrations and more supercooled water are more likely to precipitate as the ice crystals can grow to larger sizes than they might have if ice crystal concentrations were higher. As a result, lifetime of these clouds might be shorter than it would otherwise have been. Additionally, ice multiplication processes can result from the fragmentation of ice formed through primary ice nucleation processes and increase the concentration of ice crystals in clouds by several orders of magnitude (Phillips et al., 2003). The best understood of these is the Hallett-Mossop process which occurs from −3°C to −8°C (Hallett and Mossop, 1974) although other processes have also been posited (Yano and Phillips, 2011). These various processes, and others, interact in complex and generally poorly understood ways, contributing to the large uncertainty on the radiative forcing due to aerosol-cloud interactions (Field et al., 2014). These interactions between aerosol, clouds, and liquid in mixed-phase clouds need to be understood quantitatively to properly understand and assess the impact of clouds on climate and weather. This chapter focuses on experimental methods for quantifying concentrations of INPs, ways of describing the efficiency of INPs, what is known about the identity of INPs in the atmosphere, and the progress of studies into fundamental understanding of why certain substances nucleate ice efficiently.

2. EXPERIMENTAL METHODS FOR EXAMINING ICE NUCLEATION

The majority of quantitative studies of how efficiently a particular material nucleates ice have been conducted with the goal of determining what species nucleate ice in the atmosphere. The atmospheric science community has employed a wide variety of techniques. There are two overarching families of techniques for determining the immersion mode ice nucleating efficiency of nucleators. These are wet dispersion methods and dry dispersion methods (Hiranuma et al., 2015). Wet dispersion methods involve dispersion of INPs into water, which is then frozen. Dry dispersion methods involve the dispersion of aerosol particles into air, where they are then activated into water droplets before freezing. Techniques have also been divided into those which use droplets supported on the surface or suspended in oil, and those which use droplets suspended in gas (Murray et al., 2012) which are largely synonymous with wet and dry dispersion techniques, respectively. Almost invariably, raw ice nucleation data takes the form of a fraction of droplets frozen under a given set of conditions. Typical variables are temperature, cooling rate, droplet size, and nucleator identity and concentration of the nucleator in droplets.

2.1 Wet Dispersion Methods

Most wet dispersion techniques are droplet freezing experiments, also known as droplet freezing assays. These involve dividing a sample of water into multiple sub-samples and cooling these individual samples down until they freeze. For studies of heterogeneous ice nucleation a nucleator is suspended in the water prior to sub-division, or pure water droplets are placed onto a nucleating surface. The temperature at which droplets freeze is recorded, typically by simultaneous video and temperature logging. Different droplet volumes have been used, ranging from milliliters to picoliters (Murray et al., 2012; Vali, 1995). Droplets are typically either placed on hydrophobic surfaces (e.g., Lindow et al., 1982; Murray et al., 2010) or in wells or vials (e.g., Hill et al., 2014). In these cases, freezing is usually observed visually, often through a microscope. Emulsions of water droplets in oil can also be frozen, and freezing events recorded via microscope (e.g., Zolles et al., 2015) or by using a calorimeter (Michelmore and Franks, 1982). Recently, microfluidic devices have been used to create mono-disperse droplets for studying ice nucleation (Riechers et al., 2013; Stan et al., 2009).

Droplet freezing techniques typically use linear cooling rates, although isothermal experiments have also been conducted (Broadley et al., 2012; Herbert et al., 2014; Sear, 2014). Larger droplets up to milliliter volumes have typically been used for investigations of biological ice nucleators while the smallest droplets have been used for studies of homogeneous ice nucleation. The majority of studies of atmospherically relevant INPs have been conducted using smaller, nano- to picoliter-sized droplets (Murray et al., 2012).

Other techniques that use wet dispersion to produce droplets include those that freeze single droplets repeatedly many times in order to establish the variation in freezing temperature in that single droplet (Barlow and Haymet, 1995; Fu et al., 2015). Wind tunnels are similar in that they support single suspended droplets in an upward flow of air of known temperature (Diehl et al., 2002; Pitter and Pruppacher, 1973). Freezing probabilities are determined by conducting multiple experiments. Droplets are typically prepared by wet dispersion then introduced into the airflow but could also be dry dispersed. Similarly, droplets can be suspended by electrodynamic levitation (Krämer et al., 1999).

2.2 Dry Dispersion Methods

Cloud expansion chambers are large vessels in which temperature, humidity, and aerosol contents are controlled, usually with the goal of simulating clouds (Connolly et al., 2009;

Emersic et al., 2015; Niemand et al., 2012). Experiments involve pumping the chamber out to reduce temperature thereby inducing ice nucleation in the chamber. The ice nucleation efficiency of aerosols in the chamber can be determined from the appearance of ice crystals. In order to conduct experiments in the immersion mode the INPs must activate as CCN before ice nucleation occurs.

Continuous Flow Thermal Gradient Diffusion Chambers (CFDCs) flow air-containing aerosols through a space where temperature and humidity are controlled using two plates coated in ice (Garimella et al., 2016; Rogers, 1988; Stetzer et al., 2008). Typically, aerosol size distributions and concentrations are characterized going into the area of controlled supersaturation with respect to ice and the number of ice crystals coming out the other end it also determined. In this way a droplet fraction frozen can be determined. Alternatively, a pre-conditioning section can be used to ensure that all aerosol particles prior are activated as CCN prior to entry to the ice nucleation section of the instrument, thereby ensuring that all freezing is immersion mode (Lüönd et al., 2010).

3. NUCLEATION THEORY

While there is no satisfactory overarching theory for nucleation phenomena (Sear, 2012) there are various theories and descriptions used to describe ice nucleation. This section describes theories and descriptions used for describing immersion mode ice nucleation data.

3.1 Homogeneous Ice Nucleation

Homogenous nucleation is nucleation that does not involve a heterogeneous nucleator. In the atmosphere, cloud water droplets can supercool to temperatures below $-35°C$. While heterogeneous ice nucleation is probably more common in most mixed-phase clouds, homogeneous nucleation is also thought to be a factor (Sassen and Dodd, 1988) and mixed-phase clouds have been observed at sufficiently cold temperatures to support this (Choi et al., 2010; Kanitz et al., 2011). Many laboratory experiments have also investigated homogenous nucleation (Murray et al., 2010; Riechers et al., 2013; Stan et al., 2009) and it has been shown that classical nucleation theory (CNT) can describe laboratory data for homogenous nucleation well (Riechers et al., 2013).

3.1.1 Classical Description of Homogenous Ice Nucleation
The following is a derivation of CNT adapted from work by Pruppacher and Klett (1997), Mullin (2001), Debenedetti (1996), Murray et al. (2010), and Vali et al. (2015). Supercooling occurs because of a kinetic barrier to the formation of solid clusters large enough for spontaneous growth. This stems from the increasing energy cost of forming interface between ice and supercooled water as the size of a cluster grows. At the cluster size where the energy gain of adding a water molecule exceeds the energy

cost of forming an interface between the ice and supercooled water spontaneous growth will occur. This can be expressed as:

$$\Delta G = \Delta G_s + \Delta G_V \tag{1}$$

Where ΔG is the overall change in Gibbs free energy of the ice cluster, ΔG_s is the surface free energy between surface of the particle and the bulk of the supercooled water, and ΔG_V is volume excess free energy. ΔG_s and ΔG_V are competing terms, ΔG_V being negative while ΔG_s is positive. G_s can be expressed as:

$$G_s = 4\pi r^2 \gamma \tag{2}$$

where r is the radius of the solid cluster and γ is the interfacial energy between ice and water. G_v can be expressed as:

$$G_v = \frac{4\pi r^3}{3v} k_B T \ln S \tag{3}$$

where v is the volume of a water molecule in ice, k_B is the Boltzmann constant, T is the temperature, and S is the saturation ratio with respect to ice. Adding Eqs. (2) and (3) gives the total Gibbs free energy of the barrier to nucleation:

$$\Delta G = -\frac{4\pi r^3}{3v} k_B T \ln S + 4\pi r^2 \gamma \tag{4}$$

The two terms of Eq. (4) are opposing so the free energy of ice formation passes through a maximum, as shown in Fig. 1. The maximum value corresponds to the size of the critical nucleus, r_i^*.

Critical nucleus size can be calculated by differentiating Eq. (4) with respect to r_i^* and setting $d\Delta G/dr_i = 0$ before rearranging for r_i yields:

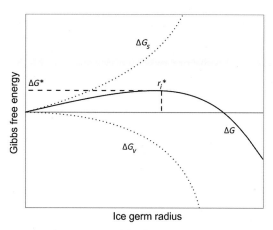

Fig. 1 Schematic of ice germ radius against Gibbs free energy.

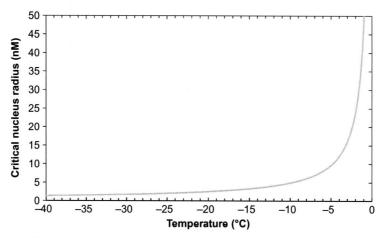

Fig. 2 Critical radius size for Ice I_{sd} as a function of temperature.

$$r_i^* = \frac{-2\gamma v}{k_B T \ln S \Delta G_v} \tag{5}$$

Eq. (5) can be used to calculate the temperature dependence of critical radius size. S can be calculated using parameterizations from Murphy and Koop (2005) along with the value for γ from Murray et al. (2010). It can be seen that the size of the critical nucleus increases sharply with rising temperature in Fig. 2.

By substituting back into Eq. (4), ΔG^* at temperature T can be calculated:

$$\Delta G^* = \frac{16\pi\gamma^3 v^2}{3(k_b T \ln S)^2} \tag{6}$$

To determine nucleation rate, the Arrhenius style Eq. (7) can be applied.

$$J_{hom} = A \exp\left(-\frac{\Delta G^*(T)}{kT}\right) \tag{7}$$

J_{hom} is the nucleation rate, A is the pre-exponential factor, and k the Boltzmann constant. Combining Eqs. (6) and (7) Eq. (8) can be written down.

$$\ln J_{hom} = \ln A - \frac{16\pi\gamma^3 v^2}{3k^3 T^3 (\ln S)^2} \tag{8}$$

Hence, a plot of $\ln J_{hom}$ against $T^{-3}(\ln S)^{-2}$ will be linear with an intercept of $\ln A$ and, over a narrow temperature range, slope:

$$m = -\frac{16\pi\gamma^3 v^2}{3k^3} \tag{9}$$

Since v is known, this allows γ to be determined from experiments determining J_{hom}.

J_{hom} has units of nucleation events $\mathrm{cm^{-3}\,s^{-1}}$. In larger volumes of water nucleation is therefore more probable. In an experiment looking at a large number of identical droplets held a constant temperature where a single nucleation event within a droplet is assumed to lead to crystallization of that droplet a freezing rate $R(t)$ can be determined. $R(t)$ is a purely experimental value that has units of events $\mathrm{s^{-1}}$. It can be determined for any droplet freezing experiment, heterogeneous or homogeneous. Application to heterogeneous experiments is discussed in the following section. $R(t)$ can be calculated using:

$$R(t) = \frac{1}{N_0 - N_F}\frac{dN_F}{dt} \tag{10}$$

where N_F is the total number of frozen droplets at time t and N_0 is the total number of droplets present, frozen or unfrozen. If V is the volume of the droplets Eq. (11) can be written down.

$$J_v = \frac{R(t)}{V} \tag{11}$$

where J_v is the volume nucleation rate. if the droplets are free of impurities so that nucleation is via the homogenous mechanism then:

$$J_{hom} = J_v = \frac{R(t)}{V} \tag{12}$$

Following on from this, for constant temperature the fraction of droplets N_L that remains unfrozen at time t can therefore be calculated using:

$$N_L = N_0 \exp\left(-J_{hom} V t\right) \tag{13}$$

In cases where droplets are constantly cooled, rather than being held at a steady temperature to small increments of it is necessary to apply Eq. (13) to small time intervals over which changes in temperature are small. In this way, $J_{hom}(T)$ can be determined.

3.2 Heterogeneous Ice Nucleation

Immersion mode heterogeneous ice nucleation takes place when an external entity lowers the energy barrier preventing ice nucleation. As a result, the probability of a nucleation event occurring at any given supercooled temperature can be far higher in the presence of a suitable heterogeneous ice nucleator. The observed outcome is that heterogeneous ice nucleation takes place at higher temperatures than homogenous ice nucleation in otherwise equivalent systems. Different nucleators nucleate ice with varying efficiency (Hoose and Möhler, 2012; Murray et al., 2012). The following sections detail methods for describing immersion mode heterogeneous ice nucleation efficiency.

3.2.1 Application of CNT to Heterogeneous Nucleation

In classical nucleation theory, the temperature dependent heterogeneous nucleation rate coefficient can be related to the energy difference by:

$$J_{het}(T) = A_{het} \exp\left(-\frac{\Delta G^* \varphi}{kT}\right) \tag{14}$$

where A_{het} is a constant and φ the factor by which the heterogeneous energy barrier to nucleation is lower than the homogenous barrier. This equation is identical to Eq. (7), except that the height of the energy barrier is lowered by a factor φ calculated using:

$$\varphi = \frac{(2 + \cos\theta)(1 - \cos\theta)^2}{4} \tag{15}$$

where θ is the contact angle between a spherical ice nucleus and a flat surface of the nucleator. It is possible to calculate contact angles if $J_{het}(T)$ is known therefore. It is not clear what contact angles mean physically although they give an indication of a material's ice nucleating ability.

3.2.2 Single Component Stochastic Models

The simplest CNT based models are a type of single component stochastic (SCS) model. $J_{het}(T)$ is usually measured per surface area of nucleator meaning it has units of events $\text{cm}^{-2}\,\text{s}^{-1}$. These models use a single nucleation rate (J_{het}) to describe a nucleator's behavior. J_{het} is in principle calculated in the same way as J_{hom} from Eq. (10) except that a rate per surface area of nucleator, J_s is used:

$$J_{het} = J_s = \frac{R(t)}{A} \tag{16}$$

J_{het} can be related to CNT as described as in the above section (e.g., Chen et al., 2008) to account for temperature dependence of J_{het} but a simple linear temperature dependence can also be used (e.g., Murray et al., 2011).

These models are not usually appropriate as they assume that all droplets in an experiment nucleate ice with the same rate. Although there are examples of nucleators which show good agreement with a single component model, notably KGa-1b kaolinite (Herbert et al., 2014; Murray et al., 2011) it is clear that this is not the case for many materials (Herbert et al., 2014; Vali, 2008, 2014). J_{het} often does not equal R/A (Herbert et al., 2014; Vali, 2008, 2014). As a result, various other models of ice nucleation have been developed. Multiple component stochastic models are an extension of single component models.

3.2.3 Multiple Component Stochastic Models

Multiple stochastic models (MCSMs) have been developed to describe the observed variation in nucleation rates between droplets. These models divide a population of droplets, or sites, into sub-populations with different single component rates. There are a number of different variations on this theme. Some use distributions of efficiencies described by CNT (Lüönd et al., 2010; Marcolli et al., 2007; Niedermeier et al., 2014; Niedermeier et al., 2011), while others use linear dependences (Broadley et al., 2012). All use multiple different curves, representing different sites, droplets, or particles and sum the freezing probabilities of all these to generate a total nucleation rate at a given temperature. Such descriptions therefore retain time dependence and account for variability between droplets.

3.2.4 Singular Models

Singular models of ice nucleation assume that each droplet in an ice nucleation experiment contains a site that induces it to freeze at a specific characteristic temperature (Vali and Stansbury, 1966). The justification for this approach is that it is typically observed that variability in freezing temperature for a single droplet frozen and thawed multiple times is much smaller than the range in freezing temperature of a population of droplets with identical nucleator content (Vali, 2008; Vali and Stansbury, 1966). The concept was originally put forward by Levine (1950). Typically, concentration of sites is related to either droplet volume or surface area of nucleator. The differential nucleus spectrum, $k(T)$, which can be calculated from the output of ice nucleation experiments using:

$$k(T) = \frac{1}{V \cdot (N_0 - N_F(T))} \cdot \frac{dN_F(T)}{dT} \tag{17}$$

where V is the droplet volume used in the experiment, N_0 is the total number of droplets in the experiment, and $N_f(T)$ is the number of droplets frozen at temperature T. By integrating this expression the cumulative nucleus spectrum, $K(T)$ can be derived:

$$K(T) = -\frac{1}{V} \cdot \ln\left(1 - \frac{N_F(T)}{N_0}\right) \tag{18}$$

$K(T)$ has dimensions of sites per volume. Recently, it has become common to determine the surface area of nucleator contained in each droplet in order to calculate the ice active site density $n_s(T)$, which is a measure of the number of sites per unit surface area of nucleator (Connolly et al., 2009). $n_s(T)$ is related to $K(T)$ by:

$$n_s(T) = \frac{K(T)}{A} \tag{19}$$

where A is the surface area of nucleator per droplet. To calculate $n_s(T)$ directly from droplet experimental data the following expression can be used:

$$n_s(T) = -\frac{1}{A} \ln\left(1 - \frac{N_F(T)}{N_0}\right) \tag{20}$$

Site-specific models of ice nucleation can also conceivably use other units besides nucleator surface area and droplet volume, for instance, the number of nucleation sites per cell or per particle can be calculated, if the number of these entities per droplet is known.

Singular models ignore time dependence. According to a site-specific model at constant temperature, no freezing will take place. This is generally not the case but it is often true that freezing does not follow the sort of exponential decay that would be predicted by a single component model (Sear, 2014).

3.2.5 The Framework for Reconciling Observable Stochastic Time-Dependence (FROST)

To overcome the difficulty that simple site-specific models do not account for time dependence, modified singular models can be used (Vali, 1994; Vali, 2008). If two identical sets of droplets (identical meaning that the two sets contain the same surface area of nucleator) are cooled at different rates a greater fraction of the droplets that are cooled more slowly will be frozen at a given relevant temperature. This is because time dependence of ice nucleation will mean that every droplet has a greater probability of freezing in the longer time interval allowed to it by the slower cooling rate, compared to the faster cooling rate. Modified singular models incorporate a factor that accounts for shifts induced by differing cooling rates into typical site-specific expressions for ice nucleation. The Framework for Reconciling Observable Stochastic Time-dependence (FROST) derived by Herbert et al. (2014) is similar to the modified singular approach, which allows ice nucleation data obtained from experiments conducted at different ramp rates, or in isothermal conditions to be reconciled. The shift in freezing temperature between two experiments conducted at cooling rates r_1 and r_2 can be calculated using:

$$\Delta T_f = \beta = \frac{1}{\lambda} \ln \left(\frac{r_1}{r_2} \right) \tag{21}$$

where β is the shift in freezing temperature caused by the change in cooling rate and λ is the slope, $-d\ln(J)/dT$, of the individual components in the MCSM of Broadley et al. (2012) and Herbert et al. (2014). This equation can be used to calculate λ from experimental fraction frozen data. A similar quantity, ω, is defined as the gradient $-d\ln(R/A)/dT$. Herbert et al. (2014) showed using computer simulations that when $\omega = \lambda$ a single component stochastic model can be applied. When $\omega \neq \lambda$ there is variation in the nucleating ability of droplets in the experiment and a MCSM must be used to account for data. λ can be regarded as a fundamental property of a nucleator.

FROST can be used to reconcile $n_s(T)$ from experiments conducted at different ramp rates by substituting fraction frozen values calculated using Eq. (21) into Eq. (20). If a standard r_1 value of 1°C min^{-1} it can be shown that:

$$\frac{N_F(T, r)}{N_0} = 1 - \exp \left(-n_s \left(T - \frac{\ln r}{-\lambda} \right) \cdot A \right) \tag{22}$$

where $N_F(T, r)$ is the number of droplets frozen at temperature T for an experiment conducted at ramp rate r. This equation is compatible with the modified singular model of Vali (1994). Typical modified singular approaches use an empirical shift from experimental data in temperature instead of λ.

By performing multiple experiments Herbert et al. (2014) showed that FROST could account for experimental data. Four sets of experiments conducted at four different ramp rates could be reconciled with single λ value for two different nucleators. For KGa-1b kaolinite this λ was equal to its ω value while for BCS 376 microcline this was not the case, meaning that a single-component model could be used to describe ice nucleation by KGa-1b, but not BCS 376.

3.2.6 Comparison and Summary of Models of Heterogeneous Nucleation

Heterogeneous ice nucleation is, in the majority of cases, a phenomenon with both site-specific and time dependent characteristics. For most freezing experiments it is likely that individual droplets contain many sites which nucleate ice more efficiently than the majority of the nucleator surface area, one of which may nucleate ice more efficiently that all others sites in the droplet, as assumed by the site-specific model. Ice nucleation at sites is likely to be stochastic, and may be well described by a single component stochastic model, possibly by classical nucleation theory with a suitably reduced free energy barrier height. As the specific mechanism of heterogeneous ice nucleation is not known it cannot be said that this is the case.

There is little reason to suppose that classical nucleation theory as applied to heterogeneous nucleation is valid for the nucleation of ice. It is generally acknowledged that the contact angle used in Eqs. (14) and (15) has no physical meaning and serves as a proxy for lowering the height of the free energy barrier calculated by CNT at a given temperature. Clearly, site-specific models are also unphysical insofar as ice nucleation is to some extent stochastic. No experiment has found that droplets repeatedly freeze at the exact same temperature.

Site-specific models account for the strong temperature dependence observed in nucleation by most nucleators while single component stochastic models account for the time dependence. They ignore time dependence and droplet to droplet variability in nucleation efficiency respectively. The various multiple component stochastic models and time dependent site-specific models seek to add the facet of the problem that the simple models do not account for.

Ultimately, none of these models of ice nucleation offer real insight into the underlying mechanism of ice nucleation (Vali, 2014). Multiple component stochastic models generally provide the best fit to experimental data, which is not surprising as they have the most degrees of freedom. They are, in a sense, fitting routines. That said, they are probably also the most physically realistic models of ice nucleation. Generally, it is convenient to use site-specific models as temperature dependence is the overriding determinant of freezing rate. Many recent studies have tended to determine n_s as a means of comparing ice nucleating species. Agreement is not universal however. For instance, efforts have been made to explain the variation in freezing rate between the individual droplets in experiments as a product of variations in the amount of material between different droplets (Alpert and Knopf, 2016).

4. PROPERTIES OF GOOD HETEROGENEOUS ICE NUCLEATORS

Ideally, it would be possible to predict the efficiency of a heterogeneous ice nucleator from knowledge of its physical and chemical properties. For comparison, it is possible to describe the activity of CCN with a single hygroscopicity parameter (Petters and Kreidenweis, 2007). At current this sort of description is not possible for ice nucleation. Indeed, it seems unlikely that it will be so straightforward. In the case of deposition mode ice nucleation there is a growing body of evidence that pore-condensation freezing is responsible for ice nucleation in many cases (Campbell et al., 2016; Marcolli, 2014), which has the potential to simplify the problem there. For the immersion mode ice nucleation relevant to mixed-phase clouds no consistent theory exists. The difficulty of understanding what makes a good INP stems from the small size of the ice critical nucleus (see Fig. 2) and the small spatial extent of the nucleation event. According to CNT, critical nuclei range in size from a 1 nm radius at $-38°C$ to 10 nm at $-4°C$. These critical nuclei are spatially rare. Whatever volume of droplet is frozen, there will usually only be a single critical nucleus present. Droplets are typically at least picometers across. No current technique is capable of locating and usefully measuring the physical properties of an event this small and rare. As a result, properties of ice nucleation have usually been inferred from experimental data.

4.1 The Traditional View of Heterogeneous Ice Nucleation

Historically, five properties were thought to be important for heterogeneous ice nucleation. These were listed and discussed by Pruppacher and Klett (1997). While these have never been regarded as hard and fast rules, discussion of the reasons for them and where they fall down are instructive. They are:

(1) The insolubility requirement: nucleators must provide an interface with water. Dissolved substances do not provide an interface and so do not nucleate ice.

(2) The size requirement: observations in the atmosphere indicate that INPs tend to be large. This requirement is somewhat vague, although it stems from the observation that larger particles in the atmosphere tend to be the ones that nucleate ice. It is also assumed that an INP must be larger than a critical nucleus.

(3) The chemical bond requirement: a nucleator must be able to bind to water in order to cause nucleation. Stronger bonding is likely to improve nucleation efficiency.

(4) The crystallographic requirement: the classic lattice matching idea first put forward by Vonnegut (1947). Substances with a similar lattice structure and spacing to ice will provide a template for a critical nucleus.

(5) The active site requirement: based on a combination of the observation that site-specific descriptions often give the best account of ice nucleation and the fact that deposition mode ice nucleation tends to occur repeatedly on specific locations on crystals. It seems likely that this is more related to vapor condensation than ice nucleation (Marcolli, 2014).

The next sections looks at how these requirements have been challenged and revised in recent years and what is known about the mechanism of heterogeneous ice nucleation from experimental studies. Computational studies of ice nucleation are then examined and the outcomes of the two approaches discussed.

4.1.1 Size and Solubility of Heterogeneous INPs

While INPs have traditionally been regarded as large and insoluble (Pruppacher and Klett, 1997) a number of counter examples are known. In recent times biological macromolecules associated with pollen that have been claimed as soluble have been shown to nucleate ice efficiently (Pummer et al., 2012, 2015). These molecules weigh from 100 to 860 kDa, which equates to a radius of less than 10 nm. This is only slightly larger than the critical nuclei they nucleate. They are perhaps 10 times smaller than the particles that Pruppacher and Klett (1997) envisaged as too small to efficiently nucleate ice on the basis of older work. Similarly, Ogawa et al. (2009) showed that solutions of poly-vinyl alcohol could nucleate ice, although only at a few degrees above homogenous nucleation temperatures.

4.1.2 Lattice Matching

The best known example of inference of ice nucleation properties is the lattice matching concept of Vonnegut (1947). The idea is that substances that have a similar crystal structure to ice, with similar lattice constants will pattern the first layer of ice. The amount of lattice mismatch, or lattice disregistry (Pruppacher and Klett, 1997; Turnbull and Vonnegut, 1952) can be readily calculated from knowledge of the crystal structure. On this basis Vonnegut (1947) identified AgI as a potentially excellent nucleator and all subsequent experimentation has shown that he was correct.

The role of lattice matching in ice nucleation by AgI has been questioned for some time. Zettlemoyer et al. (1961) argued that water likely adhered to specific sites on the surface of AgI, which may have been oxidized, rather than forming a layer over the crystal on the basis of the difference in adsorption of water and nitrogen. More recently, Finnegan and Chai (2003) postulated an alternative mechanism for ice nucleation by AgI where clustering of surface charge controls ice nucleation. There is no universal agreement on the mechanism of ice nucleation by AgI from experimentalists.

Experimental studies on BaF_2, which also has a good lattice match to ice, did not support the lattice matching argument (Conrad et al., 2005).

4.1.3 Bonding of Water to INPs

Intuitively, it seems obvious that water must be able to bind to a nucleator to induce ice nucleation and that hydrophilic surfaces will nucleate ice more efficiently than hydrophobic surfaces. There is little relevant experimental work, as studies where surfaces of differing but well understood hydrophilicity have been tested in the same system have not been conducted. It is also generally difficult to choose nucleators and systems such that all other possible variables are constrained.

Li et al. (2012) compared the ice nucleating ability of hydrophobic and hydrophilic surfaces and obtained the somewhat surprising result that the hydrophobic surface nucleated ice more efficiently. While their technique—coating of a silicon wafer with a fluoroalkylsilane—may introduce other variables, this is evidence that hydrophilic surfaces do not necessarily nucleate ice more efficiently than hydrophobic surfaces. It has been reported in the past that hydrophilic soots nucleate ice more efficiently than hydrophobic soots, although the method used leaves doubt as to the mode of ice nucleation observed (Gorbunov et al., 2001).

4.1.4 Active Sites and Topographical Effects

As discussed in the section on descriptions of heterogeneous ice nucleation, it has been argued that the surfaces of most immersion mode nucleators must have active sites on the basis of interpretations of droplet freezing experiments and repeated freezing droplets (Vali, 2008, 2014), although there are exceptions (Herbert et al., 2014). These experiments do not constitute direct observation of the nucleation sites. It has been known for some time that apparently depositional ice nucleation tends to occur on specific sites on both organic substances (Fukuta and Mason, 1963) and inorganic substances (Bryant et al., 1960). More recently it has been shown that nucleation from vapor of organic molecules probably follows a two-step process where small amounts of liquid condense prior to freezing (Campbell et al., 2013) and proposed that ice nucleation from vapor probably follows a similar route in many situations (Marcolli, 2014).

It is far easier to locate a depositional nucleation site than an immersed nucleation site as crystals grow out from point of depositional nucleation relatively slowly, whereas they grow very quickly through liquid water droplets. That said, it is possible to locate nucleation points in surfaces covered by water using a high-speed camera (Gurganus et al., 2011, 2013, 2014). Gurganus et al. (2011) conducted freezing experiments of droplets on silicon wafers and saw no tendency for nucleation to occur repeatedly on the same site, suggesting that no specific sites on their substrate nucleated ice more efficiently than any others. Their study was aimed at determining whether nucleation tended to take place at the air-water-substrate interface (contact mode, see above) or elsewhere and showed no preference for nucleation at the interface. More recent work from Gurganus et al. (2014) has shown that a 'nano-textured' silicon strand (features on a scale smaller than 100 nm) nucleates ice much more effectively than a 'micro-textured' (etched features in silicon with depths from 300 to 900 nm) silicon substrate or a smooth silicon substrate. The difference was particularly pronounced at the point of contact between the silicon strand and the droplet surface. This interesting result shows that topographical differences between chemically very similar surfaces can cause differences in ice nucleation behavior. The nature of topographical features implies that nucleation processes involving them must be to some extent "site specific."

Campbell et al. (2015) attempted to change the ice nucleating properties of silicon, glass, and mica substrates by scratching them with diamond powders ranging from 10 nm to 40–60 μm. They found that the scratching process made no significant difference to the ice nucleating efficiency of the surfaces. It might be expected that the 10 nm diamond powder would produce features on a similar scale those that Gurganus et al. (2014) observed enhancing the ice nucleating efficiency of silicon. This was not the case. There are many other differences between the two systems so it is difficult to suggest reasons for this. Other studies have also reported no impact on ice nucleation efficiency from topography on a micrometer scale (Heydari et al., 2013).

At present, it is impossible to observe ice nucleation events directly on a scale that is useful for understanding the underlying mechanisms. An approach to defining the scale problem is to use classical nucleation theory and assume that it gives a reasonable estimate for the size of a critical nucleus required for heterogeneous nucleation at temperatures above which homogenous nucleation can be observed.

The role of lattice matching in heterogeneous ice nucleation has been studies computationally. Recent molecular dynamics (MD) results have suggested that a lattice match is not the sole explanation for the efficiency of AgI and that a mechanism involving an ordering of water above the AgI surface, which then causes ice nucleation is more likely (Reinhardt and Doye, 2014; Zielke et al., 2015). Similarly, kaolinite had been thought of as good ice nucleator, and this efficiency had been attributed to a good lattice match of hydroxyl functional groups (—OH) on the basal face of kaolinite to hexagonal ice (Pruppacher and Klett, 1997). It is now seems likely that the apparent ice nucleation activity of kaolinite observed in older studies was largely due to contamination by feldspar minerals (Atkinson et al., 2013). Density Functional Theory (DFT) calculations previously questioned the validity of the lattice matching mechanism for kaolinite, instead attributing the activity to the amphoterism of the —OH groups on the surface of kaolinite which allows them to both accept and donate hydrogen bonds, favoring the formation of an overlayer of water molecules (Hu and Michaelides, 2007). Indeed, there is now a significant body of computational evidence suggesting that a simplistic lattice matching view of ice nucleation may be misleading (e.g., Cox et al., 2012, 2013; Fitzner et al., 2015).

Another variable that has been examined computationally is surface hydrophilicity. Lupi and Molinero (2014) found that simulated graphite surfaces nucleated ice less well when decorated with –OH groups to increase hydrophilicity. Recently, Fitzner et al. (2015) conducted a comprehensive, systematic MD study of the impact of crystallographic match and hydrophilicity on ice nucleation. By testing four different idealized crystal surfaces with varied lattice parameters and water interaction strengths they found three different mechanisms by which heterogeneous ice nucleation could be promoted. They name these "In-Plane Template of the First Overlayer," "Buckling of the First Overlayer," and "High Adsorption-Energy Nucleation on Compact Surfaces". It is interesting that even in this simplified system they found complex dependency on lattice parameters and interaction strength. Bi et al. (2016) also found complex dependencies of nucleation rate on the interaction between hydrophilicity and crystallinity. Computational studies have found that surface roughness on a fine scale (from several angstroms to several nanometers), with roughness of some specific periodicities found to promote nucleation better than others (Fitzner et al., 2015; Zhang et al., 2014).

Very recently, it has been shown computationally that the (100) face of the alkali feldspar microcline nucleates ice efficiently and, in the same study, that ice tends to grow out of cracks at the same angle to the exposed face of feldspar as this plane, suggesting that ice is indeed being templated by the (100) face (Kiselev et al., 2017). This breakthrough study points to more combined experimental/computational studies in the future, which eventually unpick the problem of understanding heterogeneous ice nucleation.

Overall, the picture is a complex one. It can be said with some certainty that active sites are important for ice nucleation by many nucleators. The exact properties of these active sites are much less certain. Other nucleators do not appear to have active sites. Lattice matching as a concept is well established and widely applied but increasingly questioned by both laboratory and computational studies. Hydrophilicity must play some role and simulations of ice nucleation have suggested that the relationship between hydrophilicity and lattice match can impact ice nucleation efficiency in complicated and non-intuitive ways. This may shed some light on why it has proved so hard to understand ice nucleation processes in the past; relationships between physico-chemical properties and ice nucleation efficiency are not straightforward. Immersion-mode ice nucleation micrometer scale topographical features have so far proved to play little role, although only limited numbers of experiments have been conducted to date. There are, however, hints that topography on a sufficiently small scale (atomic to nanometer scale) may play a role.

5. WHAT NUCLEATES ICE IN MIXED-PHASE CLOUDS?

Because we cannot directly predict what atmospheric aerosol species will nucleate ice well from theory, we must fall back on other methods for understanding what species might be nucleating ice in mixed-phase clouds, and how well they might be doing it.

The majority of laboratory studies of immersion-mode heterogeneous ice nucleation have been conducted with the aim of understanding and quantifying ice nucleation by substances that might nucleate ice in the atmosphere. Extensive reviews are available (Hoose and Möhler, 2012; Murray et al., 2012). The following section very briefly details these substances.

Large amounts of mineral dusts are emitted to into the atmosphere, mostly from arid regions in Africa and Asia (Prospero et al., 2002). It has, for many years, been known that snow crystals contain mineral dust residues (Kumai, 1961) and more recent work has found that mineral dusts make up a large proportion of ice crystal residues in certain cloud types (Murray et al., 2012; Pratt et al., 2009). There is a volume of older work on ice nucleation by mineral dusts (Pruppacher and Klett, 1997). In many of these cases only onset freezing temperatures are recorded and it is therefore difficult to assess the relative efficiency of freezing. Recently, n_s values for a range of natural mixed dusts have been calculated (Connolly et al., 2009; Niemand et al., 2012) as well as for proxies of natural dusts such as NX illite and Arizona test dust (ATD) (Broadley et al., 2012; Connolly et al., 2009; Marcolli et al., 2007).

Until recently it had been thought that clay minerals were responsible for the ice nucleation activity of mineral dusts (Lüönd et al., 2010; Pinti et al., 2012; Pruppacher and Klett, 1997), partially on the basis of kaolinite's lattice match to hexagonal ice (Pruppacher and Klett, 1997). Atkinson et al. (2013) have recently shown that feldspars nucleate ice far more efficiently than the other major components of mineral dusts and that they are likely to be responsible for much of the ice nucleation observed in mixed phase clouds in various regions of the world. It now also known that different polymorphs of feldspar nucleate ice differently well (Harrison et al., 2016).

The ice nucleation activities of a wide range of biological entities have been investigated in the past. The starting point for much of this work was the discovery by Schnell and Vali (1972) that decomposing leaf matter induced freezing at higher temperatures than any other nucleator they tested. It was discovered that the efficient nucleator was associated with the bacterium *Pseudomonas syringae*, (Lindow et al., 1989; Maki et al., 1974) a plant pathogen. It is generally thought that the efficient ice nucleation of *P. syringae* allows it to ingest nutrients from plants at a temperature just below the melting point of water where the plants would usually avoid frost damage by supercooling. Since the discovery of the ice nucleation activity of *P. syringae* many other bacteria have been shown to nucleate ice at high temperatures (Lee et al., 1995).

Other biological ice nucleators include fungi (Fröhlich-Nowoisky et al., 2014; O'Sullivan et al., 2014; Pouleur et al., 1992), pollen (Pummer et al., 2012, 2015), and plankton (Alpert et al., 2011; Knopf et al., 2011; Schnell, 1975). Recently it has become increasingly clear that pollen and fungi emit separable macromolecular INP of far smaller size than the pollen and fungi themselves (O'Sullivan et al., 2015, 2016; Pummer et al., 2012, 2015). It has recently been shown that small ice nucleating entities, most probably of biological origin, are present in the sea-surface microlayer and may be emitted to the atmosphere (Wilson et al., 2015).

Anthropogenic burning of fossil fuels and biomass contribute significantly to global aerosol (Bond et al., 2013). Various studies of ice nucleation by soots have been conducted (Demott, 1990; Diehl and Mitra, 1998; Gorbunov et al., 2001) as well as studies of ice nucleation by various biomass products (Petters et al., 2009). In general, it would appear that soots do not nucleate ice particularly efficiently, although the relative paucity of data makes meaningful comparison to other species challenging.

AgI and related compounds were identified as good nucleators in the early days of ice nucleation research (Passarelli et al., 1973; Vonnegut, 1947; Vonnegut and Chessin, 1971) and have been used for cloud seeding ever since. They are known to nucleate ice very efficiently (DeMott, 1995). No significant amount of AgI is likely to be present in mixed-phase clouds.

Comparison of ice nucleating species is complicated by the fact that different instruments, even those of the same type, do not always give the same answer when used to test the same INPs (Hiranuma et al., 2015). However, it is possible to say something about the general effectiveness of the different classes of nucleators. Murray et al. (2012) calculated n_s values for a wide range of immersion mode measurements. As discussed in Section 3.2.6 n_s values are probably the best metric for comparing different nucleators. Fig. 3 is a reproduction of the comparison figure from Murray et al. (2012). They are not perfect, as calculating surface areas for species with varying natures is not straightforward.

A further problem is that many nucleators have only been tested using very small-sized cloud droplets with correspondingly small amounts of nucleator surface area. As a result, there is little data at warmer temperatures for many nucleators. Conversely, biological nucleators have mostly only been tested in larger droplets, although dilution has allowed extension of the range of n_s values tested. (e.g., Wex et al., 2015). The only examples of non-biological ice nucleators tested at n_s values below $10^3 \ cm^{-2}$ in Fig. 3 are the volcanic ash tested by Fornea et al. (2009) and BCS 376 microcline (2013).

What can be seen is that *P. syringae* nucleates ice far more efficiently than any other nucleator for which n_s has been calculated, with similar site concentrations to other nucleators at much warmer temperatures. BCS 376 microcline, an alkali feldspar, was tested by Atkinson et al. (2013) and shown to nucleate ice more efficiently than the atmospherically relevant minerals they tested. BCS 376 microcline also has a higher active site density at all temperatures than all other non-biological nucleators. Other feldspar minerals have also been tested (Zolles et al., 2015; Niedermeier et al., 2015; Augustin-Bauditz et al., 2014) and are of broadly similar, or somewhat lesser activity than BCS 376 microcline. Illite, kaolinite, Arizona Test Dust (ATD) and natural dusts all appear to be rather less effective nucleators than BCS 376 microcline. It seems quite likely that nucleation by ATD and natural dusts is dominated by their feldspar content (Augustin-Bauditz et al., 2014; Atkinson et al., 2013).

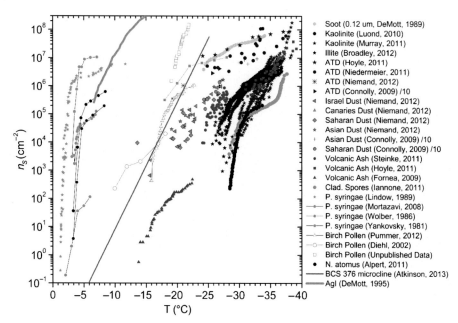

Fig. 3 An adapted version of figure 18 from Murray et al. (2012) with some additional data from subsequent studies. The figure shows ice nucleation efficiencies for a range of different nucleators. It can be seen that bacterial ice nucleators are much more effective than non-biological ice nucleators. BCS 376 microcline from Atkinson et al. (2013) nucleates ice more efficiently than other non-biological nucleators, except for AgI (DeMott, 1995).

AgI is a highly efficient nucleator (DeMott, 1995) and is better at nucleating ice than any other non-biological nucleator that has been tested, including feldspars. Birch pollen nucleates with similar efficiency to BCS-376 microcline and the plankton *N. atomus* is rather less active. Overall, biological nucleators such as *P. syringae* and fungal proteins nucleate ice more efficiently than any other species. AgI and related compounds are probably the next most efficient nucleators known but will not be present in mixed-phase clouds, unless artificially introduced. Alkali feldspar is more efficient than other minerals, volcanic ashes, and combustion products.

To assess the relative importance of different nucleator species in mixed-phase clouds both the concentration of INPs and their efficiencies must be taken into account. Several studies have attempted to do this. For instance, Hoose et al. (2010b) found that mineral dusts dominate immersion–mode ice nucleation between 0°C and −38°C globally. Different studies have come to different conclusions about the relevance of biological ice nucleators. Phillips et al. (2009) concluded that biological nucleating species may play a key role globally, while work by Hoose and co-workers has suggested otherwise (Hoose et al., 2008, 2010a). It seems likely that much of the discrepancy arises from differing assumptions about the nature and quantity of biological INPs present in the atmosphere. Further laboratory and field measurements may be needed to constrain these variables. More recently, it has been suggested that alkali feldspars can account for a large proportion of ice nucleation observed in the atmosphere (Atkinson et al., 2013) and more recently still that a combination of feldspar and marine organic sources account still better for observations, while leaving the possibility of the existence of an extra, terrestrial source of INPs (Vergara Temprado et al., 2016). Overall, while the precise identity of the most important INPs in the atmosphere is not known much progress has been made.

6. FIELD MEASUREMENTS OF ICE NUCLEATING PARTICLES

Direct measurement of the concentration of INPs in the atmosphere is important for providing inputs to and comparisons for atmospheric models, thereby contributing to understanding of ice nucleation in the atmosphere. The practice of measuring INP concentrations in the atmosphere has enjoyed something of a resurgence in recent years (DeMott et al., 2011) and the amount of data available is steadily increasing. The instruments mentioned previously can and have been used for measuring INP concentrations in the atmosphere. In the case of dry dispersion instruments, particularly CFDCs (which are often portable), aerosol collection is relatively straightforward, although much care is needed. Many studies of this sort have been carried out (e.g., Boose et al., 2016; DeMott et al., 2010; Rosinski et al., 1987, 1995). For wet-dispersion methods some approach by which atmospheric aerosol can be transferred to liquid droplets is required. Of late, this has been accomplished by collecting aerosol onto filter membranes and either washing this aerosol off the filter in order to conduct a droplet freezing assay (DeMott et al., 2016) or placing pieces of the filters directly into water before freezing (Stopelli et al., 2016). Another method is to use a micro-orifice uniform-deposit impactor (MOUDI) to collect size-separated aerosol onto glass slides. Droplet freezing experiments can then be conducted on these glass slides (Mason et al., 2016).

7. SUMMARY

Immersion mode ice nucleation must play a key role in the evolution of mixed-phase clouds. The precise nature of this role and how ice nucleation interacts with other microphysical and dynamical processes in mixed-phase clouds remains to be established. Several different methods of testing the ice nucleating efficiency of nucleators and the INP concentration in the atmosphere are available although the comparability of these methods is not entirely established at this point. What can be said is that knowledge of which aerosol species are likely to cause ice nucleation in mixed-phase clouds is improving rapidly, as is understanding of why certain substances nucleate ice efficiently. Various mathematical descriptions of ice nucleation are available and the number of measurements of INP concentrations relevant to mixed-phase clouds in the atmosphere is increasing rapidly.

ACKNOWLEDGMENTS

I would like to thank Prof. Benjamin Murray for reading a version of this chapter and providing helpful comments and Dr. Constantin Andronache for giving me the opportunity to contribute this chapter. This article has been adapted from a chapter of my PhD thesis.

REFERENCES

Alpert, P.A., Aller, J.Y., Knopf, D.A., 2011. Ice nucleation from aqueous NaCl droplets with and without marine diatoms. Atmos. Chem. Phys. 11, 5539–5555.

Alpert, P.A., Knopf, D.A., 2016. Analysis of isothermal and cooling-rate-dependent immersion freezing by a unifying stochastic ice nucleation model. Atmos. Chem. Phys. 16, 2083–2107.

Ansmann, A., Mattis, I., Müller, D., Wandinger, U., Radlach, M., Althausen, D., Damoah, R., 2005. Ice formation in Saharan dust over central Europe observed with temperature/humidity/aerosol Raman lidar. J. Geophys. Res.-Atmos. 110, D18S12.

Ansmann, A., Tesche, M., Seifert, P., Althausen, D., Engelmann, R., Fruntke, J., Wandinger, U., Mattis, I., Müller, D., 2009. Evolution of the ice phase in tropical altocumulus: SAMUM lidar observations over Cape Verde. J. Geophys. Res.-Atmos. 114, D17208.

Atkinson, J.D., Murray, B.J., Woodhouse, M.T., Whale, T.F., Baustian, K.J., Carslaw, K.S., Dobbie, S., O'Sullivan, D., Malkin, T.L., 2013. The importance of feldspar for ice nucleation by mineral dust in mixed-phase clouds. Nature 498, 355–358.

Augustin-Bauditz, S., Wex, H., Kanter, S., Ebert, M., Niedermeier, D., Stolz, F., Prager, A., Stratmann, F., 2014. The immersion mode ice nucleation behavior of mineral dusts: a comparison of different pure and surface modified dusts. Geophys. Res. Lett. 41, 7375–7382.

Barlow, T.W., Haymet, A.D.J., 1995. ALTA: An automated lag-time apparatus for studying the nucleation of supercooled liquids. Rev. Sci. Instrum. 66, 2996–3007.

Bi, Y., Cabriolu, R., Li, T., 2016. Heterogeneous ice nucleation controlled by the coupling of surface crystallinity and surface hydrophilicity. J. Phys. Chem. C 120, 1507–1514.

Bond, T.C., Doherty, S.J., Fahey, D.W., Forster, P.M., Berntsen, T., DeAngelo, B.J., Flanner, M.G., Ghan, S., Karcher, B., Koch, D., Kinne, S., Kondo, Y., Quinn, P.K., Sarofim, M.C., Schultz, M.G., Schulz, M., Venkataraman, C., Zhang, H., Zhang, S., Bellouin, N., Guttikunda, S.K., Hopke, P.K., Jacobson, M.Z., Kaiser, J.W., Klimont, Z., Lohmann, U., Schwarz, J.P., Shindell, D., Storelvmo, T., Warren, S.G., Zender, C.S., 2013. Bounding the role of black carbon in the climate system: a scientific assessment. J. Geophys. Res.-Atmos. 118, 5380–5552.

Boose, Y., Sierau, B., García, M.I., Rodríguez, S., Alastuey, A., Linke, C., Schnaiter, M., Kupiszewski, P., Kanji, Z.A., Lohmann, U., 2016. Ice nucleating particles in the Saharan air layer. Atmos. Chem. Phys. 16, 9067–9087.

Boucher, O., Randall, D., Artaxo, P., Bretherton, C., Feingold, G., Forster, P., Kerminen, V.-M., Kondo, Y., Liao, H., Lohmann, U., 2013. Clouds and aerosols. In: Climate Change 2013: The Physical Science Basis. Contribution of Working Group I to the Fifth Assessment Report of the Intergovernmental Panel on Climate Change. Cambridge University Press, Cambridge.

Broadley, S.L., Murray, B.J., Herbert, R.J., Atkinson, J.D., Dobbie, S., Malkin, T.L., Condliffe, E., Neve, L., 2012. Immersion mode heterogeneous ice nucleation by an illite rich powder representative of atmospheric mineral dust. Atmos. Chem. Phys. 12, 287–307.

Bryant, G.W., Hallett, J., Mason, B.J., 1960. The epitaxial growth of ice on single-crystalline substrates. J. Phys. Chem. Solids 12, 189, IN118.

Campbell, J.M., Meldrum, F.C., Christenson, H.K., 2013. Characterization of preferred crystal nucleation sites on Mica surfaces. Cryst. Growth Des. 13, 1915–1925.

Campbell, J.M., Meldrum, F.C., Christenson, H.K., 2015. Is ice nucleation from supercooled water insensitive to surface roughness? J. Phys. Chem. C 119, 1164–1169.

Campbell, J.M., Meldrum, F.C., Christenson, H.K., 2016. Observing the formation of ice and organic crystals in active sites. Proc. Natl. Acad. Sci. 2016. https://doi.org/10.1073/pnas.1617717114.

Chen, J.-P., Hazra, A., Levin, Z., 2008. Parameterizing ice nucleation rates using contact angle and activation energy derived from laboratory data. Atmos. Chem. Phys. 8, 7431–7449.

Choi, Y.-S., Lindzen, R.S., Ho, C.-H., Kim, J., 2010. Space observations of cold-cloud phase change. Proc. Natl. Acad. Sci. U. S. A. 107, 11211–11216.

Christenson, H.K., 2013. Two-step crystal nucleation via capillary condensation. CrystEngComm 15, 2030–2039.

Connolly, P.J., Möhler, O., Field, P.R., Saathoff, H., Burgess, R., Choularton, T., Gallagher, M., 2009. Studies of heterogeneous freezing by three different desert dust samples. Atmos. Chem. Phys. 9, 2805–2824.

Conrad, P., Ewing, G.E., Karlinsey, R.L., Sadtchenko, V., 2005. Ice nucleation on BaF2 (111). J. Chem. Phys. 122, 064709.

Cox, S.J., Kathmann, S.M., Purton, J.A., Gillan, M.J., Michaelides, A., 2012. Non-hexagonal ice at hexagonal surfaces: the role of lattice mismatch. Phys. Chem. Chem. Phys. 14, 7944–7949.

Cox, S.J., Raza, Z., Kathmann, S.M., Slater, B., Michaelides, A., 2013. The microscopic features of heterogeneous ice nucleation may affect the macroscopic morphology of atmospheric ice crystals. Faraday Discuss. 167, 389–403.

Cui, Z.Q., Carslaw, K.S., Yin, Y., Davies, S., 2006. A numerical study of aerosol effects on the dynamics and microphysics of a deep convective cloud in a continental environment. J. Geophys. Res.-Atmos. 111.

de Boer, G., Morrison, H., Shupe, M.D., Hildner, R., 2011. Evidence of liquid dependent ice nucleation in high-latitude stratiform clouds from surface remote sensors. Geophys. Res. Lett. 38L01803.

Debenedetti, P.G., 1996. Metastable Liquids: Concepts and Principles. Princeton University Press, Princeton, NJ.

Demott, P.J., 1990. An exploratory-study of ice nucleation by soot aerosols. J. Appl. Meteorol. 29, 1072–1079.

DeMott, P.J., 1995. Quantitative descriptions of ice formation mechanisms of silver iodide-type aerosols. Atmos. Res. 38, 63–99.

DeMott, P.J., Hill, T.C.J., McCluskey, C.S., Prather, K.A., Collins, D.B., Sullivan, R.C., Ruppel, M.J., Mason, R.H., Irish, V.E., Lee, T., Hwang, C.Y., Rhee, T.S., Snider, J.R., McMeeking, G.R., Dhaniyala, S., Lewis, E.R., Wentzell, J.J.B., Abbatt, J., Lee, C., Sultana, C.M., Ault, A.P., Axson, J.L., Diaz Martinez, M., Venero, I., Santos-Figueroa, G., Stokes, M.D., Deane, G.B., Mayol-Bracero, O.L., Grassian, V.H., Bertram, T.H., Bertram, A.K., Moffett, B.F., Franc, G.D., 2016. Sea spray aerosol as a unique source of ice nucleating particles. Proc. Natl. Acad. Sci. 113, 5797–5803.

DeMott, P.J., Möhler, O., Stetzer, O., Vali, G., Levin, Z., Petters, M.D., Murakami, M., Leisner, T., Bundke, U., Klein, H., 2011. Resurgence in ice nuclei measurement research. Bull. Amer. Meteorol. Soc. 92, 1623–1635.

DeMott, P.J., Prenni, A.J., Liu, X., Kreidenweis, S.M., Petters, M.D., Twohy, C.H., Richardson, M.S., Eidhammer, T., Rogers, D.C., 2010. Predicting global atmospheric ice nuclei distributions and their impacts on climate. Proc. Natl. Acad. Sci. U. S. A. 107, 11217–11222.

Denman, K.L., Brasseur, G., Chidthaisong, A., Ciais, P., Cox, P.M., Dickinson, R.E., Hauglustaine, D., Heinze, C., Holland, E., Jacob, D., Lohmann, U., Ramachandran, S., da Silva Dias, P.L., Wofsy, S.C., Zhang, X., 2007. Couplings between changes in the climate system and biogeochemistry. In: Climate Change 2007: The Physical Science Basis. Contribution of Working Group I to the Fourth Assessment Report of the Intergovernmental Panel on Climate Change. Cambridge University Press, Cambridge, UK.

Diehl, K., Matthias-Maser, S., Jaenicke, R., Mitra, S.K., 2002. The ice nucleating ability of pollen: Part II. Laboratory studies in immersion and contact freezing modes. Atmos. Res. 61, 125–133.

Diehl, K., Mitra, S.K., 1998. A laboratory study of the effects of a kerosene-burner exhaust on ice nucleation and the evaporation rate of ice crystals. Atmos. Environ. 32, 3145–3151.

Emersic, C., Connolly, P.J., Boult, S., Campana, M., Li, Z., 2015. Investigating the discrepancy between wet-suspension- and dry-dispersion-derived ice nucleation efficiency of mineral particles. Atmos. Chem. Phys. 15, 11311–11326.

Field, C.B., Barros, V.R., Mach, K.J., Mastrandrea, M.D., Aalst, M.v., Adger, W.N., Arent, D.J., Barnett, J., Betts, R., Bilir, T.E., Birkmann, J., Carmin, J., Chadee, D.D., Challinor, A.J., Chatterjee, M., Cramer, W., Davidson, D.J., Estrada, Y.O., Gattuso, J.P., Hijioka, Y., Hoegh-Guldberg, O., Huang, H.Q., Insarov, G.E., Jones, R.N., Kovats, R.S., Lankao, P.R., Larsen, J.N., Losada, I.J., Marengo, J.A., McLean, R.F., Mearns, L.O., Mechler, R., Morton, J.F., Niang, I., Oki, T., Olwoch, J.M., Opondo, M., Poloczanska, E.S., Pörtner, H.O., Redsteer, M.H., Reisinger, A., Revi, A., Schmidt, D.N., Shaw, M.R., Solecki, W., Stone, D.A., Stone, J.M.R., Strzepek, K.M., Suarez, A.G., Tschakert, P., Valentini, R., Vicuña, S., Villamizar, A., Vincent, K.E., Warren, R., White, L.L., Wilbanks, T.J., Wong, P.P., Yohe, G.W., 2014. Technical summary. In: Field, C.B., Barros, V.R., Dokken, D.J., Mach, K.J., Mastrandrea, M.D., Bilir, T.E., Chatterjee, M., Ebi, K.L., Estrada, Y.O., Genova, R.C., Girma, B., Kissel, E.S., Levy, A.N., MacCracken, S., Mastrandrea, P.R., White, L.L. (Eds.), Climate Change 2014: Impacts, Adaptation, and Vulnerability. Part A: Global and Sectoral Aspects. Contribution of Working Group II to the Fifth Assessment Report of the Intergovernmental Panel on Climate Change. Cambridge University Press, Cambridge, UK; New York.

Finnegan, W.G., Chai, S.K., 2003. A new hypothesis for the mechanism of ice nucleation on wetted AgI and AgI center dot AgCl particulate aerosols. J. Atmos. Sci. 60, 1723–1731.

Fitzner, M., Sosso, G.C., Cox, S.J., Michaelides, A., 2015. The many faces of heterogeneous ice nucleation: interplay between surface morphology and hydrophobicity. J. Am. Chem. Soc. 137, 13658–13669.

Fornea, A.P., Brooks, S.D., Dooley, J.B., Saha, A., 2009. Heterogeneous freezing of ice on atmospheric aerosols containing ash, soot, and soil. J. Geophys. Res.-Atmos. 114, D13201.

Fröhlich-Nowoisky, J., Hill, T.C.J., Pummer, B.G., Franc, G.D., Pöschl, U., 2014. Ice nucleation activity in the widespread soil fungus Mortierella alpina. Biogeosci. Discuss. 11, 12697–12731.

Fu, Q., Liu, E., Wilson, P., Chen, Z., 2015. Ice nucleation behaviour on sol–gel coatings with different surface energy and roughness. Phys. Chem. Chem. Phys. 17, 21492–21500.

Fukuta, N., Mason, B.J., 1963. Epitaxial growth of ice on organic crystals. J. Phys. Chem. Solids 24, 715–718.

Garimella, S., Kristensen, T.B., Ignatius, K., Welti, A., Voigtländer, J., Kulkarni, G.R., Sagan, F., Kok, G.L., Dorsey, J., Nichman, L., Rothenberg, D., Rösch, M., Kirchgäßner, A., Ladkin, R., Wex, H., Wilson, T.W., Ladino, L.A., Abbatt, J.P.D., Stetzer, O., Lohmann, U., Stratmann, F., Cziczo, D.J., 2016. The SPectrometer for ice nuclei (SPIN): an instrument to investigate ice nucleation. Atmos. Meas. Tech. Discuss. 2016, 1–37.

Gorbunov, B., Baklanov, A., Kakutkina, N., Windsor, H.L., Toumi, R., 2001. Ice nucleation on soot particles. J. Aerosol Sci. 32, 199–215.

Gurganus, C.W., Charnawskas, J.C., Kostinski, A.B., Shaw, R.A., 2014. Nucleation at the contact line observed on nanotextured surfaces. Phys. Rev. Lett. 113, 235701.

Gurganus, C., Kostinski, A.B., Shaw, R.A., 2011. Fast imaging of freezing drops: no preference for nucleation at the contact line. J. Phys. Chem. Lett. 2, 1449–1454.

Gurganus, C., Kostinski, A.B., Shaw, R.A., 2013. High-speed imaging of freezing drops: Still no preference for the contact line. J. Phys. Chem. 117, 6195–6200.

Hallett, J., Mossop, S.C., 1974. Production of secondary ice particles during the riming process. Nature 249, 26–28.

Harrison, A.D., Whale, T.F., Carpenter, M.A., Holden, M.A., Neve, L., O'Sullivan, D., Vergara Temprado, J., Murray, B.J., 2016. Not all feldspars are equal: a survey of ice nucleating properties across the feldspar group of minerals. Atmos. Chem. Phys. 16, 10927–10940.

Hartmann, D.L., Ockert-Bell, M.E., Michelsen, M.L., 1992. The effect of cloud type on Earth's energy balance: global analysis. J. Clim. 5, 1281–1304.

Herbert, R.J., Murray, B.J., Dobbie, S.J., Koop, T., 2015. Sensitivity of liquid clouds to homogenous freezing parameterizations. Geophys. Res. Lett. 42, 1599–1605.

Herbert, R.J., Murray, B.J., Whale, T.F., Dobbie, S.J., Atkinson, J.D., 2014. Representing time-dependent freezing behaviour in immersion mode ice nucleation. Atmos. Chem. Phys. 14, 8501–8520.

Heydari, G., Thormann, E., Järn, M., Tyrode, E., Claesson, P.M., 2013. Hydrophobic surfaces: topography effects on wetting by supercooled water and freezing delay. J. Phys. Chem. C 117, 21752–21762.

Hill, T.C., Moffett, B.F., DeMott, P.J., Georgakopoulos, D.G., Stump, W.L., Franc, G.D., 2014. Measurement of ice nucleation-active bacteria on plants and in precipitation by quantitative PCR. Appl. Environ. Microbiol. 80, 1256–1267.

Hiranuma, N., Augustin-Bauditz, S., Bingemer, H., Budke, C., Curtius, J., Danielczok, A., Diehl, K., Dreischmeier, K., Ebert, M., Frank, F., Hoffmann, N., Kandler, K., Kiselev, A., Koop, T., Leisner, T., Möhler, O., Nillius, B., Peckhaus, A., Rose, D., Weinbruch, S., Wex, H., Boose, Y., DeMott, P.J., Hader, J.D., Hill, T.C.J., Kanji, Z.A., Kulkarni, G., Levin, E.J.T., McCluskey, C.S., Murakami, M., Murray, B.J., Niedermeier, D., Petters, M.D., O'Sullivan, D., Saito, A., Schill, G.P., Tajiri, T., Tolbert, M.A., Welti, A., Whale, T.F., Wright, T.P., Yamashita, K., 2015. A comprehensive laboratory study on the immersion freezing behavior of illite NX particles: a comparison of 17 ice nucleation measurement techniques. Atmos. Chem. Phys. 15, 2489–2518.

Hoose, C., Kristjánsson, J., Burrows, S., 2010a. How important is biological ice nucleation in clouds on a global scale? Environ. Res. Lett. 5, 024009.

Hoose, C., Kristjánsson, J.E., Chen, J.-P., Hazra, A., 2010b. A classical-theory-based parameterization of heterogeneous ice nucleation by mineral dust, soot, and biological particles in a global climate model. J. Atmos. Sci. 67, 2483–2503.

Hoose, C., Lohmann, U., Erdin, R., Tegen, I., 2008. The global influence of dust mineralogical composition on heterogeneous ice nucleation in mixed-phase clouds. Environ. Res. Lett. 3, 025003.

Hoose, C., Möhler, O., 2012. Heterogeneous ice nucleation on atmospheric aerosols: a review of results from laboratory experiments. Atmos. Chem. Phys. 12, 9817–9854.

Hu, X.L., Michaelides, A., 2007. Ice formation on kaolinite: lattice match or amphoterism? Surf. Sci. 601, 5378–5381.

Kanitz, T., Seifert, P., Ansmann, A., Engelmann, R., Althausen, D., Casiccia, C., Rohwer, E.G., 2011. Contrasting the impact of aerosols at northern and southern midlatitudes on heterogeneous ice formation. Geophys. Res. Lett. 38, L17802.

Kiselev, A., Bachmann, F., Pedevilla, P., Cox, S.J., Michaelides, A., Gerthsen, D., Leisner, T., 2017. Active sites in heterogeneous ice nucleation—the example of K-rich feldspars. Science 355, 367–371.

Knopf, D.A., Alpert, P.A., Wang, B., Aller, J.Y., 2011. Stimulation of ice nucleation by marine diatoms. Nat. Geosci. 4, 88–90.

Koop, T., Murray, B.J., 2016. A physically constrained classical description of the homogeneous nucleation of ice in water. J. Chem. Phys. 145, 211915.

Kovács, T., Meldrum, F.C., Christenson, H.K., 2012. Crystal nucleation without supersaturation. J. Phys. Chem. Lett. 3, 1602–1606.

Krämer, B., Hübner, O., Vortisch, H., Wöste, L., Leisner, T., Schwell, M., Rühl, E., Baumgärtel, H., 1999. Homogeneous nucleation rates of supercooled water measured in single levitated microdroplets. J. Chem. Phys. 111, 6521–6527.

Kumai, M., 1961. Snow crystals and the identification of the nuclei in the northern United States of America. J. Meteorol. 18, 139–150.

Lee Jr., R., Warren, G.J., Gusta, L.V., 1995. Biological Ice Nucleation and Its Applications. American Phytopathological Society, St. Paul, MN.

Levine, J., 1950. CASE FiLE, 1950.

Li, K., Xu, S., Shi, W., He, M., Li, H., Li, S., Zhou, X., Wang, J., Song, Y., 2012. Investigating the effects of solid surfaces on ice nucleation. Langmuir 28, 10749–10754.

Lindow, S.E., Arny, D.C., Upper, C.D., 1982. Bacterial ice nucleation: a factor in frost injury to plants. Plant Physiol. 70, 1084–1089.

Lindow, S.E., Lahue, E., Govindarajan, A.G., Panopoulos, N.J., Gies, D., 1989. Localization of ice nucleation activity and the iceC gene product in Pseudomonas syringae and Escherichia coli. Mol. Plant-Microbe Interact. 2, 262–272.

Lohmann, U., Feichter, J., 2005. Global indirect aerosol effects: a review. Atmos. Chem. Phys. 5, 715–737.

Lüönd, F., Stetzer, O., Welti, A., Lohmann, U., 2010. Experimental study on the ice nucleation ability of size-selected kaolinite particles in the immersion mode. J. Geophys. Res. 115, D14201.

Lupi, L., Molinero, V., 2014. Does hydrophilicity of carbon particles improve their ice nucleation ability? J. Phys. Chem. A 118, 7330–7337.

Maki, L.R., Galyan, E.L., Chang-Chien, M.-M., Caldwell, D.R., 1974. Ice nucleation induced by Pseudomonas syringae. Appl. Microbiol. 28, 456–459.

Marcolli, C., 2014. Deposition nucleation viewed as homogeneous or immersion freezing in pores and cavities. Atmos. Chem. Phys. 14, 2071–2104.

Marcolli, C., Gedamke, S., Peter, T., Zobrist, B., 2007. Efficiency of immersion mode ice nucleation on surrogates of mineral dust. Atmos. Chem. Phys. 7, 5081–5091.

Mason, R.H., Si, M., Chou, C., Irish, V.E., Dickie, R., Elizondo, P., Wong, R., Brintnell, M., Elsasser, M., Lassar, W.M., Pierce, K.M., Leaitch, W.R., MacDonald, A.M., Platt, A., Toom-Sauntry, D., Sarda-Estève, R., Schiller, C.L., Suski, K.J., Hill, T.C.J., Abbatt, J.P.D., Huffman, J.A., DeMott, P.J., Bertram, A.K., 2016. Size-resolved measurements of ice-nucleating particles at six locations in North America and one in Europe. Atmos. Chem. Phys. 16, 1637–1651.

Michelmore, R.W., Franks, F., 1982. Nucleation rates of ice in undercooled water and aqueous solutions of polyethylene glycol. Cryobiology 19, 163–171.

Mullin, J.W., 2001. Crystallization. Elsevier, Oxford, UK.

Murphy, D.M., Koop, T., 2005. Review of the vapour pressures of ice and supercooled water for atmospheric applications. Q. J. R. Meteorol. Soc. 131, 1539–1565.

Murray, B.J., Broadley, S.L., Wilson, T.W., Atkinson, J.D., Wills, R.H., 2011. Heterogeneous freezing of water droplets containing kaolinite particles. Atmos. Chem. Phys. 11, 4191–4207.

Murray, B.J., Broadley, S.L., Wilson, T.W., Bull, S.J., Wills, R.H., Christenson, H.K., Murray, E.J., 2010. Kinetics of the homogeneous freezing of water. Phys. Chem. Chem. Phys. 12, 10380–10387.

Murray, B.J., O'Sullivan, D., Atkinson, J.D., Webb, M.E., 2012. Ice nucleation by particles immersed in supercooled cloud droplets. Chem. Soc. Rev. 41, 6519–6554.

Niedermeier, D., Ervens, B., Clauss, T., Voigtländer, J., Wex, H., Hartmann, S., Stratmann, F., 2014. A computationally efficient description of heterogeneous freezing: a simplified version of the Soccer ball model. Geophys. Res. Lett. 41, 736–741.

Niedermeier, D., Shaw, R., Hartmann, S., Wex, H., Clauss, T., Voigtländer, J., Stratmann, F., 2011. Heterogeneous ice nucleation: exploring the transition from stochastic to singular freezing behavior. Atmos. Chem. Phys. 11, 8767–8775.

Niedermeier, D., Augustin-Bauditz, S., Hartmann, S., Wex, H., Ignatius, K., Stratmann, F., 2015. Can we define an asymptotic value for the ice active surface site density for heterogeneous ice nucleation? J. Geophys. Res. Atmos. 120, 5036–5046.

Niemand, M., Möhler, O., Vogel, B., Vogel, H., Hoose, C., Connolly, P., Klein, H., Bingemer, H., DeMott, P.J., Skrotzki, J., Leisner, T., 2012. A particle-surface-area-based parameterization of immersion freezing on desert dust particles. J. Atmos. Sci. 69.

Ogawa, S., Koga, M., Osanai, S., 2009. Anomalous ice nucleation behavior in aqueous polyvinyl alcohol solutions. Chem. Phys. Lett. 480, 86–89.

O'Sullivan, D., Murray, B.J., Malkin, T.L., Whale, T.F., Umo, N.S., Atkinson, J.D., Price, H.C., Baustian, K.J., Browse, J., Webb, M.E., 2014. Ice nucleation by fertile soil dusts: relative importance of mineral and biogenic components. Atmos. Chem. Phys. 14, 1853–1867.

O'Sullivan, D., Murray, B.J., Ross, J., Webb, M.E., 2016. The adsorption of fungal ice-nucleating proteins on mineral dusts: a terrestrial reservoir of atmospheric ice-nucleating particles. Atmos. Chem. Phys. Discuss. 2016, 1–22.

O'Sullivan, D., Murray, B.J., Ross, J.F., Whale, T.F., Price, H.C., Atkinson, J.D., Umo, N.S., Webb, M.E., 2015. The relevance of nanoscale biological fragments for ice nucleation in clouds. Sci Rep 5.

Passarelli, R.E., Chessin, H., Vonnegut, B., 1973. Ice nucleation by solid solutions of silver-copper iodide. Science 181, 549–551.

Petters, M., Kreidenweis, S., 2007. A single parameter representation of hygroscopic growth and cloud condensation nucleus activity. Atmos. Chem. Phys. 7, 1961–1971.

Petters, M.D., Parsons, M.T., Prenni, A.J., DeMott, P.J., Kreidenweis, S.M., Carrico, C.M., Sullivan, A.P., McMeeking, G.R., Levin, E., Wold, C.E., Collett, J.L., Moosmüller, H., 2009. Ice nuclei emissions from biomass burning. J. Geophys. Res.-Atmos. 114, D07209.

Phillips, V.T.J., Andronache, C., Christner, B., Morris, C.E., Sands, D.C., Bansemer, A., Lauer, A., McNaughton, C., Seman, C., 2009. Potential impacts from biological aerosols on ensembles of continental clouds simulated numerically. Biogeosciences 6, 987–1014.

Phillips, V., Choularton, T., Illingworth, A., Hogan, R., Field, P., 2003. Simulations of the glaciation of a frontal mixed-phase cloud with the explicit microphysics model. Q. J. R. Meteorol. Soc. 129, 1351–1371.

Pinti, V., Marcolli, C., Zobrist, B., Hoyle, C.R., Peter, T., 2012. Ice nucleation efficiency of clay minerals in the immersion mode. Atmos. Chem. Phys. Discuss. 12, 3213–3261.

Pitter, R.L., Pruppacher, H.R., 1973. Wind-tunnel investigation of freezing of small water drops falling at terminal velocity in air. Q. J. R. Meteorol. Soc. 99, 540–550.

Pouleur, S., Richard, C., Martin, J.-G., Antoun, H., 1992. Ice nucleation activity in *Fusarium acuminatum* and *Fusarium avenaceum*. Appl. Environ. Microbiol. 58, 2960–2964.

Pratt, K.A., DeMott, P.J., French, J.R., Wang, Z., Westphal, D.L., Heymsfield, A.J., Twohy, C.H., Prenni, A.J., Prather, K.A., 2009. In situ detection of biological particles in cloud ice-crystals. Nat. Geosci. 2, 398–401.

Prospero, J.M., Ginoux, P., Torres, O., Nicholson, S.E., Gill, T.E., 2002. Environmental characterization of global sources of atmospheric soil dust identified with the NIMBUS 7 Total Ozone Mapping Spectrometer (TOMS) absorbing aerosol product. Rev. Geophys. 40, 1002.

Pruppacher, H.R., Klett, J.D., 1997. Microphysics of Clouds and Precipitation. Kluwer Academic Publishers, Dordrecht, The Netherlands.

Pummer, B.G., Bauer, H., Bernardi, J., Bleicher, S., Grothe, H., 2012. Suspendable macromolecules are responsible for ice nucleation activity of birch and conifer pollen. Atmos. Chem. Phys. 12, 2541–2550.

Pummer, B.G., Budke, C., Augustin-Bauditz, S., Niedermeier, D., Felgitsch, L., Kampf, C.J., Huber, R.G., Liedl, K.R., Loerting, T., Moschen, T., Schauperl, M., Tollinger, M., Morris, C.E., Wex, H., Grothe, H., Pöschl, U., Koop, T., Fröhlich-Nowoisky, J., 2015. Ice nucleation by water-soluble macromolecules. Atmos. Chem. Phys. 15, 4077–4091.

Reinhardt, A., Doye, J.P.K., 2014. Effects of surface interactions on heterogeneous ice nucleation for a monatomic water model. J. Chem. Phys. 141, 084501.

Riechers, B., Wittbracht, F., Hütten, A., Koop, T., 2013. The homogeneous ice nucleation rate of water droplets produced in a microfluidic device and the role of temperature uncertainty. Phys. Chem. Chem. Phys. 15, 5873–5887.

Rogers, D.C., 1988. Development of a continuous flow thermal gradient diffusion chamber for ice nucleation studies. Atmos. Res. 22, 149–181.

Rosinski, J., Haagenson, P.L., Nagamoto, C.T., Parungo, F., 1987. Nature of ice-forming nuclei in marine air masses. J. Aerosol Sci. 18, 291–309.

Rosinski, J., Nagamoto, C.T., Zhou, M.Y., 1995. Ice-forming nuclei over the East China Sea. Atmos. Res. 36, 95–105.

Sassen, K., Dodd, G.C., 1988. Homogeneous nucleation rate for highly supercooled cirrus cloud droplets. J. Atmos. Sci. 45, 1357–1369.

Schnell, R.C., 1975. Ice nuclei produced by laboratory cultured marine phytoplankton. Geophys. Res. Lett. 2, 500–502.

Schnell, R.C., Vali, G., 1972. Atmospheric ice nuclei from decomposing vegetation. Nature 236, 163–165.

Sear, R.P., 2012. The non-classical nucleation of crystals: microscopic mechanisms and applications to molecular crystals, ice and calcium carbonate. Int. Mater. Rev. 57, 328–356.

Sear, R.P., 2014. Quantitative studies of crystal nucleation at constant supersaturation: experimental data and models. CrystEngComm 16, 6506–6522.

Stan, C.A., Schneider, G.F., Shevkoplyas, S.S., Hashimoto, M., Ibanescu, M., Wiley, B.J., Whitesides, G.M., 2009. A microfluidic apparatus for the study of ice nucleation in supercooled water drops. Lab Chip 9, 2293–2305.

Stetzer, O., Baschek, B., Lüönd, F., Lohmann, U., 2008. The Zurich ice nucleation chamber (ZINC)-a new instrument to investigate atmospheric ice formation. Aerosol Sci. Technol. 42, 64–74.

Stopelli, E., Conen, F., Morris, C.E., Herrmann, E., Henne, S., Steinbacher, M., Alewell, C., 2016. Predicting abundance and variability of ice nucleating particles in precipitation at the high-altitude observatory Jungfraujoch. Atmos. Chem. Phys. 16, 8341–8351.

Turnbull, D., Vonnegut, B., 1952. Nucleation catalysis. Ind. Eng. Chem. 44, 1292–1298.

Vali, G., 1985. Nucleation terminology. J. Aerosol Sci. 16, 575–576.

Vali, G., 1994. Freezing rate due to heterogeneous nucleation. J. Atmos. Sci. 51, 1843–1856.

Vali, G., 1995. Principles of ice nucleation. In: Lee Jr., R., Warren, G.J., Gusta, L.V. (Eds.), Biological Ice Nucleation and Its Applications. American Phytopathological Society, St. Paul, MN.

Vali, G., 2008. Repeatability and randomness in heterogeneous freezing nucleation. Atmos. Chem. Phys. 8, 5017–5031.

Vali, G., 2014. Interpretation of freezing nucleation experiments: singular and stochastic; sites and surfaces. Atmos. Chem. Phys. 14, 5271–5294.

Vali, G., DeMott, P., Möhler, O., Whale, T., 2014. Ice nucleation terminology. Atmos. Chem. Phys. Discuss. 14, 22155–22162.

Vali, G., DeMott, P.J., Möhler, O., Whale, T.F., 2015. Technical note: a proposal for ice nucleation terminology. Atmos. Chem. Phys. 15, 10263–10270.

Vali, G., Stansbury, E.J., 1966. Time-dependent characteristics of heterogeneous nucleation of ice. Can. J. Phys. 44, 477.

Vergara Temprado, J., Wilson, T.W., O'Sullivan, D., Browse, J., Pringle, K.J., Ardon-Dryer, K., Bertram, A.K., Burrows, S.M., Ceburnis, D., DeMott, P.J., Mason, R.H., O'Dowd, C.D., Rinaldi, M., Murray, B.J., Carslaw, K.S., 2016. Contribution of feldspar and marine organic aerosols to global ice nucleating particle concentrations. Atmos. Chem. Phys. Discuss. 2016, 1–37.

Vonnegut, B., 1947. The nucleation of ice formation by silver iodide. J. Appl. Phys. 18, 593–595.

Vonnegut, B., Chessin, H., 1971. Ice nucleation by coprecipitated silver iodide and silver bromide. Science 174, 945–946.

Wex, H., Augustin-Bauditz, S., Boose, Y., Budke, C., Curtius, J., Diehl, K., Dreyer, A., Frank, F., Hartmann, S., Hiranuma, N., Jantsch, E., Kanji, Z.A., Kiselev, A., Koop, T., Möhler, O., Niedermeier, D., Nillius, B., Rösch, M., Rose, D., Schmidt, C., Steinke, I., Stratmann, F., 2015. Intercomparing different devices for the investigation of ice nucleating particles using Snomax® as test substance. Atmos. Chem. Phys. 15, 1463–1485.

Wilson, T.W., Ladino, L.A., Alpert, P.A., Breckels, M.N., Brooks, I.M., Browse, J., Burrows, S.M., Carslaw, K.S., Huffman, J.A., Judd, C., Kilthau, W.P., Mason, R.H., McFiggans, G., Miller, L.A., Najera, J.J., Polishchuk, E., Rae, S., Schiller, C.L., Si, M., Temprado, J.V., Whale, T.F., Wong, J.P.S., Wurl, O., Yakobi-Hancock, J.D., Abbatt, J.P.D., Aller, J.Y., Bertram, A.K., Knopf, D.A., Murray, B.J., 2015. A marine biogenic source of atmospheric ice-nucleating particles. Nature 525, 234–238.

Yano, J.-I., Phillips, V.T.J., 2011. Ice–ice collisions: an ice multiplication process in atmospheric clouds. J. Atmos. Sci. 68, 322–333.

Zettlemoyer, A.C., Tcheurekdjian, N., Chessick, J.J., 1961. Surface properties of silver iodide. Nature 192, 653.

Zhang, X.-X., Chen, M., Fu, M., 2014. Impact of surface nanostructure on ice nucleation. J. Chem. Phys. 141, 124709.

Zielke, S.A., Bertram, A.K., Patey, G.N., 2015. A molecular mechanism of ice nucleation on model AgI surfaces. J. Phys. Chem. B 119, 9049–9055.

Zolles, T., Burkart, J., Häusler, T., Pummer, B., Hitzenberger, R., Grothe, H., 2015. Identification of ice nucleation active sites on feldspar dust particles. J. Phys. Chem. A 119, 2692–2700.

CHAPTER 3

Detection of Mixed-Phase Clouds From Shortwave and Thermal Infrared Satellite Observations

Yoo-Jeong Noh, Steven D. Miller
Colorado State University, Fort Collins, CO, United States

Contents

1. INTRODUCTION

Clouds may contain liquid water droplets, ice particles, or a mixture of both below freezing down to -40°C (Rauber and Tokay, 1991; Cober et al., 2001). The latter, referred to as mixed-phase clouds, occur commonly in the Earth's atmosphere (e.g., Pinto, 1998; Verlinde et al., 2007; Wang et al., 2004), since water droplets often exist in the form of supercooled liquid below the freezing point. In situ measurements during field campaigns at middle and high latitudes (e.g., Fleishauer et al., 2002; Hogan et al., 2003; Zuidema et al., 2005; Niu et al., 2008; Carey et al., 2008; Shupe et al., 2008a,b; Noh et al., 2011, 2013) reveal that a large number of mid-level non-precipitating clouds contain predominately supercooled liquid droplets and a small number of ice crystals at or near cloud top, and precipitating ice virga in the middle and lower portions of the cloud. Fig. 1 shows examples from aircraft measurements of such mixed-phase clouds as observed during the Canadian CloudSat/CALIPSO Validation Programme and CLEX-10 joint field campaign (see Noh et al., 2011, 2013 for more details).

Aircraft flying through mixed-phase clouds with the presence of large supercooled liquid droplets can cause rapid ice accretion of ice (icing) on the wings and frames

Mixed-Phase Clouds
https://doi.org/10.1016/B978-0-12-810549-8.00003-9

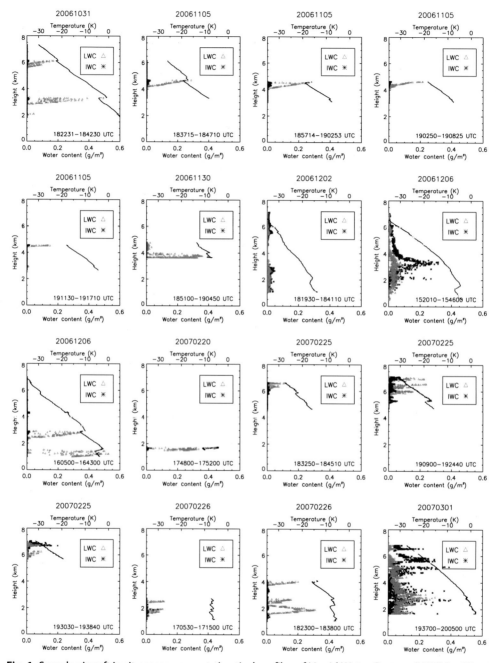

Fig. 1 Sample aircraft in situ measurements (vertical profiles of Liquid Water Content (LWC)/Ice Water Content (IWC) and temperatures) for mixed-phase clouds collected during the C3VP/CLEX-10 field campaign (Noh et al., 2013).

(Smith et al., 2012), posing a direct and serious in-flight hazard. Hence, a detailed understanding of the characteristics and microphysical properties of mixed-phase clouds is essential to improving aviation safety and reducing risk to both military operations and the civilian aviation community alike. An improved understanding of mixed-phase cloud morphology is also important for the parameterization of these ubiquitous clouds in climate and weather prediction models (e.g., Randall et al., 2007). These clouds have significant impacts on the atmospheric radiative heating profile, which feeds back to circulations occurring on both weather and climate spatial/temporal scales. In turn, changes to the local radiative heating profiles affects cloud formation and the overall radiation budget of Earth (Larson et al., 2006; Fleishauer et al., 2002).

The impact of these mixed-phase clouds is non-uniformly distributed. Hu et al. (2010) show large amounts of supercooled liquid water clouds at high latitudes, especially over the relatively pristine maritime air mass of the Southern Ocean. The Southern Ocean has the highest observed mixed-phase cloud fraction in the world (Mace et al., 2007). Trenberth and Fasullo (2010) suggest that biases in top-of-atmosphere net radiative forcing simulations in the Southern Ocean are tied to problems in cloud representation. Recent studies indicate that 40%–60% of clouds in the temperature range between 0°C and −30°C are mixed-phase and 30%–60% are supercooled liquid water clouds (e.g., Korolev et al., 2003; Mazin, 2006; Shupe et al., 2006; Zhang et al., 2010). The longevity and areal extent of these supercooled-liquid and mixed-phase clouds have a significant impact on the radiative balance (Sun and Shine, 1995; DeMott et al., 2010).

Despite their recognized importance, the properties of mixed-phase clouds are relatively unknown and remain a source of uncertainty in retrievals from satellites and numerical weather/climate models (e.g., Sun and Shine, 1994; Fowler et al., 1996; Beesley and Moritz, 1999; Harrington et al., 1999; Klein et al., 2009). Many numerical models and satellite retrieval algorithms still use simple approaches to partitioning cloud phases and microphysical distributions in temperature space, with an adjusting fraction of liquid/ice between freezing (0°C) and the homogeneous freezing point (near −40°C), with a dependence on droplet size, chemical composition, and ambient vertical velocity (Heymsfield and Miloshevich, 1993; Heymsfield et al., 2005; Swanson, 2009). Many numerical models use temperature limits (empirically based) to discriminate between liquid and ice, for instance, specifying thresholds of −23°C by Tiedtke (1993), −15°C by Smith (1990) and Boucher et al. (1995), and −9°C by Gregory and Morris (1996), as reviewed by Shupe et al. (2008a). Similar temperature thresholds are also found in various remote sensing retrieval algorithms, such as the CloudSat water content algorithm (Noh et al., 2011).

Overall, the physical mechanisms responsible for mixed-phase clouds and the frequency/scale/distribution of their occurrence globally are not well understood. Whereas in situ measurements can provide useful data points for these clouds, such observations are costly and spatially limited. Satellite remote-sensing based methods are best suited to the continuous monitoring and characterization of these mixed phase clouds. Active satellite sensor measurements such as CloudSat (Stephens et al., 2002) and CALIPSO (Winker et al., 2009) have provided a unique view of cloud vertical structures including mixed-phase clouds, but such sensors are applicable only to a very limited domain along their curtain observations. A strategy based on passive imaging radiometry would best leverage the satellite platform for observing the distribution of mixed-phase clouds and the further detailed structures such as liquid-top mixed-phase (hereafter, LTMP) clouds, globally. We propose one such approach in the discussion to follow that attempts to add a new dimension to current state of understanding.

2. CLOUD PHASE DETERMINATION FROM PASSIVE SATELLITE RADIOMETERS

Multispectral band measurements in the optical spectrum (ranging from 0.4 to 14 μm) from satellite-based radiometers have been used widely to globally determine cloud occurrence, classify cloud type, and retrieve cloud top height/pressure, integrated liquid/ice water content, cloud emissivity, and cloud top microphysics (e.g., Nakajima and King, 1990; Inoue and Ackerman, 2002; Platnick et al., 2003). These passive satellite radiometer observations also provide information about cloud top phase. Cloud phase information is often available via a combination of some infrared (IR) channels such as 8.5, 11, and 12 μm bands (Strabala et al., 1994; Baum et al., 2000; Pavolonis, 2010a). Due to the different spectral sensitivity to cloud phase (liquid vs. ice) around 8.5 μm and 11–13.5 μm, the difference in measured radiation (or brightness temperature) between an 8.5 μm channel and an 11 μm channel (or 12 μm or 13.3 μm channel) can be used for phase determination, accounting for the background conditions (e.g., surface temperature, surface emissivity, atmospheric temperature, and atmospheric water vapor) of a given cloudy scene (Pavolonis, 2010b). A key advantage of thermal-infrared-only techniques is the ability to apply the algorithms to both daytime and nighttime observations. Due to strong absorption of both liquid and ice water at these thermal infrared wavelengths (e.g., Hu and Stamnes, 1993), the phase information generally corresponds to cloud top conditions (visible optical thickness of ~1.0 into the cloud, or typically the first few hundred meters) (Pavolonis et al., 2005). Hence, the results of such retrievals are typically referred to as *cloud top properties*.

When using the thermal bands, thresholds used to determine cloud phase are relatively simple and based on in situ measurements. For example, Korolev et al. (2003) found that at temperatures below 238.0 K, the ice phase is dominant, but the relationship becomes more complicated for mixed-phase clouds. Examples from the MODerate-resolution Imaging Spectroradiometer (MODIS) (Platnick et al., 2003) IR cloud phase retrieval product (5 km × 5 km) of Collection-5 MYD06 Level-2 data (Menzel et al., 2010) are shown together with cloud top temperatures in Fig. 2. MODIS is a 36-channel

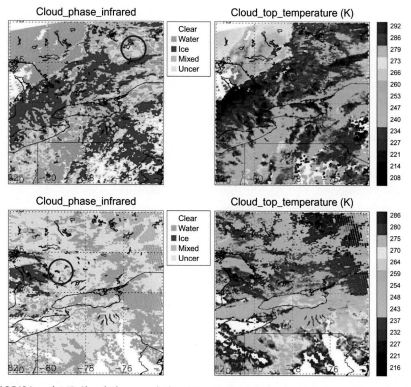

Fig. 2 MODIS Level-2 IR Cloud Phase and Cloud Top Temperature products (MYD06) for Oct. 31, 2006 and Nov. 5, 2006 with field experiment regions indicated by the *red circles* (where mixed-phase clouds were observed from the C3VP/CLEX-10 aircraft measurement).

scanning spectroradiometer with visible, near-infrared and infrared channels with a swath width of 2330 km (King et al., 2003). Brightness temperature differences between the 8.5 and 11 μm channels are compared with brightness temperatures from 11 μm to determine dominant cloud phase (liquid, ice, or mixed-phase), exploiting differences in absorption by liquid water and ice at these wavelengths. "Mixed-phase" cloud in the satellite retrieval usually means high probability of containing both liquid water and ice near cloud top (Pavolonis, 2010b). In Fig. 2, many cloudy pixels which have sub-freezing top temperatures are often classified as "uncertain." The red circles show where in situ aircraft measurement of mixed-phase clouds were collected (Noh et al., 2011). The aircraft data revealed a supercooled liquid water topped mixed-phase cloud structure (the in situ examples of these mixed phase clouds are shown in Fig. 1), and CALIPSO (not shown here) confirmed the presence of supercooled liquid water at cloud top.

For detection of supercooled liquid water clouds during the daytime, differential scattering/absorption properties between liquid and ice in the mid-wave infrared window (e.g., 3.9 μm) have been coupled with a measurement of thermal infrared window (e.g., 11.0 μm) (e.g., Ellrod, 1996; Lee et al., 1997; Ellrod and Bailey, 2007). The mid-wave IR reflectance of sunlight is higher for liquid cloud droplets, due to a higher imaginary-part of the complex index of refraction (proportional to absorption/emission) for ice at this wavelength. Thresholds imposed on this reflectance, determined conservatively from radiative transfer simulations of liquid and ice-topped clouds, are used to assign cloud top phase. Under the assumption of an optically thick cloud emitting as a blackbody in the thermal infrared band (reasonable for most liquid-phase clouds), the thermal infrared brightness temperature is a good approximation of the cloud top temperature. If the cloud top phase was determined as liquid based on the mid-wave infrared reflectance thresholds and the temperature is less than 0°C, a super-cooled liquid water classification is assigned. The classification is often referred to as mixed phase (both liquid and ice) due to uncertainties in the thresholds assumed.

3. DETERMINATION OF LIQUID-TOP MIXED-PHASE (LTMP)

Here we propose a daytime multispectral algorithm that attempts to profile and identify LTMP clouds from passive satellite radiometer observations. The previous techniques utilizing the mid-wave and thermal infrared bands (e.g., Ellrod, 1996; Lee et al., 1997; Ellrod and Bailey, 2007) would potentially classify some LTMP clouds simply as "supercooled liquid." The objective of the current algorithm (hereafter, referred to as the LTMP algorithm) is to enlist additional bands in the shortwave infrared (SIR) part of the spectrum that are capable of probing below cloud top (i.e., to deeper levels within the cloud) to identify a subset of these liquid topped clouds that may contain an ice-dominated phase below cloud top, akin to the structures shown in Fig. 1. A full description and physical basis for the LTMP algorithm is provided by Miller et al. (2014).

The LTMP algorithm makes use of reflected sunlight in narrow SIR bands at 1.6 and 2.25 μm to optically probe below liquid-topped clouds, using the unique phase-dependent behavior of these bands to infer the phase. Detection is basically predicated on differential absorption properties between liquid and ice particles for varying sun/sensor geometry and cloud optical properties. The algorithm is applied to the subset of clouds in the scene that were determined a priori, based on conventional passive radiometer phase-determination techniques described above, to be supercooled liquid top. It uses differential absorption features between liquid and ice in different atmospheric window bands in the shortwave-infrared (SIR; 1–3 μm where thermal emission signals from terrestrial and atmospheric sources are small), using reflectance measurements whose weighting functions peak below the cloud top. Comparing these measurements to those that would be expected for an entirely liquid-phase cloud (based on radiative transfer simulations), conservative thresholds are used to identify cases where a sub-cloud top ice/mixed phase layer is likely to be present.

The LTMP algorithm has been applied to spectral bands available on the Visible Infrared Imaging Radiometer Suite (VIIRS) on the Suomi National Polar-orbiting Partnership (S-NPP) and the Joint Polar Satellite System (JPSS) satellites and geostationary satellite sensors such as the Himawari-8 Advanced Himawari Imager (AHI) and the Geostationary Operational Environmental Satellite (GOES)-R series Advanced Baseline Imager (ABI; Schmit et al., 2005). A lidar/radar method (e.g., use of CALIPSO and CloudSat by Zhang et al. (2010)) would likely out-perform this simple passive radiometer algorithm, although the active sensor method is applicable only to a very limited domain along the curtain observation. Here we utilize the active sensor information as a useful tool to evaluate the passive technique.

3.1 Theoretical Basis

The LTMP algorithm is based on well established principles of atmospheric temperature and moisture soundings by passive satellite radiometers (e.g., Smith et al., 1972; Susskind et al., 2003). Conventional sounding techniques use measurements in spectral bands where the atmosphere absorbs/emits, as opposed to the clean window bands where the atmosphere is more transparent. Weighting functions, which are defined mathematically as the differential transmittance with height, describe the balance between gaseous species abundance (increasing toward the surface) and the optical path to space (increasing toward the top of the atmosphere) at a specific spectral band. For a well-mixed atmospheric gas (e.g., carbon dioxide), the weighting functions typically are bell shaped in the vertical, with the peak of the function indicating where the main measurement information comes from in the atmosphere.

The same principles of atmospheric sounding also apply to measurements in optically thick scattering/absorbing media such as clouds. In-cloud weighting function structures exist for SIR bands where clouds have non-negligible scattering characteristics. The sensitivity becomes broader in the vertical, penetrating to deeper levels within the optically thick clouds at shorter (and less absorbing) wavelengths. For example, Nakajima and King (1990) define a procedure for adjusting cloud effective radius retrievals from the Advanced Very High Resolution Radiometer (AVHRR) to arbitrary levels within the cloud to provide more representative comparisons against in situ observations (below cloud top). Miller et al. (2001) apply the same technique to compare cloud particle size retrievals from GOES-10 to instrumented aircraft observations in drizzling marine stratocumulus and similarly adjust 94 GHz radar-derived cloud particle sizes with GOES retrievals in tropical cirrus. Platnick (2000) investigates the effects of vertical photon transport in cloud microphysical retrievals for various droplet size profiles and shows strong sensitivity to the SIR observation band owing to the differential probing depths of the weighting functions. Chang and Li (2002, 2003) retrieve vertical profiles of cloud droplet effective radius using a combination of short and mid-wave IR bands.

Nakajima et al. (2010a,b) study profiles of droplet growth in warm water clouds using MODIS and CloudSat data. Nagao et al. (2013) use multi-spectral short and mid-wave IR observations to examine the vertical structures of liquid water clouds. Zhang (2013) applies a bi-spectral method using MODIS 1.6 and 2.1 μm band reflectance to study a warm-rain process associated with droplet size distribution near cloud bases. These studies assumed the cloud phase are homogeneous.

The ability to detect LTMP clouds during the daytime from passive satellite radiometers is predicated on the differential scattering/absorption properties of liquid and ice at specific bands. Fig. 3 shows a spectrum of single-scatter albedo (ω_o) for liquid and ice phases (based on Mie theory; spherical particles), showing marked differences between the spectral behaviors of ice and liquid in the SIR part of the spectrum. It is noted that ice is more absorbing than liquid at $\lambda = 1.6$ μm while liquid is more absorbing than ice at $\lambda = 2.25$ μm. Effective radii of 26 and 46 μm demonstrate how the differential scattering/absorption behavior is preserved and becomes more pronounced with increasing particle size.

While the spherical particle assumption is reasonable for the cloud top liquid layer, it may not be proper to capture the complex single-scatter behaviors of real ice particles. For the simulations to examine the spherical particle assumption for ice particles, we used the rough aggregate habit from Yang et al. (2000), which provides the scattering properties for a variety of ice crystal morphologies over a wide range of effective particle sizes

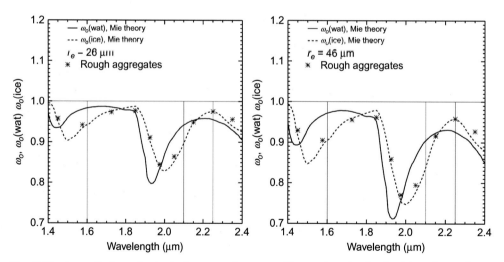

Fig. 3 Single scattering albedo (ω_o) across shortwave IR wavelengths for liquid and ice phases with effective radii of 26 μm (left) and 46 μm (right). Atmospheric window band centers for 1.6, 2.1, and 2.25 μm are shown. Note the reversal in scattering behaviors for water and ice between 1.6 and 2.25 μm. *(From Miller, S.D., Noh, Y.J., Heidinger, A.K., 2014. Liquid-top mixed-phase cloud detection from shortwave-infrared satellite radiometer observations: a physical basis. J. Geophys. Res. 119. https://doi.org/10.1002/2013JD021262.)*

across the visible to thermal infrared wavelengths. Values of spectral single-scatter albedo for these rough aggregates (asterisks at selected wavelengths) are included in Fig. 3. While some differences exist in comparison to Mie theory for ice spheres (dashed line), the comparisons exhibit an overall similar behavior in terms of both spectral and particle size variations. The general consistency and increasing strength of the signal with increasing particle size suggests that vertical variation of the lower ice layer properties will modulate but preserve reflectance-ratio signal exploited by the current algorithm. Since the ice layer is assumed to reside below an optically thick ($\tau > \sim 1.0$) scattering liquid-top layer, it will be a reasonable assumption that the global reflectance and transmittance are governed more by the single-scatter albedo and asymmetry parameter of the bulk cloud properties rather than the details of directional scattering. In other words, whereas single scatter properties (such as the phase function) may hold significant differences between various habits, the bulk optical properties in a multiple-scattering media are far less significant. For a variety of ice crystal habits and sizes considering vertical inhomogeneity, single scatter albedo ratio variation with the full complement of Yang et al. (2000)'s ice crystal habits were also compared with the spherical ice approximation (Mie theory). The majority, particularly the aggregate species that are most representative of habits found in nature (Baran et al., 2005), showed a similar bulk behavior to ice spheres (Miller et al., 2014).

Example in-cloud weighting functions for selected SIR bands (1.6, 2.1, and 2.25 μm) and the 3.9-μm mid-wave IR band are shown in Fig. 4 for different cloud particle sizes with a nadir-viewing angle and sun directly overhead. The weighting functions were calculated by perturbing the extinction coefficient at discrete levels within an idealized uniform cloud using a doubling/adding radiative transfer model (Miller et al., 2000). The SIR bands have significantly higher sensitivity to the inner portions of the cloud (thus deeper penetration below cloud top) in contrast to the mid-wave IR band (3.9 μm) reflectance, which is seen to be most sensitive to near cloud top. This cloud top sensitivity in the 3.9 μm weighting function peak is due to strong absorption of liquid water droplets. Even stronger absorption by liquid water in the thermal infrared bands (e.g., the "tri-spectral" 8.5, 10.35, and 12.3 μm thermal IR bands, which also are used for cloud top phase determination; Strabala et al., 1994; Pavolonis, 2010a) biases the determination of cloud phase to the upper-most portions of the cloud (~ 1.0 visible-wavelength optical thickness, Pavolonis et al., 2005). Hence, the shortwave IR bands provide unique insight on cloud internal structure, which the current algorithm attempts to exploit.

We examined the solar reflectance at two wavelengths (the 1.6 and 2.25 μm atmospheric window bands) whose weighting functions peak below cloud top (probing the cloud in a way that is analogous to how temperature/moisture sounding algorithms probe the clear sky atmosphere). As shown in Fig. 4, these SIR bands probe deeper into the cloud than the 3.9 μm band and provide a differential phase signal. To facilitate the assignment of a detection threshold considering the variability of pure liquid-phase

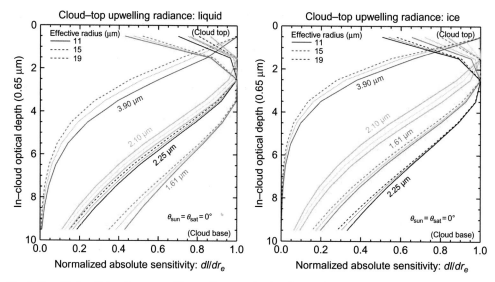

Fig. 4 Sensitivity of cloud top upwelling radiance to in-cloud perturbations at various levels (weighting function) for selected shortwave and mid-wave IR atmospheric window bands in liquid (left) and ice (right) clouds. For both clouds, the 1.6 and 2.25 μm sensitivities peak at optical thickness of ∼2–3 below cloud top. The reversal in water vs. ice cloud probing for these two bands is consistent with the single-scatter albedo properties shown in Fig. 3. *(From Miller, S.D., Noh, Y.J., Heidinger, A.K., 2014. Liquid-top mixed-phase cloud detection from shortwave-infrared satellite radiometer observations: a physical basis. J. Geophys. Res. 119. https://doi.org/10.1002/2013JD021262.)*

clouds, we compared the ratio of these measurements to the expected reflectance behavior of a pure-liquid cloud having the same total optical thickness and cloud top particle size as was retrieved for the observed cloud. If the difference between the observations and the pure-liquid baseline exceeded certain thresholds, selected conservatively, then the presence of an LTMP cloud can be inferred.

3.2 Algorithm Description

3.2.1 Construction of the Database

To simulate SIR reflectance measurements for idealized LTMP clouds, radiative transfer calculations for idealized clouds were conducted using the Santa Barbara DISORT Atmospheric Radiative Transfer (SBDART) model (Ricchiazzi et al., 1998). The radiative transfer model assumes plane-parallel cloud structure and spherical particles (Mie theory). Consideration for the effects of non-spherical ice is discussed in the sensitivity analyses to follow. Ocean/vegetation surfaces and rural/ocean aerosol models with mid-latitude winter and summer atmospheres were considered in the simulations, although it is recommended to apply the algorithm over water surfaces to minimize uncertainties.

As an approximation to the structures observed from in situ measurements shown in Fig. 1, an idealized cloud vertical structure is assumed: a simple two-layer stratified cloud with liquid upper layer and either liquid or ice phase in the lower layer. The upper (liquid) layer was placed between 5 and 5.5 km and the lower layer was placed between 3 and 5 km over the surface, based on aircraft measurements (Fleishauer et al., 2002; Carey et al., 2008; Noh et al., 2011).

Properties of the idealized clouds were varied systematically to construct look-up tables (LUT) of simulated measurements. The total cloud optical thickness (τ_{total}) was varied from 0 to 30, and the optical thickness of the top liquid layer (τ_{liquid}) component was varied from 0 to 30 with an increment of 1. To examine sensitivity to the top (liquid) layer, cloud drop effective radius of the top layer was specified as 6, 8, 10, 12, 15, and 20 μm with the lower (ice) layer held constant at 30 μm. Similarly, the effective particle radius of the lower ice layer was varied from 30, 50, 70, 100, and 120 μm with 8-μm radius for the top liquid layer. To further expand the solution space, we included several cases having the same effective radii for both top liquid and bottom ice layers, set at 10, 20, 30, and 40 μm. For each input condition, the monochromatic radiances (I_λ) were computed at 10 degrees resolution in sensor zenith angle (0–80 degrees), sun/sensor relative azimuth angle (0–170 degrees), and solar zenith angle (0–80 degrees). These values were converted to spectral reflectance (R_λ) via:

$$R_\lambda = \pi I_\lambda / (\mu_o F_{o,\lambda}) \qquad (1)$$

where μ_o is the cosine of the solar zenith angle and $F_{o,\lambda}$ is the solar spectral irradiance.

The simulated reflectance ratio LUT was built using the radiative transfer model for each input over a range of sun/satellite geometries, total optical thickness, cloud top effective radii, and surface conditions as described above. The final LUT contains over 43 million entries. The reflectance ratio between simulated 2.25 and 1.6 μm reflectance values is defined as:

$$\text{RR\_SIM} = {}^{\text{R\_sim}(2.25\,\mu\text{m})}\big/_{\text{R\_sim}(1.6\,\mu\text{m})}. \qquad (2)$$

For RR_SIM, the total optical thickness is partitioned between the upper (liquid) and lower (ice/mixed) layers for a range of values ($\tau = 0$–30), with the extreme cases being all liquid (lower layer $\tau = 0$) and all ice/mixed (top layer $\tau = 0$). By the differential absorption arguments described above, we would expect the reflectance ratio R(2.25)/R(1.6) to have a smaller denominator value when an ice/mixed lower-layer is encountered, contrary to a pure-liquid phase cloud. Thus, we would expect the same behavior to occur in our simulated reflectance ratios when considering an LTMP cloud (RR_SIM_Mixed) in comparison to a pure-liquid cloud (RR_SIM_Liquid). The RR_SIM_Liquid values are interpolated from the LUT where the total optical thickness is equal to the liquid optical thickness ($\tau_{\text{total}} = \tau_{\text{liquid}}$).

The comparisons between the pure–liquid and LTMP cloud simulations are done via the following liquid-normalized reflectance ratio.

$$\text{RR\_SIM\_COMP} = {}^{\text{RR\_SIM\_LTMP}}\!\big/_{\text{RR\_SIM\_Liquid}}. \tag{3}$$

RR_SIM_COMP exhibits higher sensitivity for larger total cloud optical thickness and a relatively optically thin liquid (top) layer. By definition, values of RR_SIM_COMP greater than 1 correspond to LTMP clouds. In the limiting case that the liquid-layer optical thickness equals the total cloud optical thickness, RR_SIM_COMP = 1.0. Per these simulations, an LTMP cloud will produce a very different (2.25 µm/1.6 µm) reflectance ratio than an all-liquid cloud. The depth of enhanced cloud top penetration by these two SIR bands depends on the optical thickness of the liquid layer near cloud top and the sun/sensor geometry.

RR_SIM_COMP values decrease with increasing cloud top (liquid) layer optical thickness. The physical basis for this decrease in the detection signal is straightforward; the weighting functions used for phase detection only penetrate so far below cloud top, such that cloud dominated by liquid phase at cloud top will produce a weaker LTMP signal. On the other hand, the detection signal is seen to increase with optically thinner cloud top liquid phase. This signal is maximized in the extreme case when the liquid-top component of the total optical depth approaches zero (i.e., the estimated signal that would be produced by a comparison between pure liquid and pure ice clouds).

For analysis of satellite SIR reflectance observations (R_sat), we define a liquid normalized reflectance ratio, RR_OBS_COMP, as:

$$\text{RR\_OBS\_COMP} = {}^{\text{RR\_OBS}}\!\big/_{\text{RR\_SIM\_Liquid}} \tag{4}$$

where, similar to Eq. (2),

$$\text{RR\_OBS} = {}^{\text{R\_sat}(2.25\,\mu m)}\!\big/_{\text{R\_sat}(1.6\,\mu m)} \tag{5}$$

and RR_SIM_Liquid is taken from the LUTs for the pure liquid case of ($\tau_{\text{total}} = \tau_{\text{liquid}}$), using the retrieved τ_{total} and cloud top effective particle size to interrogate the LUT. Normalization by RR_SIM_Liquid provides a reference point for the observations. When RR_OBS_COMP is found to exceed a specified threshold, RR_THRESH, the algorithm returns a positive detection for an LTMP cloud at that location.

3.2.2 Threshold Determination

We must specify the detection threshold (RR_THRESH) conservatively in order to avoid false alarms in LTMP detection, but not so conservative as to result in significant missed detections. As a simple first-order solution, the selection of a conservative threshold can be approached via statistical analysis of the LUT information. Here, we determined RR_THRESH by constructing cumulative distribution functions (CDFs) of the liquid-normalized reflectance ratios (RR_SIM_COMP) for all LTMP clouds in the LUT. The results indicate that 80% of all LTMP clouds were accounted for when RR_THRESH = 1.27, and 92% when RR_THRESH = 1.50. For LTMP clouds having optical thickness greater than 10 but the top (assumed liquid) layer optical thickness less than 5 (shallow liquid top) 70% of all cases are detected with RR_THRESH = 1.50, and 80% when RR_THRESH = 1.625.

A given RR_THRESH threshold may never be exceeded unless a minimum total cloud optical thickness, OT*, is present. Following Pavolonis et al. (2005), we assumed that any cloud that is classified by conventional mid-wave and thermal IR band techniques as being supercooled liquid-top will have a visible optical thickness of at least 1.0, and thus required OT* ≥ 1. The exact value of OT* changes as a function of the cloud top effective radius (via the asymmetry parameter which determines the degree of forward scatter within the medium). More information to determine the minimum total optical thickness can be found in Miller et al. (2014).

Considering all data points in the LUT, variations of OT* as a function of RR_THRESH were examined. In general, more conservative detection thresholds (i.e., higher RR_THRESH) require an optically thicker cloud to enable detection of LTMP structure. Also, relatively less absorbing and less forward-scattering media (e.g., smaller effective radii and lower values of asymmetry parameters) require optically thicker clouds to enable appreciable penetration of the cloud media and discriminate phase at these levels. Based on various sensitivity analyses, we decided a threshold RR_THRESH of 1.2 as a conservative provisional value for benchmark testing of this algorithm. In future development, a dynamic threshold based on CDFs of reflectance ratios may be worthy to explore for a wide sampling of all-liquid clouds, similar in concept to the Bayesian approach of Kummerow et al. (1996, 2001) and Noh et al. (2006). Variations on detection performance as a function of specifying more aggressive/conservative values for this threshold are considered in the case studies.

While conducting the radiative transfer simulations, liquid over drizzle cloud scenarios (e.g., drizzling marine stratocumulus) were also examined to determine whether enhanced absorption due to the presence of drizzle at lower levels of the cloud might be misconstrued for the presence of ice when applying our algorithm thresholds. The radiative transfer model calculations were performed with the same solar/sensor geometry information as before, but with the top liquid layer with effective radius set to 12 μm and bottom layers being all-liquid with each effective radii ranging from 12 to 120 μm (following Kogan and Kogan, 2001). As anticipated from the bulk scattering properties for water droplets at 1.61 and 2.25 μm (larger drizzle droplets have a lower single-scatter albedo at 2.25 μm than the smaller cloud droplets), the simulations confirmed that the presence of drizzle in fact further reduces the observable 2.25 μm/1.61 μm reflectance ratio below values of unity. It was found that over 98% liquid over drizzle cases were excluded when RR_THRESH was set to a value greater than 1.1, or less than 2% misclassified as LTMP in the detection algorithm due to the presence of drizzle.

3.2.3 Algorithm Flow

The LTMP detection algorithm operates on a pixel-by-pixel basis, following the procedure shown in Fig. 5. The first determination is whether the pixel has predominately liquid phase at cloud top and cloud top temperatures below 273 K. This determination is made via conventional passive-based cloud top phase methods mentioned in Section 2, but in principle could also be determined from an active sensor (e.g., CALIPSO or the future EarthCARE [ESA/JAXA, 2006] mission) or in situ observations including pilot reports along flight tracks for a localized cloud layer. In an operational setting, the standard Level-2 cloud products are used for this initial filtering step of the algorithm. Errors/uncertainties in this upstream assignment will be inherited by the current algorithm.

If the candidate pixel passes this initial filtering stage, it is then assessed for possible LTMP structure. Again, enlisting cloud property retrieval (Level-2) products provided from conventional satellite retrievals, the optical thickness and cloud top effective

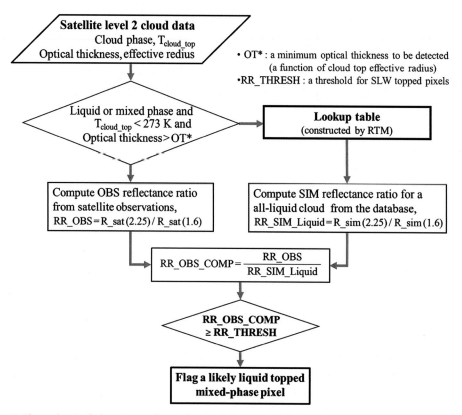

Fig. 5 Flow chart of the LTMP cloud detection algorithm. *(Modified from Miller, S.D., Noh, Y.J., Heidinger, A.K., 2014. Liquid-top mixed-phase cloud detection from shortwave-infrared satellite radiometer observations: a physical basis. J. Geophys. Res. 119. https://doi.org/10.1002/2013JD021262.)*

radius are extracted. The optical thickness for each pixel is then compared against the minimum optical thickness (OT*), prior to proceeding with the algorithm. For pixels satisfying the above criteria, we compute an observed reflectance ratio, RR_OBS (Eq. 5). Using the retrieved cloud optical thickness, cloud top effective radius, and sun/sensor geometry, we obtain from the LUTs, the simulated reflectance ratio for an all-liquid cloud (RR_SIM_Liquid, derived from the RR_SIM_Mix LUTs for the case of $\tau_{\_total} = \tau_{\_liquid}$). RR_OBS_COMP is computed following Eq. (4). If the value of RR_OBS_COMP is greater than or equal to our conservatively selected detection threshold, RR_THRESH, then the pixel is flagged as LTMP. This process is repeated until all cloudy pixels in the image have been evaluated.

4. APPLICATION TO CURRENT SATELLITE OBSERVATIONS

The idealized scenarios suggest a potential useful signal. However, it is important to evaluate the performance of the algorithm on real conditions. The VIIRS on board the S-NPP satellite (e.g., Lee et al., 2010) contains the requisite 1.61 and 2.25 μm SIR bands, which can be applied to the current algorithm. Geostationary satellite sensors such as Himawari-8 AHI and the GOES-R (now GOES-16) ABI also have the capability. Since no field campaigns providing suitable in situ cloud phase profile information co-located to those satellite overpasses were available for validation at the time of this algorithm development, we have utilized observations from a ground based observation station and other active sensor measurements from CloudSat and CALIPSO to examine the detection algorithm performance quantitatively.

4.1 S-NPP VIIRS Case Study

As a preliminary form of validation against actual observations, we considered a case study offering co-location of S-NPP VIIRS, CloudSat, and CALIPSO, all occurring near the U.S. Department of Energy (DOE) Atmospheric Radiation Measurement (ARM) Climate Research Facility on the North Slope of Alaska (NSA) at Barrow (71.323 N, 156.609 W; Verlinde et al., 2007). Ground-based observations from an upward pointing High Spectral Resolution Lidar (HSRL) with polarimetric resolution are also available at the site, which offers similar phase distinction as CALIPSO. Cloud phase discrimination from lidar data has been discussed in several studies (e.g., Sassen, 2005; Verlinde et al., 2007; Cho et al., 2008).

The satellite observations (S-NPP VIIRS, CloudSat, and CALIPSO) on Jun. 27, 2013 are shown in Fig. 6. The S-NPP and NASA A-Train ground tracks were within in-close alignment (within 1500 km, or roughly 3.5 min separation). The strong backscatter and sharp attenuation in CALIPSO is characteristic of liquid-topped clouds, and CloudSat detects the sub-cloud returns consistent with larger ice crystals typically shown in the LTMP structure. At the beginning of this co-location (2230 UTC Jun. 27, 2013

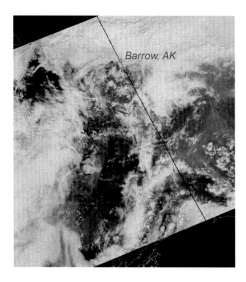
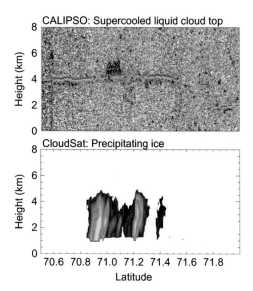

Fig. 6 Space and time co-located satellite observations over the ARM NSA area at 2230–2235 UTC on Jun. 27, 2013, (left) S-NPP VIIRS true color image with the CloudSat/CALIPSO overpass (line for the ascending node) and the nearby ARM NSA location (Barrow, AK), together with (right) CALIPSO 532-nm lidar total attenuated backscatter and CloudSat W-band radar reflectivity along the red-highlighted portion of ground track shown in the true color image. *(Modified from Miller, S.D., Noh, Y.J., Heidinger, A.K., 2014. Liquid-top mixed-phase cloud detection from shortwave-infrared satellite radiometer observations: a physical basis. J. Geophys. Res. 119. https://doi.org/10. 1002/2013JD021262.)*

through 0309 UTC Jun. 28, 2013), the satellite ground tracks passed in very close proximity to the AMR NSA site. This allowed a multi–observing system description of a complex cloud field which included supercooled and mixed-phase cloud tops, along with VIIRS data for the algorithm application. However, it should be noted that the availability of this alignment for polar-orbit satellites for the relevant meteorology condition does not occur very often. The time series of upward–pointing HSRL observations (not shown here) at the ground site during the satellite overpasses also showed the LTMP cloud structure. Combining this information with the CALIPSO observations, it was inferred that the liquid layer of the LTMP cloud was roughly 1 km thick, and precipitating ice was also observed.

We applied the LTMP detection algorithm to the VIIRS 1.61 μm (M10) and 2.25 μm (M11) SIR bands with input information about cloud type, optical depth, particle size retrieved from VIIRS standard Level-2 products, and the sun/sensor geometry information from the VIIRS geolocation data file. The algorithm was run on a variety of normalized reflectance ratio thresholds and cross-sections of these detection results were extracted along the CloudSat/CALIPSO ground track, as shown in Fig. 7. The pixels

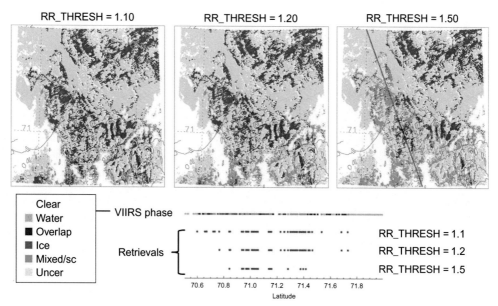

Fig. 7 Detection algorithm performance test with S-NPP VIIRS measurements for the Jun. 27, 2013 case. LTMP detection algorithm applied to the VIIRS supercooled/mixed phase pixels for various RR_THRESH as shown, with positive-detection LTMP clouds in *red*, supercooled liquid-only (negative detection) in *cyan*, and unconsidered clouds in *gray*. Shown below are cross sections of cloud phase and the corresponding detection results extracted along the CloudSat/CALIPSO ground track (*black line* in the upper-right panel). *(From Miller, S.D., Noh, Y.J., Heidinger, A.K., 2014. Liquid-top mixed-phase cloud detection from shortwave-infrared satellite radiometer observations: a physical basis. J. Geophys. Res. 119. https://doi.org/10.1002/2013JD021262.)*

with cloud top pristine ice phase were pre-screened in the algorithm. As shown in the figure, there were many areas where the algorithm flagged the presence of an LTMP structure, and at the more conservative (higher) reflectance ratio threshold of 1.5, the algorithm still retains significant patches of LTMP areas, including an area very close to the CloudSat/CALIPSO/ARM NSA matchup observations. Although this case study is not enough to represent a vigorous validation for the very complex LTMP scenes, the result suggested that the algorithm was performing as anticipated.

4.2 Himawari-8 AHI Case Study

The detection algorithm has been applied to a geostationary sensor, Himawari-8/9 AHI (e.g., Bessho et al., 2016; Miller et al., 2016) launched in Oct. 2014 and operated by the Japan Meteorological Agency. It is the first of the next generation of GEO satellites including the GOES-R (now GOES-16) carrying the ABI and the Geostationary Korea

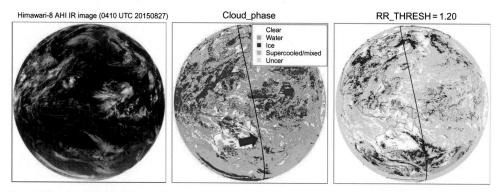

Fig. 8 Himawari-8 AHI IR image (10.35 μm), Cloud Phase, and LTMP detection result at the threshold value of 1.2 with the CloudSat/CALIPSO overpass (*black line* with a *red* colored target area in the middle panel for the evaluation) for the Aug. 27, 2015 case (0410 UTC).

Multipurpose Satellite 2A (Geo-KOMPSAT-2A) with the Advanced Meteorological Imager (AMI) operated by the Korea Meteorological Administration. Those sensors have very similar bands.

Fig. 8 shows Himawari-8 AHI IR image (10.35 μm), Cloud Phase, and LTMP detection result at the threshold value of 1.2 with the CloudSat/CALIPSO overpass (black line with a red colored target area in the middle panel for the evaluation) for the Aug. 27, 2015 case (0410 UTC). Since the operational Level-2 cloud products are not released in public yet, cloud retrieval products for the algorithm input were processed in the Clouds from Advanced Very High Resolution Radiometer (AVHRR) Extended (CLAVR-x) system with AHI Level-1 data, which is the NOAA operational cloud algorithm processing framework for the AVHRR on the NOAA—POES satellite and many other satellite sensors including GOES-R ABI (Walther and Heidinger, 2012).

Detection algorithm performance test results based on varying RR_THRESH values of 1.1, 1.2, and 1.5 are shown in Fig. 9, along with CloudSat (radar reflectivity) and CALIPSO (532 nm total attenuated backscatter and depolarization ratio) observations. The upper right panel shows cross-sections of cloud phase and the corresponding detection results extracted along the CloudSat/CALIPSO ground track (a red arrow indicated area in the previous figure). The subset of clouds determined to be all supercooled liquid water are shown in cyan colors, while clouds determined as possibly LTMP are shown in red. Vertical cross sections through the cloudy scenes over the eastern Australia region show positive detection of LTMP. The number of flagged pixels decreases as RR_THRESH increases. The over passing CloudSat and CALIPSO observations reveal the presence of LTMP clouds, which shows the current algorithm may be able to detect the presence of the below-cloud top ice phase. However, it should be noted that it cannot provide the exact information on whether that phase existed in a pristine state or is possibly mixed-phase. Performance based on these case studies offer only a hint into

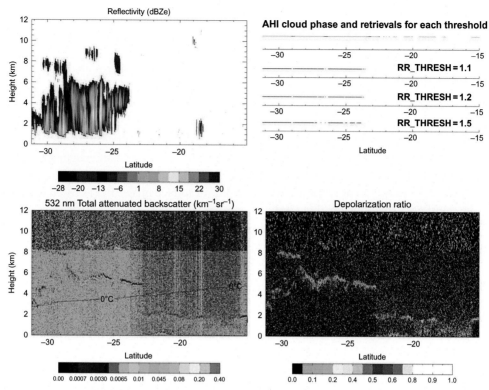

Fig. 9 Detection algorithm performance test results with CloudSat (radar reflectivity) and CALIPSO (532 nm total attenuated backscatter and depolarization ratio) observations. The upper-right panel shows cross sections of cloud phase and the corresponding detection results extracted along the CloudSat/CALIPSO ground track (a *red-arrow*-indicated area in the previous figure).

expected performance for various cloud scenes. A more rigorous validation of the current algorithm under a broader variety of conditions, and including in situ observations, will be required to help identify and characterize in a comprehensive way the actual limits of the algorithm.

5. DISCUSSION AND CONCLUSION

Multispectral band measurements (visible to infrared) from satellite radiometers have been widely used to determine cloud occurrence and retrieve cloud optical properties. These passive satellite radiometer observations offer information about cloud phase, utilizing differences in absorption by liquid water and ice at these wavelengths. Cloud phase is often determined via a combination of infrared channels but with sensitivity biased strongly toward cloud top. Proper assignment of mixed-phase clouds remains challenging with many uncertainties. Here, an algorithm targeting the daytime detection of clouds exhibiting liquid-top and mixed-phase or pristine ice below cloud top has been introduced.

The LTMP algorithm (Miller et al., 2014) utilizes additional bands in the SIR part of the spectrum that can probe the sub-cloud-phase below cloud top, particularly for LTMP clouds. It begins with upstream information about cloud mask, cloud top phase, cloud top particle size (effective radius), and cloud optical thickness. The algorithm takes advantage of differential optical properties of liquid and ice phase cloud particles using SIR bands whose weighting functions peak below cloud top and below levels of sensitivity for conventional cloud top phase discrimination techniques, thus enabling probing of the optically thick media. To develop the algorithm a simple two-layer cloud model was assumed, composed of variable fractions of liquid and ice phase. LUTs based on radiative transfer calculations for this idealized scenario span the practical solution space of cloud optical thickness, cloud top effective radius, and sun/sensor geometry. The departure of reflectance ratios between the observed cloud and an idealized all-liquid cloud having the same bulk properties (total cloud optical depth, cloud top effective particle size, under the assumption of similar sun/sensor geometry) as those retrieved is used to gauge the likelihood of the LTMP condition based on a conservatively selected and cloud-property-dependent threshold value. The outcomes of the LTMP algorithm are a flag for positive identification of LTMP, and as a by-product, an estimate of the liquid-top cloud optical thickness.

The algorithm holds promise for assisting researchers in inferring the distribution and nature of these clouds and there also exist practical applications for aviation communities. This algorithm was designed for the next generation GEO satellites such as Himawari-8/9 with AHI and GOES-R with ABI. The ABI and AHI have approximately 1.58–1.64 μm (1.61 μm nominal band center) and 2.225–2.275 μm (2.25 μm) narrow spectral bands, useful for the current approach. It is also applicable to selected polar-orbiting sensors such as VIIRS onboard S-NPP and JPSS satellites. While the S-NPP VIIRS and Himawari-8 AHI case studies are promising in terms of basic performance, we anticipate that some LTMP clouds may hold additional challenges to detection (e.g., optically thin cloud layers). Whereas positive detection of some LTMP clouds and correct exclusion of pure-liquid phase clouds was achieved, performance was tied heavily to the reflectance ratio threshold selected. Unfortunately, the algorithm cannot provide a definitive statement on whether the sub-cloud top layer is pure ice vs. mixed phase—the presence of sufficient ice phase within the sub-cloud top optical thickness region of SIR band sensitivity will trigger detection.

As with any algorithm developed on idealized assumptions, there exist a host of theoretical and practical limitations and several areas for potential improvement. Foremost among them, clouds in nature will not exhibit the pristine bifurcation of phase that is assumed in the simulations. The sub-cloud top layer may be a complex mixed-phase distribution with a variety of particle sizes and habits. The current technique is not equipped to make a detailed assessment, but the bulk absorption properties offer a chance of inferring the presence of this mixed phase when the observed reflectance ratio exceeds critical thresholds. Given its usage of solar reflectance information, the daytime-only nature of the algorithm precludes its ability to provide diurnal information and eliminates its utility during the polar winter where many of the LTMP clouds of interest may reside.

The plane-parallel assumptions of the simulated reflectance ratios may fail when confronted with sub-pixel heterogeneity, particularly over land surfaces where heterogeneity can be significant. Higher spatial resolution observations can mitigate these issues in part, but for GOES-R (GOES-16) and its 2 km IR pixels this can be an issue. Our plan when implementing the LTMP algorithm on GOES-R data is to examine the 0.5 km visible band as a way of assessing sub-pixel heterogeneity and potentially filtering these pixels from consideration. Being a passive measurement technique, the current algorithm is not applicable in cases of multi-layered clouds. Although the upstream cloud phase/type retrievals may provide information about multi-layer clouds (e.g., Pavolonis and Heidinger, 2004), the limitations of passive sensors to identify multi-layered clouds are well understood, and it is reported the portion of multi-layer (or overlap) flagged pixels is not significant (Noh et al., 2017). Finally, it should be noted that errors in the upstream cloud products used as input will be inherited by the current algorithm, which could lead to false detections of LTMP in the algorithm. The reflectance ratio signal will continue to increase with decreasing cloud top liquid phase, so applying our current algorithm to such clouds could yield detections of LTMP when in fact these clouds may not have a pure liquid-top phase. As such, proper quality control on the upstream cloud top phase retrieval would be an essential first step.

In future work, the algorithm will be tested and validated against more cases (such as LTMP clouds in the Southern Ocean) to improve quantitative uncertainty estimates and to refine dynamically-specified reflectance ratio detection thresholds. Although a threshold (RR_THRESH) on the reflectance ratio of 1.2 was found to have the best compromise between detection and avoidance of false alarms for the current analysis, for a given sun/viewing geometry and set of retrieved cloud optical properties, a critical threshold may be more effectively derived as opposed to a conservatively selected CDF-based value. Additional constraints, using various channel combinations and potentially the incorporation of lidar depolarization and backscatter intensity information (which provides phase discrimination; e.g., Hu et al., 2010) will be examined to delineate more clearly between liquid-top and mixed-phase top clouds. More robust validation against in situ observations will be essential to building confidence in the algorithm and improving detection thresholds. VIIRS and AHI (and ABI and AMI sensors) matchups with CALIPSO and/or the future EarthCARE (ESA/JAXA, 2006) active sensors, together with coordinated under-flights by instrumented aircraft (expected to coincide with satellite sensor overpass) or intensive ground measurements upon the availability will be examined.

REFERENCES

Baran, A.J., Shcherbakov, V.N., Baker, B.A., Gayet, J.F., Lawson, R.P., 2005. On the scattering phase function of non-symmetric ice-crystals. Q. J. R. Meteorol. Soc. 131, 2609–2616.

Baum, B.A., Soulen, P.F., Strabala, K.I., King, M.D., Ackerman, S.A., Menzel, W.P., et al., 2000. Remote sensing of cloud properties using MODIS Airborne Simulator imagery during SUCCESS. II. Cloud thermodynamic phase. J. Geophys. Res. 105, 11781–11792.

Beesley, J.A., Moritz, R.E., 1999. Toward an explanation of the annual cycle of cloudiness over the Arctic Ocean. J. Clim. 12, 395–415.

Bessho, K., et al., 2016. An introduction to Himawari-8/9—Japan's new-generation geostationary meteorological satellites. J. Meteor. Soc. Jpn. 94, 151–183. https://doi.org/10.2151/jmsj.2016-009.

Boucher, O., Le Treut, H., Baker, M.B., 1995. Precipitation and radiation modeling in a general circulation model: Introduction of cloud microphysical processes. J. Geophys. Res. 100, 16395–16414.

Carey, L.D., Niu, J., Yang, P., Kankiewicz, J.A., Larson, V.E., Vonder Haar, T.H., 2008. The vertical profile of liquid and ice water content in mid-latitude mixed-phase altocumulus clouds. J. Appl. Meteorol. Climatol. 47, 2487–2495.

Chang, F. L., Li, Z., 2002. Estimating the vertical variation of cloud droplet effective radius using multispectral near-infrared satellite measurements. J. Geophys. Res. 107(D15). https://doi.org/10.10292/2001JD000766.

Chang, F.-L., Li, Z., 2003. Retrieving vertical profiles of water-cloud droplet effective radius: algorithm modification and preliminary application. J. Geophys. Res. 108(D24). https://doi.org/10.1029/2003JD003906.

Cho, H.M., Yang, P., Kattawar, G.W., Nasiri, S.L., Hu, Y., Minnis, P., et al., 2008. Depolarization ratio and attenuated backscatter for nine cloud types: analyses based on collocated CALIPSO LIDAR and MODIS measurements. Opt. Express 16, 3931–3948.

Cober, S.G., Isaac, G.A., Korolev, A.V., Strapp, J.W., 2001. Assessing cloud-phase condition. J. Appl. Meteorol. 40, 1967–1983.

DeMott, P.J., Prenni, A.J., Liu, X., Kreidenweis, S.M., Petters, M.D., Twomey, C.H., et al., 2010. Predicting global atmospheric ice nuclei distributions and their impacts on climate. Proc. Natl. Acad. Sci. 107, 11217–11222.

Ellrod, G.P., 1996. The use of GOES-8 multispectral imagery for the detection of aircraft icing regions. Preprints, In: Eighth Conf. Satellite Meteorology and Oceanography, Atlanta, GA. American Meteorological Society, pp. 168–171.

Ellrod, G.P., Bailey, A.A., 2007. Assessment of aircraft icing potential and maximum icing altitude from geostationary meteorological satellite data. Weather Forecast. 22, 160–174.

ESA/JAXA, 2006. EarthCARE Mission Requirement Document Version 5, EC-RS-ESA-SY-012. Available at http://esamultimedia.esa.int/docs/EarthObservation/EarthCARE_MRD_v5.pdf.

Fleishauer, R.P., Larson, V.E., Vonder Haar, T.H., 2002. Observed microphysical structure of mid-level, mixed-phase clouds. J. Atmos. Sci. 59, 1779–1804.

Fowler, L.D., Randall, D.A., Rutledge, S.A., 1996. Liquid and ice cloud microphysics in the CSU general circulation model. Part I: model description and simulated microphysical processes. J. Clim. 9, 489–529.

Gregory, D., Morris, D., 1996. The sensitivity of climate simulation to the specification of mixed-phase cloud. Climate Dyn. 12, 641–651.

Harrington, J.Y., Reisin, T., Cotton, W.R., Kreidenweis, S.M., 1999. Cloud resolving simulations of Arctic stratus—Part II: transition-season clouds. J. Atmos. Res. 51, 45–75.

Heymsfield, A.J., Miloshevich, L.M., 1993. Homogeneous ice nucleation and supercooled liquid water in orographic wave clouds. J. Atmos. Sci. 50, 2335–2353.

Heymsfield, A.J., Miloshevich, L.M., Schmitt, C., Bansemer, A., Twohy, C., Poellot, M.R., et al., 2005. Homogeneous ice nucleation in subtropical convection and its influence on cirrus anvil microphysics. J. Atmos. Sci. 62, 41–64.

Hogan, A., Illingworth, J., O'Connor, E.J., Poiares Baptista, J.P.V., 2003. Characteristics of mixed-phase clouds. II: a climatology from ground-based lidar. Q. J. R. Meteor. Soc. 129, 2117–2134.

Hu, Y.-X., Stamnes, K., 1993. An accurate parameterization of the radiative properties of water clouds suitable for use in climate models. J. Clim. 6 (4), 728–742.

Hu, Y., Rodier, S., Xu, K., Sun, W., Huang, J., Lin, B., et al., 2010. Occurrence, liquid water content, and fraction of supercooled water clouds from combined CALIOP/IIR/MODIS measurements. J. Geophys. Res. 115, D00H34. https://doi.org/10.1029/2009JD012384.

Inoue, T., Ackerman, S.A., 2002. Radiative effects of various cloud types as classified by the split window technique over the eastern sub-tropical Pacific derived from collocated ERBE and AVHRR data. J. Meteor. Soc. Jpn. 80 (6), 1383–1394.

King, M.D., Menzel, W.P., Kaufman, Y.J., Tanré, D., Gao, B.-C., Platnick, S., et al., 2003. Cloud and aerosol properties, precipitable water and profiles of temperature and water vapor from MODIS. IEEE Trans. Geosci. Remote Sens. 41, 442–458.

Klein, S.A., et al., 2009. Intercomparison of model simulations of mixed-phase clouds observed during the ARM Mixed-Phase Arctic Cloud Experiment I: single layer cloud. Q. J. R. Meteorol. Soc. 135 (641). https://doi.org/10.1002/qj.416.

Kogan, Z.N., Kogan, Y.L., 2001. Parameterization of drop effective radius for drizzling marine stratocumulus. J. Geophys. Res. 106, 9757–9764.

Korolev, A.V., Isaac, G.A., Cober, S.G., Strapp, J.W., Hallett, J., 2003. Microphysical characterization of mixed-phase clouds. Q. J. R. Meteor. Soc. 129, 39–65.

Kummerow, C., Olson, W.S., Giglio, L., 1996. A simplified scheme for obtaining precipitation and vertical hydrometeor profiles from passive microwave sensors. IEEE Trans. Geosci. Remote Sens. 34, 1213–1232.

Kummerow, C., Hong, Y., Olson, W.S., Yang, S., Adler, R.F., McCollum, J., et al., 2001. The evolution of the Goddard Profiling Algorithm (GPROF) for rainfall estimation from passive microwave sensors. J. Appl. Meteorol. 40, 1801–1820.

Larson, V.E., Smith, A.J., Faulk, M.J., Kotenberg, K.E., Golaz, J.-C., 2006. What determines altocumulus dissipation time? J. Geophys. Res. 111, D19207. https://doi.org/10.1029/2005JD007002.

Lee, T.F., Turk, F.J., Richardson, K., 1997. Stratus and fog products using GOES-8–9 3.9-μm data. Weather Forecast. 12, 664–677.

Lee, T.F., et al., 2010. NPOESS: next-generation operational global Earth observations. Bull. Am. Meteorol. Soc. 91, 727–740.

Mace, G.G., Marchand, R., Stephens, G.L., 2007. Global hydrometeor occurrence as observed by CloudSat; initial observations from summer 2006. Geophys. Res. Lett. 34, L09808. https://doi.org/10.1029/2006GL029017.

Mazin, I.P., 2006. Cloud phase structure: experimental data analysis and parameterization. J. Atmos. Sci. 63, 667–681.

Menzel, W.P., Frey, R.A., Baum, B.A., 2010. Cloud top properties and cloud phase algorithm theoretical basis document, MODIS Algorithm theoretical basis document. 55 pp, Available online at http://modis-atmos.gsfc.nasa.gov/_docs/CTP_ATBD_oct10.pdf.

Miller, S.D., Stephens, G.L., Drummond, C.K., Heidinger, A.K., Partain, P.T., 2000. A multisensor diagnostic cloud property retrieval scheme. J. Geophys. Res. 105 (D15), 19955–19971.

Miller, S.D., Stephens, G.L., Austin, R.T., 2001. GOES 10 cloud optical property retrievals in the context of vertically varying microphysics. J. Geophys. Res. 106 (D16), 17891–17995.

Miller, S.D., Noh, Y.J., Heidinger, A.K., 2014. Liquid-top mixed-phase cloud detection from shortwave-infrared satellite radiometer observations: a physical basis. J. Geophys. Res. 119. https://doi.org/10.1002/2013JD021262.

Miller, S.D., Schmit, T.L., Seaman, C.J., Lindsey, D.T., Gunshor, M.M., Kohrs, R.A., et al., 2016. A sight for sore eyes: the return of true color to geostationary satellites. Bull. Am. Meteorol. Soc. 97 (10), 1803–1816.

Nagao, T.M., Suzuki, K., Nakajima, T.Y., 2013. Interpretation of multiwavelength-retrieved droplet effective radii for warm water clouds in terms of in-cloud vertical inhomogeneity by using a spectral bin microphysics cloud model. J. Atmos. Sci. 70, 2376–2392.

Nakajima, T., King, M.D., 1990. Determination of the optical thickness and effective particle radius of clouds from reflected solar radiation measurements. Part I: theory. J. Atmos. Sci. 47 (15), 1878–1893.

Nakajima, T.Y., Suzuki, K., Stephens, G.L., 2010a. Droplet growth in warm water clouds observed by the A-Train. Part I: Sensitivity analysis of the MODIS-derived cloud droplet sizes. J. Atmos. Sci. 67, 1884–1896.

Nakajima, T.Y., Suzuki, K., Stephens, G.L., 2010b. Droplet growth in warm water clouds observed by the A-Train. Part II: a multi-sensor view. J. Atmos. Sci. 67, 1897–1907.

Niu, J., Carey, L.D., Yang, P., Vonder Haar, T.H., 2008. Optical properties of a vertically inhomogeneous mid-latitude mid-level mixed-phase altocumulus in the infrared region. Atmos. Res. 88, 234–242.

Noh, Y.J., Liu, G., Seo, E.K., Wang, J.R., Aonashi, K., 2006. Development of a snowfall retrieval algorithm at high microwave frequencies. J. Geophys. Res. 111, D22216. https://doi.org/10.1029/2005JD006826.

Noh, Y.J., Seaman, C.J., Vonder Haar, T.H., Hudak, D.R., Rodriguez, P., 2011. Comparisons and analyses of wintertime mixed-phase clouds using satellite and aircraft observations. J. Geophys. Res. 116. https://doi.org/10.1029/2010JD015420.

Noh, Y.J., Seaman, C.J., Vonder Haar, T.H., Liu, G., 2013. In situ aircraft measurements of water content profiles in various midlatitude mixed-phase clouds. J. Appl. Meteorol. Climatol. D110202. https://doi.org/10.1175/JAMC-D-11-0202.1.

Noh, Y.J., Forsythe, J.M., Miller, S.D., Seaman, C.J., Li, Y., Heidinger, A.K., et al., 2017. Cloud base height estimation from VIIRS. Part II: a statistical algorithm based on A-Train satellite data. J. Atmos. Ocean. Technol. https://doi.org/10.1175/JTECH-D-16-0110.1.

Pavolonis, M.J., 2010a. Advances in extracting cloud composition information from spaceborne infrared radiances—a robust alternative to brightness temperatures. Part I: theory. J. Appl. Meteorol. Climatol. 49, 1992–2012.

Pavolonis, M.J., 2010b. GOES-R Advanced Baseline Imager (ABI) Algorithm Theoretical Basis Document For Cloud Type and Cloud Phase. Available at http://www.star.nesdis.noaa.gov/goesr/docs/ATBD/Cloud_Phase.pdf.

Pavolonis, M.J., Heidinger, A.K., 2004. Daytime cloud overlap detection from AVHRR and VIIRS. J. Appl. Meteorol. 43 (5), 762–778.

Pavolonis, M.J., Heidinger, A.K., Uttal, T., 2005. Daytime global cloud typing from AVHRR and VIIRS: algorithm description, validation, and comparisons. J. Appl. Meteorol. 44, 804–826.

Pinto, J.O., 1998. Autumnal mixed-phase cloudy boundary layers in the arctic. J. Atmos. Sci. 55, 2016–2037.

Platnick, S., 2000. Vertical photon transport in cloud remote sensing problems. J. Geophys. Res. 105, 22919–22935.

Platnick, S., King, M.D., Ackerman, S.A., Menzel, W.P., Baum, B.A., Riedi, J.C., et al., 2003. The MODIS cloud products: algorithms and examples from Terra. IEEE Trans. Geosci. Remote Sens. 41 (2), 459–473.

Randall, D.A., et al., 2007. Climate models and their evaluation. In: Solomon, S. et al., (Eds.), Climate Change 2007: The Physical Science Basis. Contribution of Working Group I to the Fourth Assessment Report of the Intergovernmental Panel on Climate Change. Cambridge University Press, Cambridge, UK; New York.

Rauber, R.M., Tokay, A., 1991. An explanation for the existence of supercooled water at the tops of cold clouds. J. Atmos. Sci. 48, 1005–1023.

Ricchiazzi, P., Yang, S., Gautier, C., Sowle, D., 1998. SBDART: a research and teaching software tool for plane parallel radiative transfer in the Earth's atmosphere. Bull. Am. Meteorol. Soc. 79, 2101–2114.

Sassen, K., 2005. Polarization in lidar. In: Weitkamp, C. (Ed.), Lidar: Range-Resolved Optical Remote Sensing of the Atmosphere. Springer Science and Business Media Inc., New York, NY, USA, 455 pp.

Schmit, T.J., Gunshor, M.M., Menzel, W.P., Gurka, J.J., Li, J., Bachmeier, A.S., 2005. Introducing the next generation advanced baseline imager on GOES-R. Bull. Am. Meteorol. Soc. 86, 1079–1096.

Shupe, M.D., Matrosov, S.Y., Uttal, T., 2006. Arctic mixed-phase cloud properties derived from surface-based sensors at SHEBA. J. Atmos. Sci. 63, 697–711.

Shupe, M.D., Daniel, J.S., De Boer, G., Eloranta, E.W., Kollias, P., Long, C.N., et al., 2008a. A focus on mixed-phase clouds: the status of ground-based observational methods. Bull. Am. Meteorol. Soc. 89, 1549–1562.

Shupe, M.D., Kollias, P., Ola, P., Persson, G., McFarquhar, G.M., 2008b. Vertical motions in arctic mixed-phase stratiform clouds. J. Atmos. Sci. 65, 1304–1322.

Smith, R.N.B., 1990. A scheme for predicting layer clouds and their water content in a general circulation model. Q. J. R. Meteorol. Soc. 116, 435–460.

Smith, W.L., Woolf, H.M., Fleming, H.E., 1972. Retrieval of atmospheric temperature profiles from satellite measurements for dynamical forecasting. J. Appl. Meteorol. 11, 113–122.

Smith Jr., W.L., Minnis, P., Fleeger, C., Spangenberg, D., Palikonda, R., Nguyen, L., 2012. Determining the flight icing threat to aircraft with single-layer cloud parameters derived from operational satellite data. J. Appl. Meteorol. Climatol. 51, 1794–1810. https://doi.org/10.1175/JAMC-D-12-057.1.

Stephens, G.K., et al., 2002. The CLOUDSAT Mission and the A-Train—a new dimension of space-based observations of clouds and precipitation. Bull. Am. Meteorol. Soc. 83, 1771–1790.

Strabala, K.I., Ackerman, S.A., Menzel, W.P., 1994. Cloud properties inferred from 8-12 μm data. J. Appl. Meteorol. 33, 212–229.

Sun, Z., Shine, K.P., 1994. Studies of radiative properties of ice and mixed phase clouds. Q. J. R. Meteorol. Soc. 120, 111–137.

Sun, Z., Shine, K.P., 1995. Parameterization of ice cloud radiative properties and its application to the potential climatic importance of mixed-phase clouds. J. Clim. 8, 1874–1888.

Susskind, J., Barnet, C.D., Blaisdell, J.M., 2003. Retrieval of atmospheric and surface parameters from AIRS/AMSU/HSB data in the presence of clouds. IEEE Trans. Geosci. Remote Sens. 41, 390–409.

Swanson, B.D., 2009. How well does water activity determine homogeneous ice nucleation temperature in aqueous sulfuric acid and ammonium sulfate droplets? J. Atmos. Sci. 66, 741–754.

Tiedtke, M., 1993. Representation of clouds in large-scale models. Mon. Weather Rev. 121, 3040–3061.

Trenberth, K.E., Fasullo, J.T., 2010. Simulation of present-day and twenty-first-century energy budgets of the Southern Oceans. J. Clim. 23, 440–454.

Verlinde, J., et al., 2007. The mixed-phase Arctic cloud experiment. Bull. Am. Meteorol. Soc. 88 (2), 205–221.

Walther, A., Heidinger, A.K., 2012. Implementation of the daytime cloud optical and microphysical properties algorithm (DCOMP) in PATMOS-x. J. Appl. Meteorol. Climatol. 51, 1371–1390.

Wang, Z., Sassen, K., Whiteman, D.N., Demoz, B.B., 2004. Studying altocumulus with ice virga using ground-based active and passive remote sensors. J. Appl. Meteorol. 43, 449–460.

Winker, D.M., Vaughan, M.A., Omar, A.H., Hu, Y., Powell, K.A., Liu, Z., et al., 2009. Overview of the CALIPSO Mission and CALIOP data processing algorithms. J. Atmos. Ocean. Technol. 26, 2310–2323. https://doi.org/10.1175/2009JTECHA1281.1.

Yang, P., Liou, K.N., Wyser, K., Mitchell, D., 2000. Parameterization of the scattering and absorption properties of individual ice crystals. J. Geophys. Res. 105 (D4), 4699–4718.

Zhang, Z., 2013. On the sensitivity of cloud effective radius retrieval based on spectral method to bi-modal droplet size distribution: a semi-analytical model. J. Quant. Spectrosc. Radiat. Transf. 129, 79–88.

Zhang, D., Wang, Z., Liu, D., 2010. A global view of mid-level liquid-layer topped stratiform cloud distribution and phase partition from CALIPSO and CloudSat measurements. J. Geophys. Res. 115D00H13 https://doi.org/10.1029/2009JD012143.

Zuidema, P., Baker, B., Han, Y., Intrieri, J., Key, J., Lawson, P., et al., 2005. An Arctic sprintime mixed-phase cloudy boundary layer observed during SHEBA. J. Atmos. Sci. 62, 160–176.

CHAPTER 4

Microphysical Properties of Convectively Forced Mixed-Phase Clouds

Robert Jackson*, Jeffrey French†, Joseph Finlon‡
*Argonne National Laboratory, Environmental Sciences Division, Lemont, IL, United States
†University of Wyoming, Laramie, WY, United States
‡University of Illinois at Urbana–Champaign, Urbana, IL, United States

Contents

1. CONVECTIVELY FORCED MIXED-PHASE CLOUDS

This chapter covers clouds that have formed via convection and grown to subfreezing temperatures where both supercooled water and ice can coexist in the same cloud. Although it is unknown exactly what percentage of convective clouds are mixed phase, 30%–60% of precipitating clouds in the midlatitudes and the poles have been found to be mixed phase (Mülmenstädt et al., 2015). Therefore, it is important to understand the growth and evolution of such clouds due to their ubiquity throughout the higher latitudes. They are also more complex to study than liquid convective clouds as there are many more microphysical pathways that lead to precipitation than in liquid–only clouds.

Mixed-Phase Clouds
https://doi.org/10.1016/B978-0-12-810549-8.00004-0

There are also more hydrometeor types, which can include graupel, hail, snow, and cloud ice. These added complexities make it more difficult to make quantitative precipitation forecasts, and complicate remote sensing retrievals of such clouds.

Supercooled water in convective clouds is constrained to temperatures T from 0°C to −40°C, but portions of these clouds can extend to colder ($T < -40$°C) or warmer ($T > 0$°C) temperatures. Such types of clouds have been encountered in the tropics with cloud bases near 20°C (Lawson et al., 2015) during the Ice in Clouds—Tropical experiment or from 0°C to 10°C in the midlatitudes (Blyth and Latham, 1993; Leon et al., 2016). Some convectively forced mixed-phase clouds have cloud bases at $T < 0$°C which are more frequently observed in the midlatitudes (i.e., Rangno and Hobbs, 1991).

Convectively forced mixed-phase clouds can have varying depths, sizes, and modes of organization. Shallow to moderate cumuli that have depths of around 3 km have been observed in midlatitude regions (i.e., Leon et al., 2016). They have also been observed as elevated convection (Murphy et al., 2017) and can also manifest as deep convection such as supercells and mesoscale convective systems (Jensen et al., 2015). This wide variability in convective modes results in an equally wide variability in the observed microphysical properties. These properties depend greatly on environmental conditions such as temperature, aerosol concentration, updraft velocity, stage of convective development, proximity to neighboring convection, and synoptic and mesoscale forcing. More information on how some of these factors can influence the microphysical properties of convectively forced mixed- phase clouds can be found in the papers listed in the bibliography of this chapter.

Several microphysical growth processes can occur in convectively forced mixed-phase clouds. For example, liquid droplets can grow via condensation up to a maximum dimension D of around 40 μm. In order for the drop to grow to $D > \sim 40$ μm, growth by collision-coalescence is needed to form drizzle and raindrops. These raindrops can then form into ice precipitation via freezing. Ice crystals can grow by vapor deposition, aggregation, and accretion. Microphysical process rates are often expressed in terms of measurable quantities such as number concentration, and liquid and ice water content. For accurate model representation of convective clouds—which is essential for accurate weather and climate forecasting—it is important that the rates at which these processes occur be known.

2. QUANTITIES TYPICALLY MEASURED

There are many different measurable quantities in which microphysical process rates are quantified. One such quantity is called the number distribution function $N(D)$ which provides information of the number concentration of particles present in the cloud as a function of particle diameter. The number distribution function is defined as:

$$N(D) = dN/dD \tag{1}$$

where dN is the number concentration of particles measured in a given size range with width dD.

The major quantities of interest that are derived from $N(D)$ are the moments of the size distribution. In order to find the nth moment of the size distribution in a given range of particle maximum dimension D_{min} to D_{max}, $N(D)$ is integrated over D_{min} to D_{max}:

$$M_n = \int_{D_{min}}^{D_{max}} N(D)D^n dD \tag{2}$$

These moments are typically derived separately for liquid droplets and ice crystals. Defining D for ice particles is difficult, as there are several different definitions of D for an ice particle. Certain moments of the size distribution can provide relevant information about the microphysical properties of the cloud. For example, the 0th moment is the number concentration, which provides the total number of liquid droplets or ice crystals per unit volume. This quantity is important for studying the production and nucleation of liquid and ice crystals. The 1st moment of the distribution is also better correlated to the size of the hydrometeors than the 0th moment and therefore can provide a metric for approximating the average size of the hydrometeors in cloud. The 2nd moment is related to the cross-sectional area of the liquid and ice particles. This quantity is important as it greatly impacts the radiative properties of mixed-phase clouds and the terminal velocity. The 3rd moment of the distribution corresponds to the volume of spherical particles, important for studying growth processes and mass fluxes. The 6th moment is related to the radar reflectivity of the liquid and ice particles, relevant for developing active remote sensing retrievals of the microphysical properties of convectively forced mixed-phase clouds.

3. FIELD EXPERIMENTS

In order to obtain direct observations of the microphysical properties of clouds, field experiments are conducted using aircraft with particle measurement probes attached to the outside of the aircraft, often on the wings or along the fuselage. The aircraft then flies into the cloud recording the sizes, shapes, and counts of particles as well as the total mass content of the cloud in situ. Table 1 shows the various aircraft that have been used in some field experiments sampling convectively forced mixed-phase clouds. This list is not exhaustive, but shows examples of where observations have been conducted and the multitude of aircraft that have been used.

There are also indirect methods for retrieving the microphysical properties of convective clouds using airborne and ground-based radars and spaceborne radars. The benefit of such indirect techniques is that they provide information of the microphysical properties of a greater volume of the cloud than what aircraft can achieve. However,

Table 1 Aircraft used in each field experiment

Aircraft	Field experiment	Location
University of Wyoming King Air	Cooperative Convective Precipitation Experiment (CCOPE) (Knight, 1982) Convective Precipitation Experiment (COPE) (Leon et al., 2016)	Montana (CCOPE), England (COPE)
National Center for Atmospheric Research C-130	ICE in clouds—Tropical (ICE-T) (Heymsfield and Willis, 2014; Johnson et al., 2014; Lawson et al., 2015) Profiling of Winter Storms (PLOWS) (Murphy et al., 2017)	St. Croix (ICE-T), Midwest US (PLOWS), Colorado (Heymsfield)
University of Washington Convair C-131A	Improvement of Microphysical Parameterization through Observational Verification Experiment (IMPROVE-1/2) (Hobbs and Rangno, 1990; Ikeda et al., 2007; Rangno and Hobbs, 1991)	Washington coast
Facility for Airborne Atmospheric Measurements (FAAM) Bae-146	Convective and Orographically-induced Precipitation Study (COPS) (Huang et al., 2011), the ICE and Precipitation Initiation in Cumulus (ICEPIC) (Huang et al., 2008)	Black Forest in Germany (COPS) UK (ICEPIC)
Stratton Park Engineering Company Learjet	ICE-T	St. Croix
Weather Modification Inc. Learjet	Rosenfeld and Woodley (2000)	Texas
NCAR King Air	Blyth and Latham (1993)	New Mexico
Aero Commander 690A	Cloud Aerosol Interaction and Precipitation Enhancement Experiment (Patade et al., 2016)	India

the resolution of such measurements are much more coarse. Furthermore, the empirical relationships derived from aircraft microphysical observations and assumptions about the particle shapes are required to convert quantities retrieved from radar into quantities such as number concentration. The most direct measurements of the microphysical properties of convective clouds come from particle measurement probes installed on aircraft and are therefore the focus of this chapter.

4. IN SITU MEASUREMENT METHODS

In this section, the types of instrumentation that are installed on aircraft in order to measure the microphysical properties of convectively forced mixed-phase clouds are outlined. There are two major classes of airborne particle measurement probes that measure the microphysical properties of clouds. The first major class of probes, bulk measurement probes, directly measures a moment of $N(D)$ such as LWC, listed in Section 4.1. The second class of probes, single particle instruments, provide estimates of $N(D)$ when sampling over a period of time that depends on both the number concentration of the particles as well as the sample volume of the probe. Only a brief explanation of these probes and techniques are given here. For a more detailed explanation of all of the methodologies that are used, see Baumgardner et al. (2017) and McFarquhar et al. (2017).

4.1 Bulk Instruments

The bulk instruments listed in this section directly measure a moment of $N(D)$ such as LWC. Many classes of probes do this by holding a heated wire at a constant temperature and then measuring the amount of heat that needs to be added to hold the wire at that temperature as cloud particles evaporate on the wire. The heat is dissipated both by the dry air passing over the wire via convective cooling and by water evaporating on the wire. There exist several methods for estimating the convective cooling over the wire, and those estimates are required to estimate the amount of water evaporated. The heat dissipated by convective cooling can be much larger than the heat dissipated by the evaporation of water on the wire at aircraft speeds, creating an uncertainty in the baseline LWC. Such examples include the King probe (King et al., 1978) and the Johnson-Williams probe.

A special kind of hotwire probe, the Nevzorov total water content probe (Korolev et al., 2013a), includes two separate sensors for LWC and total water content. The hotwire is located inside a cone that captures the liquid water and ice that enters the cone. In addition, there is a reference sensor on the Nevzorov probe that records the baseline LWC apart from the liquid with a sensor that is isolated from the flow of ice and water particles coming into the probe. This allows for improved corrections of the heat dissipated by convective cooling compared to other hotwire probes.

The Cloud Virtual Impactor (CVI) measures total water content by introducing particles through an inlet and into a counterflow. The counterflow ensures that only particles greater than a certain cut size are able to pass through the probe in order to isolate cloud droplets and precipitation size particles from aerosols and water vapor (Twohy et al., 1997). The cloud droplets and precipitation are then evaporated. A hygrometer inside the probe then measures the total water content from the evaporated cloud and precipitation particles. However, the decreased collection efficiency of droplets with

$D < 30$ μm can cause the CVI to underestimate *LWC*. The Particle Volume Monitor (PVM-100) uses the amount of light scattered from the targets in the sample volume to derive an estimate of *LWC*, and is typically best at deriving LWC from particles of D from 4 to about 45 μm, with less sensitivity to particles outside of that size range (Gerber et al., 1994). Therefore, while these probes can directly measure the mass content of the cloud, they require careful considerations in the interpretations of the data they provide.

Another important piece of information is locating where regions of supercooled water are present. In order to do this, the Rosemount Icing Detector (RID) has been used to detect the presence of supercooled water (Cober et al., 2001; Jackson et al., 2012, 2014; Jackson and McFarquhar, 2014; Plummer et al., 2014). This device detects supercooled water by allowing supercooled water to collect on a metal cylinder and freeze on impact. This freezing changes the vibration frequency of the sensor, which is detected by a piezoelectric sensor. As soon as the sensor accumulates a set amount of ice, the sensor is heated to a temperature that melts the ice on the sensor.

4.2 Single Particle Probes

In this section, airborne particle measurement probes that measure properties of single particles are discussed. Forward scattering probes focus on measuring small hydrometeors of $D < 50$ μm. The three most commonly used forward scattering probes are the Forward Scattering Spectrometer Probe, Cloud Droplet Probe (CDP), and the Cloud and Aerosol Spectrometer. These have been used in most of the field experiments in Table 1.

These probes size particles by shining a laser beam through an inlet, shroud, or open path where the particles enter through. When the particle enters the laser beam, light is scattered in the forward direction. The particles are then sized based on the intensity of scattered light. However, converting the intensity of light to a size requires an assumption to be made about the particle shape. Usually, it is assumed that these particles are spherical, which means that particle sizes from ice particles are more uncertain than in liquid clouds. Furthermore, until the last few years, calibrations of particle size have been made using glass beads, for which the relationship between the scattered intensity of light and water particle size is uncertain due to the different index of refraction of glass compared to water.

Measurements of $N(D)$ in ice and mixed-phase clouds from scattering probes have also been significantly contaminated by particles shattering on the probe inlets and shrouds before entering the sample volume (McFarquhar et al., 2007; Korolev et al., 2012, 2013b). However, modifications to scattering probes have been designed to sweep away shattered artifacts from the sample volume (Korolev et al., 2012). Also, processing algorithms (Field et al., 2003, 2006) have been designed to mitigate this issue. Therefore, it is important to consider possible factors such as probe calibration and ice shattering when interpreting the microphysical data from single particle probes.

Another class of single particle probe measures the difference in scattered light as a function of direction in order to obtain information about the shape of the particles. Three of these probe types have been made so far: Small Ice Detector (SID), SID-2 (Cotton et al., 2010), and SID-3 (Ulanowski et al., 2014). The SID, like the scattering probe, also focuses on measuring smaller hydrometeors, typically <0.1 mm in D. Unlike the scattering probes, the SID-2 is able to discriminate between spherical and irregular particles, making it useful for determining the relative contributions of liquid and ice for particles smaller than 0.1 mm. However, it is also prone to the over-identification of irregular particles due to the coincidence of two or more spherical particles in the SID-2 sample volume. This can give a scattering pattern similar to that of an irregular particle and therefore cause a misidentification (Johnson et al., 2014). When studying the initiation of ice particles in a liquid cloud, ice, if present, will be minute. Therefore, it is important to consider this limitation of the SID in order to determine if the ice seen is really there.

4.3 Imaging Probes

The previous section detailed airborne measurement probes that primarily focused on particles with $D < 50$ μm. However, there is also the need to sample particles of $D > 50$ μm as raindrops and hail can have $D > 1$ mm or even 1 cm. Therefore, most field experiments combine measurements from scattering probes and SIDs with those from optical array probes (OAPs) which are designed to measure particles of $D > 50$ μm. These probes use a laser shining onto a photodiode array to acquire two-dimensional black and white images of particles. Since the size of the pixels in the image is known, it is possible to acquire statistics such as the particle area, perimeter, and maximum dimension from the particle image. Information such as the particle habit and the IWC from $N(D)$ can therefore also be derived from optical array probes using the techniques discussed in Section 5.2.

Two classes of OAPs exist: cloud imaging probes and precipitation imaging probes. Cloud imaging probes such as the 2D Cloud Probe, Cloud Imaging Probe, and the 2D Stereo Probe sample particles of D from 25 μm to 1.6 mm at a 10–25 μm resolution. Precipitation imaging probes such as the 2D Precipitation Probe and the High Volume Precipitation Sampler sample particles with D of 0.8 mm to 1 cm at a resolution of 200 μm.

There are also limitations associated with optical array probes. Like with FSSPs, ice particle size distributions collected from OAPs have also been significantly contaminated by shattered artifacts (Jackson et al., 2014; Jackson and McFarquhar, 2014; Korolev et al., 2013a; Lawson, 2011). The same methods listed in Section 4.1 to correct for shattered artifacts in FSSPs are also used to correct for them in OAPs. Another uncertainty in $N(D)$

from OAPs is due to the poorly characterized depth of field of OAPs which increases the uncertainty of $N(D < 100 \ \mu m)$ (Baumgardner and Korolev, 1997). Therefore, a major challenge in sampling $N(D)$ from clouds is adequately sampling $N(50 \ \mu m < D < 100 \ \mu m)$, since this is the size range that is not sampled by most scattering probes and not adequately sampled by most OAPs.

While optical array probes have been used to collect shadow images of particles, newer probes have been developed to capture high quality images of cloud and ice particles using a charged-coupled (CCD) array. One such probe is the Cloud Particle Imager (CPI) (Lawson et al., 2001). While habit identification is less ambiguous with the CPI than with OAPs due to the greatly improved quality of the crystal images, the meager sample volume of the CPI prohibits the collection of a representative sample of particles on timescales less than 60 s, or about 6 km of flight. This probe has also been integrated with the 2D Stereo Probe, an optical array probe, to provide simultaneous sampling of particles by an OAP and a CCD array. Typically, the CPI is only used for habit identification, and algorithms such as Um and McFarquhar (2011) and McFarquhar et al. (2013) have been developed to identify the habits of particles from derived geometric properties of the particle images.

5. PROPERTIES OF BULK WATER MASS

The previous section demonstrated the methodologies that have been employed in the measurement of the microphysical properties of convectively forced mixed-phase clouds. In the following sections, examples of observations made by the probes listed in Section 4 for two different field experiments are shown. In one field experiment, the COnvective Precipitation Experiment (COPE), the University of Wyoming King Air (UWKA) penetrated within 1 km of the tops of developing cumulus over southwest England on Jul. 29, 2013. The clouds sampled were organized into a convergence line caused by the interaction of the sea breeze with the topography of southwest England along with synoptic scale lift provided by a 1000 hPa low pressure system to the northwest of the United Kingdom. The approximate cloud bases as derived from ground-based sounding launches were around 10°C, which is warm enough for liquid droplets to develop at the cloud base. Cloud tops reached -15°C, which is cold enough for ice formation. Data from COPE are then shown alongside data collected from other field experiments listed in Table 1 in order to provide a broader overview of the microphysical properties of convectively forced mixed-phase clouds

5.1 Liquid Water Content

Fig. 1 shows the median and quartiles of LWC, the amount of liquid water per unit volume, for given T ranges on the y-axis observed on the Jul. 29 case during COPE. Here the LWC is calculated from the size distributions from the CDP and therefore only

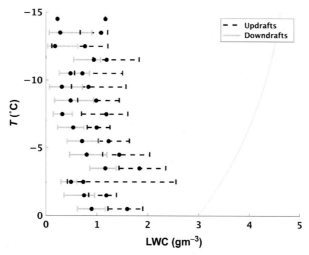

Fig. 1 Liquid water contents as a function of temperature for Jul. 29. *Solid line* is adiabatic. *Solid error bars* are ranges observed in downdraft, and *dashed error bars* represent the spread between the quartiles of *LWC* observed in updrafts and downdrafts.

includes contributions from particles with $D < 50$ μm. The different shading of lines in Fig. 1 represent whether the data were taken in updrafts ($w > 1$ m s^{-1}), or downdrafts ($w < 1$ m s^{-1}). The curve in the figure shows the LWC that would be generated if a parcel were lifted psuedoadiabatically from the approximate cloud base, the adiabatic LWC. Adiabatic LWC increases as the parcel ascends from cloud base because the droplets are assumed to grow by condensation as they ascend. Therefore, LWC can be a strong function of distance above cloud base in convectively forced mixed-phase clouds.

In general, LWC in Fig. 1 is under 50% of the adiabatic value. One reason why the LWC is subadiabatic in Fig. 1 is due to the presence of precipitation inside the cloud, as the falling precipitation collects and depletes the liquid cloud droplets. Another reason for the subadiabatic LWC is that supercooled water could be depleted by ice particles collecting supercooled water and growing via accretion. Another cause of these subadiabatic LWCs could be due to the entrainment of dry air from outside of the cloud mixing with the cloudy air, which can decrease LWC by totally or partially evaporating some of the drops.

Fig. 1 shows that LWC is about twice as high in updrafts as in downdrafts. There are many possible explanations for why this can occur in convectively forced mixed-phase clouds. First, since precipitation is more likely to fall in the downdraft region, precipitation scavenging of liquid particles could be reducing LWC in downdrafts. Second, there could be stronger dry air entrainment occurring in the downdrafts of these clouds. Finally, the enhanced supersaturations in the updrafts of these clouds could also be providing an environment more favorable for growth by condensation in the updrafts.

Fig. 1 also shows that LWC in the clouds sampled during COPE exhibited wide variability, showing values from 0.05 to 2.5 g m^{-3}. Table 2 shows the range of LWC that have been observed in other field projects sampling convectively forced mixed-phase clouds. When looking at examples from a few other field experiments in Table 1, this same conclusion also holds. For example, Lawson et al. (2015) have reported LWCs up to 4.5 g m^{-3} in the updrafts of tropical convection they observed. Heymsfield and Hjelmfelt (1984) have reported LWCs greater than 6 g m^{-3} in the hailstorm that they

Table 2 LWC values from studies sampling convectively forced mixed-phase clouds

Paper	LWC range	Cloud type
Heymsfield and Willis (2014)	0–1 g m^{-3}	Tropical deep convection growing to -17°C
Heymsfield and Hjelmfelt (1984)	0–3 g m^{-3} in congestus, up to 6.28 in hail storm, subadiabatic	Squall lines, cumulus congestus, one isolated hail storm
Lawson et al. (2015)	3–5 g m^{-3}	Tropical deep convection, developing to −22°C
Rangno and Hobbs (2005)	0.9–3.2 g m^{-3}, 10%–115% of adiabatic, generally subadiabatic	Tops of 9°C to −10°C
Rosenfeld and Woodley (2000)	0.9–2.4 g m^{-3}	Lubbock, TX cumulonimbus growing to −37.5°C
Murphy et al. (2017)	0–0.1 g m^{-3}	Elevated convection in winter cyclone
Lin and Colle (2009)	0.5–2 g m^{-3}	Elevated convection in winter cyclone
Koenig (1963)	1–2 g m^{-3}	Developing cumulus
Stith et al. (2004)	0–1 g m^{-3}	Tropical convection (monsoonal/oceanic), tops to −45°C
Huang et al. (2008)	0.6–0.8 g m^{-3}	Mid-level convection in UK (convergence line)
Huang et al. (2011)	0–1.2 g m^{-3}	Mid-level convection in Germany (convergence line)
Ackerman (1963)	0.5–3.2 g m^{-3}	Convective regions of hurricanes
MacCready and Takeuchi (1968)	0–1.5 g m^{-3}, subadiabatic, increasing with height	Cumulus bases 6°C–8°C, tops −6°C to −8°C
Patade et al. (2016)	0.1–4 g m^{-3}	Monsoonal convection over India, developing tops −10°C
Heymsfield et al. (2009)	0.001–2 g m^{-3}	Tropical isolated and deep convection from −20°C to −60°C
Cooper and Lawson (1984)	0–1.25 g m^{-3}	Cumulus, bases −5°C to 10°C

sampled, indicating that higher *LWC*s have been observed in deeper convection than sampled during COPE. Therefore, the *LWC* can be a function of the intensity of convection. This shows that it is important to consider such factors when interpreting *LWC* data in convection. Table 2 also shows that whenever sufficient data were available to determine the adiabatic *LWC*, such as in Heymsfield and Hjelmfelt (1984), Rangno and Hobbs (2005), and MacCready and Takeuchi (1968), *LWC* was generally observed to be subadiabatic. This is consistent with precipitation scavenging and entrainment reducing *LWC*.

Table 2 indicates that many of the observations of *LWC* have been made in developing midlatitude and tropical cumulus clouds in order to determine the evolution of the microphysical properties as these clouds develop. However, fewer studies have been conducted in the elevated convection embedded in a winter cyclone such as in Murphy et al. (2017). Since such observations are important for quantifying microphysical process rates used in numerical weather prediction models, a greater number of observations in elevated convection embedded in winter cyclones is warranted for improved winter storm forecasting.

Furthermore, likely due to safety concerns, few observations have been made in convection intense enough to produce hail, or in the convective cores of midlatitude mesoscale convective systems (MCS) and supercells. Field experiments sampling MCSes have made observations of microphysical properties in their stratiform regions (i.e., McFarquhar et al., 2007; Giangrande et al., 2016). An unfortunate limitation of making in situ observations of the microphysical properties of convection is that the convection needs to be safe enough for the crew to fly through, limiting such observations to weaker convection. Observations of microphysical properties in such strong convection are typically retrieved by radar. However, since methodologies for radar-based retrieval of microphysical properties incorporate information from in situ observations, the lack of in situ observations in such regions adds to the uncertainty in the retrieval of microphysical properties in stronger convection.

5.2 Ice Water Contents

The previous section focused on the liquid component of the bulk mass of mixed-phase clouds. In this section, the focus will be on the ice component of bulk mass. To do this, *IWC* values from several studies sampling convectively forced mixed-phase clouds are shown. Here, *IWC* is typically estimated from OAP measurements of particles of $D > 100$ μm, or from the Nevzorov probe, and therefore includes contributions from ice particles such as aggregates and graupel.

IWC can be calculated from OAPs using many different methodologies, which strongly depend on particle habit. The most commonly used methodology is to estimate the mass m of the particles by using an empirical relationship between m and D in the form of $m = aD^b$. There are many examples of these relationships that are developed for specific habits like those listed in Brown and Francis (1995) and Mitchell (1996). In order to calculate IWC using these relationships, the habits of particles must be first determined by identifying the habits of particles from images from an OAP or CPI. Then, for a given size bin D_k, the fraction of particles $f(D_k)$, is determined. The IWC using this methodology, employed by studies such as Jackson and McFarquhar (2014) is then:

$$IWC_{Habit} = \sum_j \sum_k f_k(D_j) \alpha_k D_j^{b_k} N(D_j) \Delta D_j$$

where $N(D_j)$ is the discrete number distribution function with midpoints D_j and widths ΔD_j, and a_k and b_k are the a and b coefficients corresponding to differing habit categories. Some have also proposed using the relationship between m and the cross-sectional area A of the particle (Baker and Lawson, 2006). Jackson and McFarquhar (2014) have shown that the difference in IWC calculated from these two methodologies can exceed 50%. This shows that determining IWC from airborne measurement is much more difficult than determining LWC. Therefore, reducing the uncertainty in IWC from airborne cloud microphysical measurements due to the unknown relationship between m and D for ice particles remains an active area of research.

Due to the difficulties in quantifying ice mass content using airborne microphysical measurements, fewer studies exist that attempt to quantify the IWC in convectively forced mixed-phase clouds than for the other quantities listed in this chapter. Table 3 shows significantly fewer observations of IWC than for LWC listed in Table 2, showing that there is a great need for IWC measurements in convectively forced mixed-phase clouds. For the studies in Fig. 2 sampling such clouds, there is a wide variability between the different types of cumulus that have been sampled, with variability in IWC ranging from 0.001 to 7.6 g m^{-3} depending on whether you are in elevated or deep convection and depending on what temperature the sample originated from. For example,

Table 3 IWC values from studies sampling convectively forced mixed-phase clouds

Paper	IWC range	Cloud type
Lawson et al. (2015)	0.02 g m^{-3} at −8°C, up to 7.6 g m^{-3} at higher levels	Tropical deep convection, developing to −22°C, St. Croix
Murphy et al. (2017)	0–1 g m^{-3}	Elevated convection in winter cyclone
Lin and Colle (2009)	0–0.4 g m^{-3}	Elevated convection in winter cyclone
Heymsfield et al. (2009)	0.001–2 g m^{-3}	Tropical isolated and deep convection from −20°C to −60°C
Hobbs et al. (1980)	0.001 in developing Cu to 6 g m^{-3} in complexes	Summertime cumulus in Montana

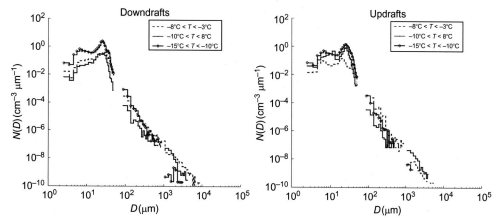

Fig. 2 Particle size distributions as a function of temperature for Jul. 29. Updrafts and downdrafts are as in Fig. 1.

Murphy et al. (2017) observed IWC up to 0.4 g m^{-3} in elevated convection embedded in a winter cyclone. However, Lawson et al. (2015) observed values in excess of 7 g m^{-3} in the tropical convection that they studied. In general, a greater number of observations of IWC in convectively forced mixed-phase clouds are needed in order to quantify mass flux rates within convection that are used in model simulations.

6. PARTICLE SIZE DISTRIBUTIONS

The previous sections outlined examples of mass content in convectively forced mixed-phase clouds. The following sections will focus on the properties of individual particles sampled during the Jul. 29 case during COPE. In this section, the size distributions of particles from the Jul. 29 case during COPE are shown. Fig. 2 shows the mean $N(D)$, sampled by a Cloud Droplet Probe, a Cloud Imaging Probe and a 2D Precipitation Probe, for given temperature ranges on Jul. 29 collected in downdrafts and in updrafts. In Fig. 2, the $N(D)$ are separated by whether the UWKA was penetrating updrafts or downdrafts as defined in Fig. 1, as well as by T. The $N(D)$ curves in Fig. 2 show that particle sizes span 8–10 orders of magnitude in convective clouds, with cloud droplets as small as 2 µm in D present alongside graupel particles up to 1 cm in D.

A wide variability in particle sizes is present for different T ranges and updraft velocities in Fig. 2. For example, inside the downdrafts particles up to 1 cm in D are present, but particles only have sizes up to 3 mm in D in the updrafts. As T decreases, there is a factor of 2 increase in the $N(0.1 < D < 1$ mm) as well as $N(D > 1$ mm). Particle sizes in convectively forced mixed-phase clouds in other field experiments have exhibited wide variability as well. Table 4 shows the maximum D that has been observed in various field experiments. For example, Lawson et al. (2015) observed a similar range of particle sizes

Table 4 Summary of observed precipitation particle sizes in convectively forced mixed-phase clouds

Paper	Largest D observed	Cloud type
Heymsfield and Willis (2014)	1 cm	Tropical deep convection in St. Croix and Cape Verde (NAMMA) growing to −17°C
Lawson et al. (2015)	3 mm	Tropical deep convection, developing to −22°C, St. Croix
Rangno and Hobbs (2005)	3 mm	Tops of 9 to −10°C
Murphy et al. (2017)	2 mm	Elevated convection in winter cyclone
Lin and Colle (2009)	8 mm	Elevated convection in winter cyclone
Taylor et al. (2016)	3 mm	Mid-level convection in UK
Patade et al. (2016)	4 mm	Monsoonal convection over India, developing tops −10°C
Heymsfield et al. (2009)	1 cm	Tropical isolated and deep convection from −20 to −60°C

compared to Fig. 5 in the tropical convection sampled at −20°C to −12°C. In general, Table 4 shows that many of the maximum D observed in various field projects is around 3 cm. However, Heymsfield and Willis (2014) and Heymsfield et al. (2009) have observed particles as large as a centimeter in the tropical deep convection that they sampled. Therefore, sizes of the particles observed can vary greatly depending on the environmental conditions that the particle was subjected to throughout its growth history.

7. LIQUID PARTICLE CONCENTRATIONS

This section focuses on the numbers of particles with $D < 50$ μm which include cloud droplets and small ice particles. Fig. 3 shows the quartiles and median of liquid droplet number concentration of particles with $D < 50$ μm $N_{<50}$ derived from the CDP equipped on the UWKA during COPE the observed for given temperature T levels on the y-axis. Fig. 3 shows that $N_{<50}$ for this case ranges from 30 to 250 cm^{-3}. As we ascend to a T of −10°C, Fig. 3 shows that the median $N_{<50}$ decreases from 150 cm^{-3} to 60 cm^{-3}. This shows that $N_{<50}$ can vary for different sets of conditions. During the Aug. 2 mission during COPE, values of $N_{<50}$ of up to 600 cm^{-3} were observed at T of 0 to −3°C. Therefore, observations of $N_{<50}$ can vary greatly in convection.

Many factors can influence $N_{<50}$ in convectively forced mixed-phase clouds. One such factor is the vertical velocity inside the updraft. If the vertical velocity is increased,

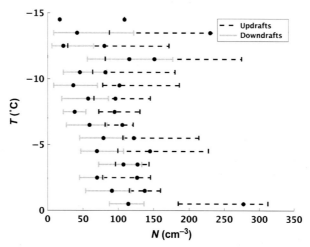

Fig. 3 Liquid number droplet concentration as a function of temperature for Jul. 29. *Error bars* are as in Fig. 1.

this increases the buoyancy of the parcel and causes the parcel to rise to colder temperatures before the water vapor in excess of water saturation can be completely condensed. This brief moment of enhanced supersaturation can lead to a greater number of cloud particles nucleating inside the updraft. The amount of cloud condensation nuclei (CCN) available in the updraft can also determine how many liquid particles nucleate as they rise in the updraft. Another factor that may influence $N_{<50}$ is entrainment of dry air from outside of the cloud by turbulent eddies that form at the edges of the cloud. The mixing of dry air into regions of supersaturated air can then result in either the partial or total evaporation of some of the droplets inside the cloud, causing a decrease in $N_{<50}$ closer to the edges of the cloud. Therefore, $N_{<50}$ can strongly depend on the distance of the sampled region from the edge of the cloud.

Finally, the presence of precipitation particles may also have a drastic influence on $N_{<50}$ as the precipitation falling through the cloud scavenges the smaller drops as they either coalesce or rime onto the particle. In general, Fig. 3 shows that N is about twice as high in updrafts as in downdrafts, consistent with greater dry air entrainment and precipitating scavenging in downdrafts compared to updrafts. Therefore, the liquid droplet concentration inside convectively forced mixed-phase clouds can be drastically affected by a multitude of factors, making the task of discerning what microphysical processes are occurring in convection from in situ observations a challenge.

Table 5 shows the $N_{<50}$ that has been observed in other field projects sampling convectively forced mixed- phase clouds. Table 5 shows a wide variability in $N_{<50}$ in other field projects as well. For example, in the updraft cores of tropical maritime convection sampled by Lawson et al. (2015), $N_{<50}$ was generally less than 200 cm^{-3}. However, in the

Table 5 Summary of observed $N_{<50}$ in convectively forced mixed-phase clouds

Paper	Range of $N_{<50}$	Cloud type
Heymsfield and Willis (2014)	0.1–500 cm^{-3}	Tropical deep convection in St. Croix and Cape Verde (NAMMA) growing to −17°C
Heymsfield and Hjelmfelt (1984)	20–500 cm^{-3}	Squall lines, cumulus congestus, one isolated hail storm in Oklahoma
Lawson et al. (2015)	15–160 cm^{-3}	Tropical deep convection, developing to −22°C, St. Croix
Rangno and Hobbs (2005)	60–105 cm^{-3}	Tops of −10 to −9°C
Rosenfeld and Woodley (2000)	200–1000 cm^{-3}	Lubbock, TX cumulonimbus growing to −37.5°C
Stith et al. (2004)	1–200 cm^{-3}	Tropical convection (monsoonal/oceanic), tops to −45°C
Huang et al. (2008)	80–120 cm^{-3}	Mid-level convection in UK (convergence line)
Heymsfield et al. (1979)	700–1300 cm^{-3}	Unmixed cumulus updraft cores in NE Colorado to −24°C

continental convection sampled by Blyth and Latham (1993), $N_{<50}$ ranged from 80 to 700 cm^{-3}, showing a stark contrast from the observations of Lawson et al. (2015) and higher than those in Fig. 3. The largest values of $N_{<50}$ in Table 5 are from the studies of Rosenfeld and Woodley (2000) and Heymsfield et al. (1979) that were collected in continental convective clouds. Therefore, some of the wide variability in $N_{<50}$ could be attributed to the greater availability of CCN over the continents compared to maritime environments. However, factors such as the updraft velocity, position in cloud, and age of the cloud sampled can also play a role in determining $N_{<50}$.

8. PRECIPITATION PARTICLE CONCENTRATIONS AND SHAPES

Fig. 4 shows the average number concentration of particles of $D > 100$ μm $N_{>100}$ derived from a combination of the greyscale Cloud Imaging Probe (CIP-Grey) and 2D Precipitation Probe (2DP) for the Jul. 29 case during COPE. Particles in this size range can be drizzle, raindrops, or ice particles such as graupel and aggregates. Fig. 4 shows that $N_{>100}$ increases with decreasing T from 1 L^{-1} to up to 200 L^{-1}. This could be due to the growth of precipitation as the cloud ascends to colder temperatures. At T of −13°C to −11°C, Fig. 4 shows that $N_{>100}$ is an order of magnitude higher in downdrafts than in updrafts, consistent with the notion of the greater presence of precipitation in the downdraft at this level.

There are many different processes that can control $N_{>100}$. Some are related to the formation of drizzle from cloud droplets. There are multiple pathways in which cloud

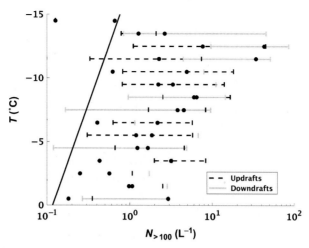

Fig. 4 $N_{>100}$ as a function of temperature for Jul. 29. *Error bars* are as in Fig. 1

droplets grow to a broad enough distribution of sizes for collision-coalescence to occur to form drizzle. One such way is through the entrainment and mixing into the cloud producing droplets of varying sizes. The variability in the trajectories and resultant growth histories of the droplets can produce this broad spectra of droplets (Cooper et al., 2013). Also, giant cloud condensation nuclei (GCCN) that have $D > 10\ \mu m$ can also nucleate into drizzle size droplets and therefore initiate the warm rain process (Feingold et al., 1999; Johnson, 1982).

These liquid drops can then nucleate into ice particles through a process called primary ice nucleation. There are two different pathways in which primary nucleation can occur:

(1) Homogeneous nucleation of liquid drops. Water drops freeze into ice particles without the presence of an ice nucleating particle (INP). This process only occurs at temperatures less than $-40°C$.

(2) Heterogeneous nucleation of liquid drops. In this process, a water droplet will not freeze without the presence of an INP. This is the only primary nucleation mechanism that can operate at temperatures greater than $-40°C$.

There are four different heterogeneous nucleation mechanisms as defined by Vali et al. (2015):

(1) Deposition nucleation is when an INP nucleates directly from water vapor depositing on the INP.

(2) Immersion freezing is when nucleation initiates by INP immersed within liquid.

(3) Contact freezing is when INP comes into contact with a liquid particle and initiates freezing.

(4) Condensation freezing is when a liquid drop initially grows on a CCN and then freezes.

Therefore, many different processes can contribute to the amount of ice particles present. Several empirical relationships have been developed between the amount of INP as a function of a given temperature in order to estimate the number concentration of ice particles that form by primary ice nucleation alone. Such relationships include Fletcher (1962), Meyers et al. (1992), and DeMott et al. (2010). In addition to these, DeMott et al. (2010) have also developed a relationship between the number concentration of INP between T and aerosol concentration in order to account for the amount of available aerosols to act as INP.

The black line in Fig. 4 shows the relationship between INP and T from DeMott et al. (2010). Fig. 4 shows that the $N_{>100}$ are several orders of magnitude above the predicted number of INP from DeMott et al. (2010) and that there are three orders of magnitude variability in the observed $N_{>100}$. Therefore, in the clouds sampled during COPE, this would indicate that other processes in addition to primary ice nucleation must also be occurring.

In addition to primary nucleation, several secondary ice production mechanisms that occur in convection can contribute to the wide variability in ice number concentrations observed. One such mechanism is the Hallett-Mossop mechanism (Hallett and Mossop, 1974). In this process, as graupel collects supercooled water droplets and grows by accretion, some of the droplets form splinters as they rime onto the graupel particle. However, this process only occurs in a specific set of conditions:

(1) T is in between $-8°C$ and $-3°C$
(2) Number concentration of cloud droplets $D > 24$ μm $N_{>24}$ > few drops per cm^{-3}
(3) Number concentration of cloud droplets $D > 13$ μm $N_{>13}/N_{>24} > 0.1$
(4) Graupel is present

Typically, the presence of needles, columns, drizzle, and graupel at temperatures from $-8°C$ and $-3°C$ are good indicators that the Hallett-Mossop process could be occurring. Therefore, while this mechanism could explain the formation of ice in excess of INP concentrations $-8°C$ and $-3°C$, there have been many instance of such observations at temperatures too cold for the Hallett-Mossop to occur in the updrafts of developing cumulus (Lawson et al., 2015).

In order to explain the production of ice in excess of INP concentrations at colder temperatures, additional secondary ice production mechanisms have been proposed. One such mechanism is through the fragmentation of frozen droplets (Leisner et al., 2014). These drops can also form spicules in which air bubbles can eject from the particle and freeze into secondary ice particles. While the greatest amount of evidence of this mechanism has been observed in the laboratory, the observations of Lawson et al. (2015) suggest that this mechanism produced the ice in the tropical convection they observed. Lawson et al. (2015) suggests that the presence of drizzle in an updraft at the $-6°C$ level increases the likelihood of secondary ice production by frozen drop fragmentation. Another mechanism that can produce secondary ice particles at temperatures too cold for the Hallett-Mossop process to operate is through collisions between graupel particles producing ice fragments (Yano and Phillips, 2010).

Table 6 shows the observed concentration of precipitation size particles in studies sampling convectively forced mixed-phase clouds. The definition of precipitation size particles varies throughout these studies due to the availability of instrumentation throughout the different field experiments. Table 6 shows a wide variability in the number concentration of precipitation particles in the different projects. For example, Blyth and Latham (1993) have found ice particle concentrations up to 1300 L^{-1}, and drizzle concentrations up to 10 L^{-1}. Lawson et al. (2015) have observed ice crystal concentrations anywhere from 120 to 1600 L^{-1} in the tropical convection they sampled. Factors such as distance from cloud edge, temperature, vertical velocity, or whether the developing cloud grew into a preceding cloud layer can all play a role in determining the precipitation number concentration.

In general, Table 6 shows that the observed concentrations of precipitation were above the observed or predicted concentrations of Ice Nuclei (IN), showing that primary nucleation alone cannot explain the observed concentrations of precipitation. However, it is worth noting that a majority of the observations in Table 6 were made using probes that did not have any modifications to account for shattering artifacts. As Korolev et al. (2013a) and Jackson and McFarquhar (2014) have shown that $N_{>100}$ can be significantly contaminated by shattered artifacts, it is possible that the many of values given here are overestimates of the actual concentration of ice particles. Therefore, more studies should focus on sampling such convection using probes that have shatter mitigating tips and use shattered artifact removal algorithms, such as what was done by Lawson et al. (2015) and Taylor et al. (2016). Such data will extend the generality of the conclusion that ice particle concentrations generally exceed IN concentrations to studies that account for shattered artifacts.

Furthermore, since differing studies have arrived at differing conclusions about the relative impact of the differing secondary ice production process in convection, the study of the relative impact of such processes remains an active area of research. Elucidating the relative impact of such processes will guide the improvement of model parameterizations of ice processes in convection that are associated with heavy rainfall events.

Fig. 5 shows representative particle images recorded by the CIP-Grey probe onboard the UWKA on Jul. 29 during COPE at different temperature ranges and in regions sampled in updrafts and downdrafts. In the updrafts, the images generally show the presence of spherical particles in updrafts at the $-2°C$ level with sizes ranging from 0.2 to ~1 mm. As the temperature decreases to $-8°C$, Fig. 5 shows that the particles present in the updrafts are generally columns and graupel. This mixture of particle habits is an indicator that the Hallett-Mossop process could be occurring in this temperature range in the updraft. At temperatures of $-11°C$, the particles are generally a mixture of drizzle, columns, and graupel in Fig. 5. The presence of graupel indicates that these particles are

Table 6 Summary of observed concentrations of precipitation in convectively forced mixed-phase clouds

Paper	Observed precipitation concentrations	Cloud type
Blyth and Latham (1993)	10 L^{-1}	Cumulonimbus in new Mexico
Heymsfield and Willis (2014)	$30–200 \text{ L}^{-1}$	Tropical deep convection in St. Croix and Cape Verde growing to $-17°C$
Hobbs and Rangno (1990)	100 L^{-1}, $1–20 \text{ L}^{-1}$ in chimney clouds	Cumulus off coast of Washington
Lawson et al. (2015)	Ice $1–100 \text{ L}^{-1}$ in first ice, $129–1630 \text{ L}^{-1}$ in glaciated region	Tropical deep convection, developing to $-22°C$, St. Croix
Hallet et al. (1978)	New towers $<0.1 \text{ L}^{-1}$, old $>10 \text{ L}^{-1}$	Florida cumulus
Rangno and Hobbs (2005)	$10–100 \text{ L}^{-1}$, up to 600 L^{-1}	Tops of 9 to $-10°C$
Murphy et al. (2017)	$1–40 \text{ L}^{-1}$	Elevated convection in winter cyclone
Taylor et al. (2016)	350 L^{-1}	Mid-level convection in UK
Koenig (1963)	$1–50 \text{ L}^{-1}$	Developing cumulus
Stith et al. (2002)	$50–600 \text{ L}^{-1}$	Tropical convection
Huang et al. (2008)	$3–70 \text{ L}^{-1}$	Mid-level convection in UK (convergence line)
Huang et al. (2011)	$10–400 \text{ L}^{-1}$	Mid-level convection in Germany (convergence line)
Black and Hallett (1986)	Ice 200 L^{-1}, $20–30 \text{ L}^{-1}$ graupel in downdrafts	Convective regions of hurricanes
Murgatroyd and Garrod (1960)	$>1 \text{ L}^{-1}$	Cumulus 1 km thick
Patade et al. (2016)	$1–400 \text{ L}^{-1}$	Monsoonal convection over India, developing tops $-10°C$
Heymsfield et al. (1979)	$0.1–1 \text{ L}^{-1}$ ice, $0.1–10 \text{ L}^{-1}$ in mixed regions	Unmixed cumulus updraft cores to $-24°C$
Hobbs et al. (1980)	Ice $1–10 \text{ L}^{-1}$	Summertime cumulus in Montana
Cooper and Lawson (1984)	1000 L^{-1}	Cumulus, bases $-5°C$ to $10°C$

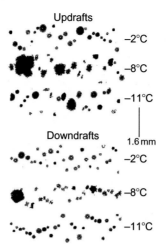

Fig. 5 Example CIP-Grey images taken at three temperature levels inside updrafts and downdrafts.

growing by the accretion of supercooled water droplets. The presence of columns also indicates growth by diffusion. Therefore, this shows that liquid and ice can both be present in convection at temperatures well below freezing.

Looking at the sample of particles from the downdrafts, Fig. 5 shows that spherical particles are generally present at −2°C with $D < 1$ mm. Some of the particles could be frozen droplets, but it is difficult to determine if these particles are frozen droplets due to the limited resolution of the CIP-Grey. At temperatures of −8°C, graupel and irregular particles with $D < 1$ mm are present in the downdrafts. At temperatures of −11°C, a mixture of drizzle and graupel can be seen in the CIP-Grey images sampled in a downdraft region. Therefore, a wide variability in particle shapes were sampled in the mixed-phase convection during COPE.

Table 7 shows the particle habits that have been observed by other studies sampling convectively forced mixed-phase clouds. A majority of the studies in Table 7 have observed graupel, which would be expected when sampling clouds with supercooled water available for particles to grow by accretion. However, many of the studies in Table 7 have also observed frozen drops, sometimes with ice spicules as observed by Lawson et al. (2015). This is indicative that there is an active warm rain process, which has been proposed to be crucial for the development of graupel and the initiation of secondary production of ice (Koenig, 1963; Leon et al., 2016). Indeed, in many of the studies, columns are observed alongside graupel, which would be expected if secondary ice production were occurring.

However, not all of the studies in Table 7 have observed graupel. For example, Blyth and Latham (1993) observed a mixture of lightly rimed aggregates and irregular particles.

Table 7 Summary of observed concentrations of precipitation in convectively forced mixed-phase clouds

Paper	Particle shape	Cloud type
Heymsfield and Willis (2014)	Columns, frozen drops, graupel	Tropical deep convection in St. Croix and Cape Verde (NAMMA) growing to −17°C
Heymsfield and Hjelmfelt (1984)	planar crystals, graupel, aggregates	Squall lines, cumulus congestus, one isolated hail storm in Oklahoma
Lawson et al. (2015)	Graupel	Tropical deep convection, developing to −22°C, St. Croix
Rangno and Hobbs (2005)	Graupel (−12°C), columns, spicules, frozen drops (−3°C to −10°C)	Tops of 9 to −10°C
Murphy et al. (2017)	Aggregates, columns, rimed dendrites	Elevated convection in winter cyclone
Lin and Colle (2009)	Dendrites, aggregates to 8 mm	Elevated convection in winter cyclone
Taylor et al. (2016)	Graupel, columns to −12°C	Mid-level convection in UK
Koenig (1963)	Graupel, frozen drops	Developing cumulus
Stith et al. (2002)	Aggregates, frozen drops	Tropical convection
Stith et al. (2004)	Chain aggregates, frozen drops	Tropical convection (monsoonal/oceanic), tops to −45°C
Huang et al. (2008)	Pristine and rimed columns	Mid-level convection in UK (convergence line)
Huang et al. (2011)	Columns, rimed columns	Mid-level convection in Germany (convergence line)
Black and Hallett (1986)	Graupel in downdrafts, columns	Convective regions of hurricanes
Braham (1963)	Snow pellets, ice pellets, frozen drops	Cumulus to −22°C
Cannon et al. (1974)	Rimed columns, rimed plates, rimed branched crystals	NE Colorado cumulus clouds
MacCready and Takeuchi (1968)	Graupel outside updrafts	Cumulus bases −6°C to 8°C, tops −6°C to −8°C
Patade et al. (2016)	Graupel, aggregates	Monsoonal convection over India, developing tops −10°C
Johnson et al. (2014)	Irregular ice particles	Tropical deep convection
Heymsfield et al. (2009)	Graupel, columns, aggregates	Tropical isolated and deep convection from −20°C to −60°C
Cooper and Lawson (1984)	Graupel in unseeded cloud, aggregates in seeded clouds	Cumulus, bases −5°C to 10°C

Stith et al. (2004) have observed chain aggregates at −8°C, which were confined to temperatures below −43°C in Stith et al. (2002). Lin and Colle (2009) and Murphy et al. (2017) have also observed dendrites in elevated convection. The wide differences in the particle habits between the studies are likely due to the fact that the particles were subjected to vastly different temperatures and relative humidities throughout their growth history. Therefore, a wide variability of particle shapes can exist in convectively forced mixed-phase clouds.

9. SUMMARY

This chapter summarizes the methodologies used and field experiments conducted to sample mixed-phase convectively forced clouds, and various examples of the microphysical properties of convectively forced mixed-phase clouds. The earlier sections showed that there are uncertainties in determining the liquid and ice microphysical properties of convectively forced mixed-phase clouds related to uncertainties in the measurements due to calibration errors in forward scattering spectrometer probes, shattering of ice particles on probe tips and inlets, and difficulties in distinguishing smaller ice particles from liquid particles due to the limited resolution of many of the imaging probes that have been used in field experiments sampling convection. Estimates of ice water content are further complicated by the difficulty in converting two-dimensional images of ice particles from optical array probes to mass due to the unknown relationship between mass and particle maximum dimension.

Examples of liquid and ice water contents, particle concentrations, and shapes from two field experiments sampling convective clouds were then shown. In general, the microphysical properties of convective clouds can be widely variable due to numerous factors that include, but are not limited to: temperature, position in cloud, vertical velocity, strength of entrainment, and the amount of cloud condensation nuclei loaded into the cloud. It is therefore extremely important to consider the environmental conditions the particles were sampled in to properly interpret microphysical data taken in convectively forced mixed-phase clouds. The observations in this chapter show the need to improve quantifications of IWC from in situ measurements as well as to sample a greater number of convectively forced mixed-phase clouds using modern probes that mitigate shattered artifacts.

ACKNOWLEDGMENTS

The collection of the data during COPE was supported by the National Science Foundation grant number AGS-1230203. The authors would like to acknowledge Greg McFarquhar and David Plummer for their useful feedback on this manuscript. We would also like to acknowledge Greg McFarquhar and Wei Wu of University of Illinois for the use of the University of Illinois Optical Array Processing Software for processing of OAP data.

REFERENCES

Ackerman, B., 1963. Some observations of water contents in hurricanes. J. Atmos. Sci. 20, 288–298. https://doi.org/10.1175/1520-0469(1963)020<0288:SOOWCI>2.0.CO%3B2.

Baker, B., Lawson, R.P., 2006. Improvement in determination of ice water content from two-dimensional particle imagery. Part I: image-to-mass relationships. J. Appl. Meteorol. Climatol. 45, 1282–1290. https://doi.org/10.1175/JAM2398.1.

Baumgardner, D., Korolev, A., 1997. Airspeed corrections for optical array probe sample volumes. J. Atmos. Ocean. Technol. 14, 1224–1229. https://doi.org/10.1175/1520-0426(1997)014<1224:ACFOAP>2.0.CO;2.

Baumgardner, D., Abel, S.J., Axisa, D., Cotton, R., Crosier, J., Field, P., Gurganus, C., Heymsfield, A., Korolev, A., Krämer, M., Lawson, P., McFarquhar, G., Ulanowski, Z., Um, J., 2017. Cloud ice properties: in situ measurement challenges. Meteor. Monogr. 58, 9.1–9.23. https://doi.org/10.1175/AMSMONOGRAPHS-D-16-0011.1.

Black, R.A., Hallett, J., 1986. The distribution of ice in hurricanes. J. Atmos. Sci. 43 (8), 802–822. https://doi.org/10.1175/1520-0469(1986)043<0802:OOTDOI>2.0.CO;2.

Blyth, A.M., Latham, J., 1993. Development of ice and precipitation in New Mexican summertime cumulus clouds. Q. J. R. Meteorol. Soc. 119, 91–120. https://doi.org/10.1002/qj.49711950905.

Braham Jr., R.R., 1963. Some measurements of snow pellet bulk densities. J. Appl. Meteorol. 2, 498–500. https://doi.org/10.1175/1520-0450(1963)002<0498:SMOSPB>2.0.CO%3B2.

Brown, P.R., Francis, P.N., 1995. Improved measurements of the ice water content in cirrus using a total-water probe. J. Atmos. Oceanic Technol. 12, 410–414. https://doi.org/10.1175/1520-0426(1995)012<0410:IMOTIW>2.0.CO;2.

Cannon, T.W., Dye, J.E., Toutenhoodf, V., 1974. The mechanism of precipitation formation in northeastern Colorado Cumulus II. Sailplane measurements. J. Atmos. Sci. 31, 2148–2174. https://doi.org/10.1175/1520-0469(1974)031<2148%3ATMOPFI>2.0.CO%3B2.

Cober, S.G., Isaac, G.A., Korolev, A.V., 2001. Assessing the Rosemount icing detector with in situ measurements. J. Atmos. Ocean. Technol. 18, 515–528. https://doi.org/10.1175/1520-0426(2001)018<0515:ATRIDW>2.0.CO;2.

Cooper, W., Lawson, R., 1984. Physical interpretation of results from the HIPLEX-1 experiment. J. Clim. Appl. Meteorol. 23, 523–540. https://doi.org/10.1175/1520-0450(1984)023<0523:PIORFT>2.0.CO;2.

Cooper, W.A., Lasher-Trapp, S.G., Blyth, A.M., 2013. The influence of entrainment and mixing on the initial formation of rain in a warm cumulus cloud. J. Atmos. Sci. 70, 1727–1743. https://doi.org/10.1175/JAS-D-12-0128.1.

Cotton, R., Osborne, S., Ulanowski, Z., Hirst, E., Kaye, P.H., Greenaway, R.S., 2010. The ability of the small ice detector (SID-2) to characterize cloud particle and aerosol morphologies obtained during flights of the FAAM BAe-146 research aircraft. J. Atmos. Ocean. Technol. 27, 290–303. https://doi.org/10.1175/2009JTECHA1282.1.

DeMott, P.J., Prenni, A.J., Liu, X., Kreidenweis, S.M., Petters, M.D., Twohy, C.H., et al., 2010. Predicting global atmospheric ice nuclei distributions and their impacts on climate. Proc. Natl. Acad. Sci. 107, 11217–11222. https://doi.org/10.1073/pnas.0910818107.

Feingold, G., Cotton, W.R., Kreidenweis, S.M., Davis, J.T., 1999. The impact of giant cloud condensation nuclei on drizzle formation in stratocumulus: implications for cloud radiative properties. J. Atmos. Sci. 56, 4100–4117. https://doi.org/10.1175/1520-0469(1999)056<4100:TIOGCC>2.0.CO;2.

Field, P.R., Wood, R., Brown, P.R.A., Kaye, P.H., Hirst, E., Greenaway, R., et al., 2003. Ice particle inter-arrival times measured with a fast FSSP. J. Atmos. Ocean. Technol. 20, 249–261. https://doi.org/10.1175/1520-0426(2003)020<0249:IPITMW>2.0.CO;2.

Field, P.R., Heymsfield, A.J., Bansemer, A., 2006. Shattering and particle interarrival times measured by optical array probes in ice clouds. J. Atmos. Ocean. Technol. 23, 1357–1371. https://doi.org/10.1175/JTECH1922.1.

Fletcher, N.H., 1962. The Physics of Rainbands. Cambridge University Press, p. 386.

Gerber, H., Arends, B.G., Ackerman, A.S., 1994. New microphysics sensor for aircraft use. Atmos. Res. 31, 235–252. https://doi.org/10.1016/0169-8095(94)90001-9.

Giangrande, S.E., Toto, T., Bansemer, A., Kumjian, M.R., Mishra, S., Ryzhkov, A.V., 2016. Insights into riming and aggregation processes as revealed by aircraft, radar, and disdrometer observations for a 27 April 2011 widespread precipitation event. J. Geophys. Res. Atmos. 121, 5846–5863. https://doi.org/10.1002/2015JD024537.

Hallet, J., Sax, R.I., Lamb, D., Murty, A.S.R., 1978. Aircraft measurements of ice in Florida cumuli. Q. J. Roy. Metor. Soc. 104, 631–650. https://doi.org/10.1002/qj.49710444108.

Hallett, J., Mossop, S.C., 1974. Production of secondary ice particles during the riming process. Nature 249, 26–28. https://doi.org/10.1038/249026a0.

Heymsfield, A.J., Hjelmfelt, M.R., 1984. Processes of hydrometeor development in Oklahoma convective clouds. J. Atmos. Sci. 41, 2811–2835. https://doi.org/10.1175/1520-0469(1984)041<2811:POHDIO>2.0.CO;2.

Heymsfield, A., Willis, P., 2014. Cloud conditions favoring secondary ice particle production in tropical maritime convection. J. Atmos. Sci. 71, 4500–4526. https://doi.org/10.1175/JAS-D-14-0093.1.

Heymsfield, A., Knight, C., Dye, J., 1979. Ice initiation in unmixed updraft cores in northeast colorado cumulus congestus clouds. J. Atmos. Sci. 36, 2216–2229. https://doi.org/10.1175/1520-0469(1979)036<2216:IIIUUC>2.0.CO;2.

Heymsfield, A.J., Bansemer, A., Heymsfield, G., Fierro, A.O., 2009. Microphysics of maritime tropical convective updrafts at temperatures from −20° to −60°. J. Atmos. Sci. 66, 3530–3562. https://doi.org/10.1175/2009JAS3107.1.

Hobbs, P.V., Rangno, A.L., 1990. Rapid development of high ice particle concentrations in small polar maritime cumuliform clouds. J. Atmos. Sci. 47, 2710–2722. https://doi.org/10.1175/1520-0469(1990)047<2710:RDOHIP>2.0.CO;2.

Hobbs, P., Politovich, M., Radke, L., 1980. The structures of summer convective clouds in eastern Montana. I: natural clouds. J. Appl. Meteorol. 19, 645–663. https://doi.org/10.1175/1520-0450(1980)019<0645:TSOSCC>2.0.CO;2.

Huang, Y., Blyth, A.M., Brown, P.R.A., Choularton, T.W., Connolly, P., Gadian, A.M., et al., 2008. The development of ice in a cumulus cloud over southwest England. New J. Phys. 10, 105021. https://doi.org/10.1088/1367-2630/10/10/105021.

Huang, Y., Blyth, A.M., Brown, P.R.A., Cotton, R., Crosier, J., Bower, K.N., et al., 2011. Development of ice particles in convective clouds observed over the Black Forest mountains during COPS. Q. J. R. Meteorol. Soc. 137, 275–286. https://doi.org/10.1002/qj.749.

Ikeda, K., Rasmussen, R.M., Hall, W.D., Thompson, G., 2007. Observations of freezing drizzle in extratropical cyclonic storms during IMPROVE-2. J. Atmos. Sci. 64, 3016–3043. https://doi.org/10.1175/JAS3999.1.

Jackson, R.C., McFarquhar, G.M., 2014. An assessment of the impact of antishattering tips and artifact removal techniques on bulk cloud ice microphysical and optical properties measured by the 2D cloud probe. J. Atmos. Ocean. Technol. 31, 2131–2144. https://doi.org/10.1175/JTECH-D-14-00018.1.

Jackson, R.C., McFarquhar, G.M., Korolev, A.V., Earle, M.E., Liu, P.S.K., Lawson, R.P., et al., 2012. The dependence of ice microphysics on aerosol concentration in arctic mixed-phase stratus clouds during ISDAC and M-PACE. J. Geophys. Res. Atmos. 117D15207. https://doi.org/10.1029/2012JD017668.

Jackson, R.C., McFarquhar, G.M., Stith, J., Beals, M., Shaw, R.A., Jensen, J., et al., 2014. An assessment of the impact of antishattering tips and artifact removal techniques on cloud ice size distributions measured by the 2D cloud probe. J. Atmos. Ocean. Technol. 31, 2567–2590. https://doi.org/10.1175/JTECH-D-13-00239.1.

Jensen, M.P., Petersen, W.A., Bansemer, A., Bharadwaj, N., Carey, L.D., Cecil, D.J., et al., 2015. The midlatitude continental convective clouds experiment (MC3E). Bull. Am. Meteorol. Soc. https://doi.org/10.1175/BAMS-D-14-00228.1.

Johnson, D.B., 1982. The role of giant and ultragiant aerosol particles in warm rain initiation. J. Atmos. Sci. 39, 448–460. https://doi.org/10.1175/1520-0469(1982)039<0448:TROGAU>2.0.CO;2.

Johnson, A., Lasher-Trapp, S., Bansemer, A., Ulanowski, Z., Heymsfield, A.J., 2014. Difficulties in early ice detection with the small ice detector-2 HIAPER (SID-2H) in maritime cumuli. J. Atmos. Ocean. Technol. 31, 1263–1275. https://doi.org/10.1175/JTECH-D-13-00079.1.

King, W.D., Parkin, D.A., Handsworth, R.J., 1978. A hot-wire liquid water device having fully calculable response characteristics. J. Appl. Meteorol. 17, 1809–1813. https://doi.org/10.1175/1520-0450(1978)017<1809:AHWLWD>2.0.CO;2.

Knight, C.A., 1982. The cooperative convective precipitation experiment (CCOPE), 18 May–7 August 1981. Bull. Am. Meteorol. Soc. 63, 386–398. https://doi.org/10.1175/1520-0477(1982)063<0386:TCCPEM>2.0.CO;2.

Koenig, L.R., 1963. The glaciating behavior of small cumulonimbus clouds. J. Atmos. Sci. 20, 29–47. https://doi.org/10.1175/1520-0469(1963)020<0029:TGBOSC>2.0.CO;2.

Korolev, A., Emery, E., Creelman, K., 2012. Modification and tests of particle probe tips to mitigate effects of ice shattering. J. Atmos. Ocean. Technol. 30, 690–708. https://doi.org/10.1175/JTECH-D-12-00142.1.

Korolev, A., Strapp, J.W., Isaac, G.A., Emery, E., 2013a. Improved airborne hot-wire measurements of ice water content in clouds. J. Atmos. Ocean. Technol. 30, 2121–2131. https://doi.org/10.1175/JTECH-D-13-00007.1.

Korolev, A.V., Emery, E.F., Strapp, J.W., Cober, S.G., Isaac, G.A., 2013b. Quantification of the effects of shattering on airborne ice particle measurements. J. Atmos. Ocean. Technol. 30, 2527–2553. https://doi.org/10.1175/JTECH-D-13-00115.1.

Lawson, R.P., 2011. Effects of ice particles shattering on the 2D-S probe. Atmos. Meas. Tech. 4, 1361–1381. https://doi.org/10.5194/amt-4-1361-2011.

Lawson, R.P., Baker, B.A., Schmitt, C.G., Jensen, T.L., 2001. An overview of microphysical properties of Arctic clouds observed in May and July 1998 during FIRE ACE. J. Geophys. Res.-Atmos. 106, 14989–15014. https://doi.org/10.1029/2000JD900789.

Lawson, R.P., Woods, S., Morrison, H., 2015. The microphysics of ice and precipitation development in tropical cumulus clouds. J. Atmos. Sci. 72, 2429–2445. https://doi.org/10.1175/JAS-D-14-0274.1.

Leisner, T., Pander, T., Handmann, P., Kiselev, A., 2014. Secondary ice processes upon heterogeneous freezing of cloud droplets. In: Presented at the 14th Conference on Cloud Physics, Boston, MA.

Leon, D.C., French, J.R., Lasher-Trapp, S., Blyth, A.M., Abel, S.J., Ballard, S., Barrett, A., Bennett, L.J., Bower, K., Brooks, B., Brown, P., Charlton-Perez, C., Choularton, T., Clark, P., Collier, C., Crosier, J., Cui, Z., Dey, S., Dufton, D., Eagle, C., Flynn, M.J., Gallagher, M., Halliwell, C., Hanley, K., Hawkness-Smith, L., Huang, Y., Kelly, G., Kitchen, M., Korolev, A., Lean, H., Liu, Z., Marsham, J., Moser, D., Nicol, J., Norton, E.G., Plummer, D., Price, J., Ricketts, H., Roberts, N., Rosenberg, P.D., Simonin, D., Taylor, J.W., Warren, R., Williams, P.I., Young, G., 2016. The Convective Precipitation Experiment (COPE): investigating the origins of heavy precipitation in the Southwestern United Kingdom. Bull. Amer. Meteor. Soc. 97, 1003–1020. https://doi.org/10.1175/BAMS-D-14-00157.1.

Lin, Y., Colle, B.A., 2009. The 4–5 December 2001 IMPROVE-2 event: observed microphysics and comparisons with the weather research and forecasting model. Mon. Weather Rev. 137, 1372–1392. https://doi.org/10.1175/2008MWR2653.1.

MacCready, P., Takeuchi, D., 1968. Precipitation initiation mechanisms and droplet characteristics of some convective cloud cores. J. Appl. Meteorol. 7, 591–602. https://doi.org/10.1175/1520-0450(1968)007<0591:PIMADC>2.0.CO;2.

McFarquhar, G.M., Um, J., Freer, M., Baumgardner, D., Kok, G.L., Mace, G., 2007. Importance of small ice crystals to cirrus properties: observations from the tropical warm pool international cloud experiment (TWP-ICE). Geophys. Res. Lett. 34L13803. https://doi.org/10.1029/2007GL029865.

McFarquhar, G.M., Um, J., Jackson, R., 2013. Small cloud particle shapes in mixed-phase clouds. J. Appl. Meteorol. Climatol. 52, 1277–1293. https://doi.org/10.1175/JAMC-D-12-0114.1.

McFarquhar, G.M., Baumgardner, D., Bansemer, A., Abel, S.J., Crosier, J., French, J., et al., 2017. Processing of cloud in-situ data collected by bulk water, scattering and imaging probes: fundamentals, uncertainties and efforts towards consistency. Amer. Meteor. Soc. Monographs, under review.

Meyers, M.P., DeMott, P.J., Cotton, W.R., 1992. New primary ice-nucleation parameterizations in an explicit cloud model. J. Appl. Meteorol. 31, 708–721. https://doi.org/10.1175/1520-0450(1992)031<0708:NPINPI>2.0.CO;2.

Mitchell, D.L., 1996. Use of mass- and area-dimensional power laws for determining precipitation particle terminal velocities. J. Atmos. Sci. 53, 1710–1723. https://doi.org/10.1175/1520-0469(1996)053<1710: UOMAAD>2.0.CO;2.

Mülmenstädt, J., Sourdeval, O., Delanoë, J., Quaas, J., 2015. Frequency of occurrence of rain from liquid-, mixed-, and ice-phase clouds derived from A-Train satellite retrievals. Geophys. Res. Lett. 42, 6502–6509. https://doi.org/10.1002/2015GL064604.

Murgatroyd, R.J., Garrod, M.P., 1960. Observations of precipitation elements in cumulus clouds. Q. J. R. Meteorol. Soc. 86, 167–175. https://doi.org/10.1002/qj.49708636805.

Murphy, A.M., Rauber, R.M., McFarquhar, G.M., Finlon, J.A., Plummer, D.M., Rosenow, A.A., et al., 2017. A microphysical analysis of elevated convection in the comma head region of continental winter cyclones. J. Atmos. Sci. https://doi.org/10.1175/JAS-D-16-0204.1.

Patade, S., Shete, S., Malap, N., Kulkarni, G., Prabha, T.V., 2016. Observational and simulated cloud microphysical features of rain formation in the mixed-phase clouds observed during CAIPEEX. Atmos. Res. A 169, 32–45. https://doi.org/10.1016/j.atmosres.2015.09.018.

Plummer, D.M., McFarquhar, G.M., Rauber, R.M., Jewett, B.F., Leon, D.C., 2014. Structure and statistical analysis of the microphysical properties of generating cells in the comma-head region of continental winter cyclones. J. Atmos. Sci. 71, 4181–4203. https://doi.org/10.1175/JAS-D-14-0100.1.

Rangno, A.L., Hobbs, P.V., 1991. Ice particle concentrations and precipitation development in small polar maritime cumuliform clouds. Q. J. R. Meteorol. Soc. 117, 207–241. https://doi.org/10.1002/qj.49711749710.

Rangno, A.L., Hobbs, P.V., 2005. Microstructures and precipitation development in cumulus and small cumulonimbus clouds over the warm pool of the tropical Pacific Ocean. Q. J. R. Meteorol. Soc. 131, 639–673. https://doi.org/10.1256/qj.04.13.

Rosenfeld, D., Woodley, W.L., 2000. Deep convective clouds with sustained supercooled liquid water down to -37.5°C. Nature 405, 440–442. https://doi.org/10.1038/35013030.

Stith, J.L., Dye, J.E., Bansemer, A., Heymsfield, A.J., Grainger, C.A., Petersen, W.A., et al., 2002. Microphysical observations of tropical clouds. J. Appl. Meteorol. 41, 97–117.

Stith, J.L., Haggerty, J.A., Heymsfield, A., Grainger, C.A., 2004. Microphysical characteristics of tropical updrafts in clean conditions. J. Appl. Meteorol. 43, 779–794. https://doi.org/10.1175/2104.1.

Taylor, J.W., et al., 2016. Observations of cloud microphysics and ice formation during COPE. Atmos. Chem. Phys. 16, 799–826. https://doi.org/10.5194/acp-16-799-2016.

Twohy, C.H., Schanot, A.J., Cooper, W.A., 1997. Measurement of condensed water content in liquid and ice clouds using an airborne counterflow virtual impactor. J. Atmos. Ocean. Technol. 14, 197–202. https://doi.org/10.1175/1520-0426(1997)014<0197:MOCWCI>2.0.CO;2.

Ulanowski, Z., Kaye, P.H., Hirst, E., Greenaway, R.S., Cotton, R.J., Hesse, E., et al., 2014. Incidence of rough and irregular atmospheric ice particles from Small Ice Detector 3 measurements. Atmos. Chem. Phys. 14, 1649–1662. https://doi.org/10.5194/acp-14-1649-2014.

Um, J., McFarquhar, G.M., 2011. Dependence of the single-scattering properties of small ice crystals on idealized shape models. Atmos. Chem. Phys. 11, 3159–3171. https://doi.org/10.5194/acp-11-3159-2011.

Vali, G., DeMott, P.J., Möhler, O., Whale, T.F., 2015. Technical note: a proposal for ice nucleation terminology. Atmos. Chem. Phys. 15, 10263–10270. https://doi.org/10.5194/acp-15-10263-2015.

Yano, J.-I., Phillips, V.T.J., 2010. Ice–ice collisions: an ice multiplication process in atmospheric clouds. J. Atmos. Sci. 68, 322–333. https://doi.org/10.1175/2010JAS3607.1.

FURTHER READING

Brenguier, J.L., 1989. Coincidence and dead-time corrections for particles counters. Part II: high concentration measurements with an FSSP. J. Atmos. Ocean. Technol. 6, 585–598. https://doi.org/10.1175/1520-0426(1989)006<0585:CADTCF>2.0.CO;2.

Crosier, J., Choularton, T.W., Westbrook, C.D., Blyth, A.M., Bower, K.N., Connolly, P.J., et al., 2014. Microphysical properties of cold frontal rainbands. Q. J. R. Meteorol. Soc. 140, 1257–1268. https://doi.org/10.1002/qj.2206.

Fritsch, J.M., Carbone, R.E., 2004. Improving quantitative precipitation forecasts in the warm season: a USWRP research and development strategy. Bull. Am. Meteorol. Soc. 85, 955–965. https://doi.org/10.1175/BAMS-85-7-955.

Gardiner, B.A., Hallett, J., 1985. Degradation of in-cloud forward scattering spectrometer probe measurements in the presence of ice particles. J. Atmos. Ocean. Technol. 2, 171–180. https://doi.org/10.1175/1520-0426(1985)002<0171:DOICFS>2.0.CO;2.

Hirst, E., Kaye, P.H., 1996. Experimental and theoretical light scattering profiles from spherical and non-spherical particles. J. Geophys. Res. Atmos. 101, 19231–19235. https://doi.org/10.1029/95JD02343.

Holroyd, E.W., 1987. Some techniques and uses of 2D-C habit classification software for snow particles. J. Atmos. Ocean. Technol. 4, 498–511. https://doi.org/10.1175/1520-0426(1987)004<0498:STAUOC>2.0.CO;2.

Koenig, L.R., 1965. Drop freezing through drop breakup. J. Atmos. Sci. 22, 448–451. https://doi.org/10.1175/1520-0469(1965)022<0448:DFTDB>2.0.CO;2.

Korolev, A., Khain, A., Pinsky, M., French, J., 2016. Theoretical study of mixing in liquid clouds—Part 1: classical concepts. Atmos. Chem. Phys. 16, 9235–9254. https://doi.org/10.5194/acp-16-9235-2016.

Lance, S., Brock, C.A., Rogers, D., Gordon, J.A., 2010. Water droplet calibration of the Cloud Droplet Probe (CDP) and in-flight performance in liquid, ice and mixed-phase clouds during ARCPAC. Atmos. Meas. Tech. 3, 1683–1706. https://doi.org/10.5194/amt-3-1683-2010.

Locatelli, J.D., Hobbs, P.V., 1974. Fall speeds and masses of solid precipitation particles. J. Geophys. Res. 79, 2185–2197. https://doi.org/10.1029/JC079i015p02185.

Lu, C., Liu, Y., Yum, S.S., Niu, S., Endo, S., 2012. A new approach for estimating entrainment rate in cumulus clouds. Geophys. Res. Lett. 39L04802. https://doi.org/10.1029/2011GL050546.

Rangno, A.L., Hobbs, P.V., 1994. Ice particle concentrations and rainfall development in small continental cumuliform clouds. Q. J. R. Meteorol. Soc. 120, 573–601. https://doi.org/10.1002/qj.49712051705.

Rosenow, A.A., Plummer, D.M., Rauber, R.M., McFarquhar, G.M., Jewett, B.F., Leon, D., 2014. Vertical velocity and physical structure of generating cells and convection in the comma head region of continental winter cyclones. J. Atmos. Sci. 71, 1538–1558. https://doi.org/10.1175/JAS-D-13-0249.1.

Shupe, M.D., Matrosov, S.Y., Uttal, T., 2006. Arctic mixed-phase cloud properties derived from surface-based sensors at SHEBA. J. Atmos. Sci. 63, 697–711. https://doi.org/10.1175/JAS3659.1.

CHAPTER 5

Characterization of Mixed-Phase Clouds: Contributions From the Field Campaigns and Ground Based Networks

Constantin Andronache
Boston College, Chestnut Hill, MA, United States

Contents

1. INTRODUCTION

The characterization of mixed–phase clouds (MPCs) is part of a greater challenge: how to improve cloud observations and their representation in atmospheric numerical models. Ultimately, the goal is to increase the accuracy of the weather forecast and reduce uncertainties in climate projections. Over the last decades, the cloud observing systems, including ground observations, aircraft measurements, and spaceborne remote sensing, aided by

Mixed-Phase Clouds
https://doi.org/10.1016/B978-0-12-810549-8.00005-2

numerical modeling, have contributed to a better, more detailed representation of global cloudiness. Throughout this process, the focus has been to solve various parts of the problem considered to be of immediate importance. For example, the prediction of precipitation quantity and type at mid-latitudes has been of considerable interest over many decades. Starting with the Wegener-Bergeron-Findeisen (WBF) process (Findeisen, 1938), it has been noted that supercooled and MPCs can play a substantial role in the formation of precipitation. Later, aircraft experiments showed that in certain atmospheric conditions, supercooled water droplets form a layer at the top of ice clouds. These observations were confirmed in various parts of the world (Rauber, 1987; Heymsfield et al., 1991; Heymsfield and Miloshevich, 1993; Gayet et al., 2002; Hogan et al., 2002, 2003; Shupe et al., 2008a,b). Particular attention was given to the Arctic region, a vast area that influences the climate at northern latitudes and plays a significant role in the climate system. During a series of field experiments, it has been confirmed that stratiform MPCs are persistent, frequent during many months of the year, and cover large regions.

A significant step forward was the establishment of the Atmospheric Radiation Measurement (ARM) program by the US Department of Energy (DOE), which started in the 1990s and evolved towards a fully equipped network to conduct cloud and aerosol observations (Stokes and Schwartz, 1994; Stokes, 2016). Additional efforts to understand the conditions of aircraft icing, improved the characterization of supercooled and mixed-phase clouds at mid-latitudes (Cober et al., 2001; Isaac et al., 2001, 2005; Köhler and Görsdorf, 2014). In Europe, sustained research to improve weather forecast resulted in a continuous remote sensing of cloud properties under the Cloudnet program (Illingworth et al., 2007). While these efforts delivered essential, continuous data at several ground-based sites, the need for global observations resulted in satellites such as CloudSat and Cloud-Aerosol Lidar and Infrared Pathfinder Satellite Observations (CALIPSO), which were instrumented to observe MPC characteristics at the global scale and provide much needed observational references for models (Stephens et al., 2002, 2008; Zhang et al., 2010). For a better understanding of the MPCs problems and their implications, we start with a short description of the observational aspects, followed by a review of the main applications.

1.1 Observations

Observations of MPCs are based on measurements taken using in situ aircraft sampling, ground-based networks, and remote sensing from satellite platforms. In some instances, such as in aircraft icing and cloud seeding studies, the observations were specifically designed to focus on supercooled MPCs. In a more general perspective, the MPC studies are part of comprehensive observations of all cloud systems, such as in the ARM and Cloudnet programs, which are particularly important for weather and climate modeling.

1.1.1 Earlier Observations

Rauber and Tokay (1991) reviewed some of the earlier investigations of cold cloud microstructure, which reported the existence of a thin layer of supercooled water droplets at the tops of both stratiform and convective clouds. Cooper and Vali (1981) made measurements in wintertime stratiform and convective clouds over Wyoming's Elk Mountain as well as over the San Juan Mountains of Southwest Colorado and found the top liquid layer with a depth of about 30 m. The results were confirmed by additional aircraft measurements in both stratiform and convective clouds (Hobbs and Rangno, 1985; Rauber and Grant, 1986). The liquid layer was common in orographic cloud systems that developed over the mountains of Northern Colorado, and this was observed at the top of clouds as cold as $-31°C$ (Rauber and Grant, 1986). The importance of the cloud top liquid layer, and its characteristics in the formation and subsequent ice evolution in orographic cloud systems, was discussed by Rauber (1987). Observations that followed showed that MPCs could be found in other forms: some in multiple layers, and others consisting of regions dominated by homogeneous supercooled cloud droplets next to zones of predominant ice particles.

1.1.2 Types of Mixed-Phase Clouds

There are many reported observations of different types of MPCs, including low-level stratiform (frequently found in Arctic clouds), mid-level stratiform (commonly found at mid-latitudes), and cumulus (often connected with clouds in winter storms at mid-latitudes). MPC conditions are associated with frontal systems (Hogan et al., 2002), and orography (Politovich and Vali, 1983; Rauber, 1987; Heymsfield and Miloshevich, 1993). MPC observations have been reported for various parts of the world: Arctic region (Shupe et al., 2005, 2006, 2008a,b; Shupe, 2007; McFarquhar et al., 2007; Ehrlich et al., 2008a, 2009; Lampert et al., 2010), Europe (Hogan et al., 2003; Field et al., 2004), North America (Heymsfield et al., 1991; Fleishauer et al., 2002; Noh et al., 2013; Campos et al., 2014), Asia (Gayet et al., 2002), Australia (Platt, 1977), and Antarctica (Lubin, 2004). Many studies have focused on the high-latitudes stratiform clouds, due to their frequency and spatial coverage.

Low-level mixed-phase stratiform clouds contribute significantly to the radiative budget of the Arctic, potentially reducing wintertime net surface cooling by about $40–50\ Wm^{-2}$ (Curry et al., 1996). Summaries of cloud climatology show that low-altitude stratus frequency in the Arctic is of up to 70% during transitional seasons. High-latitude observations from the Mixed-Phase Arctic Clouds Experiment (M-PACE) (Verlinde et al., 2007; Morrison et al., 2012) reveal long-lived mixed-phase layers, with continuous cloud coverage lasting up to several days at a time.

Shallow mixed-phase cloud layers like altocumulus, altostratus, or stratocumulus, have been used by different research groups to study aerosol-cloud-dynamics interaction under ambient conditions (Fleishauer et al., 2002; Sassen and Khvorostyanov, 2007;

Bühl et al., 2013, 2016; Schmidt et al., 2015). Altostratus can be irregularly shaped and spaced, and often cover the sky over an area of several thousand square kilometers. Observations show that within an altostratus cloud with large vertical extent, a heterogeneous particulate composition might be present: (1) an upper part, dominated by ice crystals; (2) a middle part, made of a mixture supercooled water droplets, ice, and snowflakes; (3) a lower part, dominated by supercooled or ordinary water droplets (Bühl et al., 2016).

These observations of MPCs have been instrumental in the progress to solve a series of practical problems. Some of the main applications with particular attention to MPCs are: (a) aircraft icing, a problem of great importance for aviation safety; (b) cloud seeding, weather modification, rain enhancement, and hail suppression; (c) improvement of numerical weather prediction, particularly of precipitation intensity and type; (d) reduction of uncertainties in climate model projections.

1.2 Aircraft Icing

Aircraft icing is the formation of ice that can occur on aircraft while on the ground or in flight. Since ice nuclei are often present in a small concentration in the atmosphere (Phillips et al., 2008), for a large temperature range $-38°C < T < 0°C$ supercooled liquid water (SLW) can exist for a long time. The surface of an aircraft flying through such a cloud or rain acts as an ice nucleus. Ice accretion in flight may affect the aerodynamic characteristics and engine performance. In this context, ice accretion is the process by which a layer of ice forms on objects exposed to freezing precipitation, supercooled cloud droplets, or fog. There are several types of airframe icing related to supercooled and MPC conditions. The first has the aspect of a white opaque deposit that forms in clouds of low water content, containing small SLW droplets, at temperatures well below 0°C. The second consists of a coating of clear ice that forms in clouds of high liquid water content consisting of large SLW droplets in the form of drizzle or rainfall on aircraft, with a temperature near or below 0°C. The third is a mixed ice or cloudy ice that occurs in clouds with a mix of ice crystals, cloud droplets, and snowflakes. These icing scenarios involve the presence of SLW droplets, a characteristic of MPCs, and have been studied extensively (Cober et al., 2001; Cober and Isaac, 2012; Isaac et al., 2001, 2005; Köhler and Görsdorf, 2014).

1.3 Cloud Seeding

Another instance where supercooled and mixed-phase clouds are studied extensively is related to cloud seeding and weather modification. Cloud seeding involves the addition of aerosol, such as silver iodide aerosol, that modifies the phase and size distribution of hydrometeors. The goal of cloud seeding is to alter the natural development of the cloud to enhance precipitation, suppress hail, dissipate fog, or reduce lightning. Various cloud seeding techniques are employed, as particles are released from rockets, aircraft, or ground. The seeding of ice-phase clouds can induce the phase transition from a super-cooled water cloud to one composed of ice. In the case of dynamic cloud seeding, the purpose is to stimulate vertical air motions through increased buoyancy caused by the release of latent heat of freezing (Hobbs, 1975; Cotton and Pielke, 1995).

1.4 Precipitation Forecast

Since the early finding that the WBF process is important in rain formation, the prediction of precipitation quantity and type has presented challenges, particularly in complex terrain, and during extreme rainfall events. Operational weather forecast models have a typical spatial resolution of the order of 10 km horizontally, with a trend toward using finer resolutions. Besides the shortcomings in input observations, data assimilation, and model resolution, there are limitations in the representation of clouds in these models, particularly those of MPCs. Comparison between Cloudnet high-resolution observational data and operational weather numerical models showed significant discrepancies between observed and predicted cloud fields (Hogan et al., 2001; Illingworth et al., 2007; Bouniol et al., 2010). Meanwhile, these studies provided valuable hints on how to improve the models and cloud observations. Given the importance of precipitation forecast in many practical meteorological and hydrological applications, a better representation of MPCs in such models remains a priority.

1.5 Climate Modeling

Numerical climate models are essential tools to understand possible climate changes. Accurate representation of clouds and aerosols in these models is one of the current research challenges and requires a better parameterization of physical processes in numerical schemes (Klein et al., 2009). In this effort, the representation of MPCs is of particular importance (Furtado et al., 2016; McCoy et al., 2016; Tan et al., 2016). Progress in the treatment of MPCs in general circulation models (GCMs) benefited greatly from studies at a smaller scale. Thus, the results from Large Eddy Simulations (LES) and Weather Research and Forecasting (WRF) modeling provide useful hints of what parameterization might work, moving from a detailed description to a more schematic cloud representation at larger grid models. Modeling approaches for the treatment of MPCs can be grouped in: (a) Theoretical and simplified models (Korolev and Field, 2008; Field et al., 2014); (b) LES (Marsham et al., 2006; Hill et al., 2014; Field et al., 2014); (c) WRF (Morrison and Pinto, 2005; Morrison et al., 2009), and (d) Global models (Lohmann and Hoose, 2009; Choi et al., 2014; Komurcu et al., 2014; McCoy et al., 2016).

This chapter presents the development of MPC characterization of interest in numerical weather forecast and climate models. The focus is on the observational aspects of MPCs from field campaigns and ground-based networks. The satellite remote sensing of MPCs from space to achieve continuous global monitoring of cloud properties is treated in other chapters. Section 2 describes a suite of field experiments, many of them designed to support the development of permanent ground stations in the ARM program. Section 3 presents the ARM and Cloudnet networks and their main accomplishments. Finally, Section 4 gives a summary of the main achievements and some of the outstanding issues facing the research of MPCs.

2. OBSERVATIONS OF MIXED-PHASE CLOUDS DURING FIELD CAMPAIGNS

This section reviews several field experiments in which MPCs were observed, with a focus on middle and high latitude regions. These experiments are listed in chronological order. One exception is related to aircraft icing experiments, which are presented at the end of this section. The list is not comprehensive. Particularly, the convective MPCs are treated in detail in this volume by Jackson et al. (2017). The primary objectives, results, and related references are given. Many of these field campaigns contributed to the development of ground-based observation systems, as in the case of the ARM North Slope of Alaska (NSA) site. These campaigns show rapid progress in improving instrumentation and increasing the ability to address complex scientific problems concerning the role of clouds in the climate system. A series of airborne experiments, combined with ground-based observations in the Arctic region, documented the single-layer and multilayer MPCs. The typical single-layer MPC has a single supercooled liquid layer located at cloud top. Below this liquid layer, ice crystals are present, falling to the surface (Shupe et al., 2005; Verlinde et al., 2007; Gayet et al., 2009a; Mioche et al., 2015). Many of the field campaigns were conducted in the North American Arctic and the European Arctic regions. These regions have distinct climate and meteorological features with impact on MPC properties and their climatology.

2.1 Beaufort and Arctic Storms Experiment (BASE)

The Beaufort and Arctic Storms Experiment (BASE), part of the Canadian Global Energy and Water Cycle Experiment (GEWEX), was conducted off the Canadian Beaufort coast, during Sep. and Oct. 1994. The goal of BASE was to understand the storms that occur in this region and their impact on local hydrology and sea ice conditions. The project provided the characteristics of the cloudy boundary layer observed from a research aircraft. It made available the data needed to understand the evolution of the atmospheric boundary layer (BL) and sea ice properties during the autumnal freezing of the Beaufort Sea. Observations from 12 research flights during the period Sep. 21 through Oct. 25, 1994, were taken when the underlying surface made a transition from late summer melt conditions to winter conditions. The data set includes measurements of radiative fluxes, cloud microphysical properties, aerosols and condensation nuclei, as well as high-resolution meteorological variables. The location of the ice edge changed, the multiyear ice made a transition to wintertime conditions as the melt ponds and leads froze, and the surface became increasingly snow covered. Most of the observed BL clouds had a cloud base lower than 100 m and a cloud top below 1 km. They were associated with stable temperature profiles, with temperature inversions above or within the cloud deck. Analysis of the cloud phase indicated a transition from primarily liquid phase clouds at the beginning of the experiment to predominantly ice clouds at the end of the experiment. MPCs were observed in a temperature range of $-8°C$ to $-15°C$. Data analysis revealed that BL clouds are strongly influenced by the underlying surface and large-scale atmospheric dynamics (Curry et al., 1997).

2.2 The First International Satellite Cloud Climatology Project (ISCCP) Regional Experiment Arctic Clouds Experiment (FIRE-ACE)

FIRE-ACE was conducted Apr. to Jul. 1998 using aircraft to fly over surface-based observational sites in the Arctic Ocean and at Barrow, Alaska. The principal goal was to examine the impact of Arctic clouds on the exchange of radiation between the surface and atmosphere, and to study how the surface influences the development of the BL clouds under spring and summer conditions. The observations were designed to: (a) improve climate model parameterizations of cloud and radiation processes; (b) develop satellite remote sensing of cloud and surface characteristics, and (c) advance the understanding of cloud radiation feedbacks. One issue addressed was to understand the transition from liquid to ice phase BL clouds, its dependence on temperature and aerosol characteristics, and seasonal characteristics of such processes. FIRE-ACE interacted closely with the Surface Heat Budget of the Arctic (SHEBA) and the ARM program.

Observations showed a persistent humidity inversion above BL cloud tops correlated with the static stability of the arctic environment. These conditions contribute to the homogeneity and persistence of the cloud by inhibiting evaporative cooling due to entrainment mixing at cloud top. Data showed significant variations in the relationship between cloud temperature and phase, influenced by the presence of ice nuclei, seeding of the cloud by ice particles falling from above, and the size of the liquid drops. The spatial inhomogeneity of the surface physical and optical characteristics is substantial, especially during the summer melt season. The experiment proved that a combination of visible, near-infrared, sub-millimeter, and microwave wavelengths sensors has considerable skill in discriminating the characteristics of the surface and clouds under many conditions (Curry et al., 2000).

2.3 The Surface Heat Budget of the Arctic (SHEBA) Program

SHEBA occurred from Oct. 1997 through Oct. 1998 in the Beaufort Sea, and from the ARM Program's measurements at the North Slope of Alaska (NSA) site in Barrow,

Alaska. Measurements from these two sites give a comprehensive view of Arctic cloudiness in all seasons and a large set of MPC observations. SHEBA provided information of the trends and magnitude of MPCs over a full year in Arctic environment. Some of the major conclusions concerning the macro- and microphysical properties based on the measurements from SHEBA showed that:

- MPCs occur about 41% of the time in the Arctic, predominantly during the spring and fall transition seasons, at temperatures ranging from $-25°C$ to $-5°C$, have a thickness of $\sim$0.5–3 km, with a cloud base near the surface. On average, such clouds persisted for about 12 h, while in many instances the persistence occurred for many days.
- The liquid fraction (the ratio of liquid water to total condensed water) increases with temperature. The annual average relationship shows a transition from full glaciation at $-24°C$ to complete liquid water at $-14°C$ (Shupe et al., 2006).

The ground-based instruments at SHEBA for the identification and characterization of MPCs were: (a) the millimeter cloud radar, which provided profiles of radar reflectivity, Doppler velocity, and Doppler spectrum width; (b) the microwave radiometer (MWR), which provided estimates of the column integrated liquid water path and water vapor amounts; (c) a depolarization lidar, which provided information on cloud phase; and (d) radiosondes, which provided profiles of temperature and humidity. Also, the cloud-type classification was produced using ground-based sensors.

Clouds above the SHEBA ice camp were categorized as being all ice, all liquid, mixed-phase, or precipitating based on data from the above instruments and surface observations. MPCs are defined as cloud layers that contain both liquid and ice. Many MPCs also contained embedded regions of liquid, as identified by the depolarization lidar. While all measurements were combined to determine a cloud classification, the following criteria were typically used to identify MPCs: (1) a positive liquid water path (LWP) derived from the MWR measurements; (2) cloud temperatures $<0°C$ from the radiosonde data; (3) radar reflectivity typically >-15 dBZ; and (4) radar Doppler velocity typically >0.5 m s^{-1} at some height in the cloud. Also, surface-observer records were used to distinguish periods of snowfall (Shupe et al., 2006).

2.4 The Mixed-Phase Arctic Cloud Experiment (M-PACE)

The Mixed-Phase Arctic Cloud Experiment (M-PACE) took place from Sep. 27 to Oct. 22, 2004, during the fall season in the area of the North Slope of Alaska (NSA) (Fig. 1). The primary objective was to collect observations needed to advance understanding of the cloud microphysics, dynamics, radiative properties, and evolution of Arctic MPCs (Verlinde et al., 2007). Observations showed that MPCs tend to be long-lived, with liquid tops that continually precipitate ice, and they dominate the low-cloud fraction within the Arctic during the coldest periods of the year (McFarquhar et al., 2007). M-PACE addressed a set of outstanding scientific questions designed to understand how

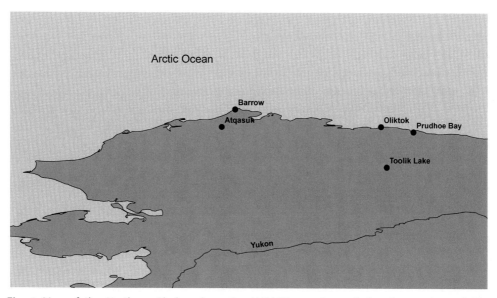

Fig. 1 Map of the Northern Alaska where the M-PACE experimental domain was located. The operation center was situated in Prudhoe Bay, southeast of Oliktok Point. Shown are four surface sites (Barrow, Atqasuk, Oliktok Point, and Toolik Lake) where radiosonde launches were conducted. For details, see Verlinde et al. (2007).

MPCs microphysics, radiation, and cloud dynamics are related. With the advances in ground-based remote sensing instruments available during the campaign, another question was how well they can characterize the macro- and microscopic properties of MPCs. Expanding on previous studies, the mission also explored the characteristics of Arctic midlevel clouds.

One significant achievement of the program was the MPC classification using remote sensing at ARM site in the NSA area. A series of innovative techniques involving the ARM millimeter cloud radar (MMCR) and the lidar made possible very detailed measurements of MPCs. The method used is a multisensor approach, which exploits phase-specific signatures from radar, lidar, MWR, and radiosonde measurements to discriminate between phases (Shupe, 2007). The criteria utilized to detect MCPs were quite similar to those employed during SHEBA, with some improvements, particularly in the use of the low lidar depolarization ratio (<0.1) to detect the presence of the liquid water droplets. Some of the M-PACE observations, based on aircraft flights underneath and coincident with satellite overpasses, contributed as a basis for the new algorithms to identify MPCs from satellite measurements. Another task of the mission was to obtain measurements in an area close to the ARM NSA site to be used in single-column-modeling (SCM) validation.

The M-PACE documented the microphysical structure of Arctic MPCs, with multiple in situ profiles, collected in both single- and multilayer clouds over the two ground-based remote sensing sites at Barrow and Oliktok Point. The liquid phase was found in clouds with cloud-top temperatures as low as $-30°C$, in clouds forming as the result of strong surface forcing, and also in weakly forced stratiform clouds. This result confirmed the SHEBA finding that MPCs are common in the Arctic region at very low temperatures, showing it is also true in the coastal regions of the North Slope (Verlinde et al., 2007).

2.5 The Arctic Study of Tropospheric Cloud, Aerosol and Radiation (ASTAR)

The experimental campaign ASTAR was carried out in the European sector of the Arctic region in two phases: (1) between May 15 and Jun. 19, 2004, and (2) Apr. 7–9, 2007. The primary objective of ASTAR was the measurement of aerosol and cloud properties in the polar troposphere using a research aircraft. During the 2004 campaign, airborne measurements in slightly supercooled Arctic BL stratocumulus have been performed in Spitsbergen on May 29. The flight was executed over the Greenland Sea in the area of the west coast of the Svalbard Archipelago (Fig. 2). In situ cloud measurements, in both warm and

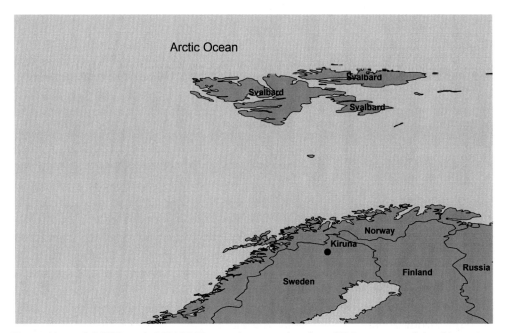

Fig. 2 Map of ASTAR and POLARCAT experiments related to MPCs presented in this review: ASTAR 2004 experiment was carried out from Svalbard from May 10 to Jun. 11, 2004; ASTAR 2007 (Apr. 7–9) had measurements over the open Greenland Sea; POLARCAT had data collected during Apr. 1, 2008, north of Kiruna, Sweden.

cold sectors of a cold front passing the observation area, displayed a north-south gradient in freezing properties, and evidence of significant differences in the cloud microstructure (Gayet et al., 2009a).

During the Apr. 7–9, 2007 campaign, a synoptic case was characterized by a cold air outbreak. The northerly winds initiated extended boundary-layer cloud fields over the open Greenland Sea. The convection above the relatively warm open sea maintained the coexistence of ice and liquid water particles in the clouds. In addition to the predominating MPCs, pure ice, and pure liquid water clouds were observed during this period of ASTAR 2007, providing the opportunity to test cloud phase identification methods. The thermodynamic phase of the detected clouds was estimated by measurements of the reflected solar radiation (for details, see Ehrlich et al., 2008b). For a case study on Apr. 9, 2007, the combination of CALIPSO/CloudSat data with collocated in situ observations provided new insights on mixed-phase layer clouds in the Arctic region. The MPC on Apr. 9, 2007 exhibited a cloud top layer dominated by liquid water in which ice precipitation was produced. This event confirmed the common feature observed in Arctic mixed-phase stratocumulus clouds, even for cloud top temperatures down to −25°C during ASTAR (Gayet et al., 2009b).

2.6 The Polar Study Using Aircraft, Remote Sensing Surface Measurements and Models of Climate, Chemistry, Aerosols and Transport (POLARCAT)

During the POLARCAT campaign in 2008, the airborne radar-lidar instrument RALI was used to study cloud processes and evaluate satellite products. The general goal of the POLARCAT campaign was to identify the impact of trace gases and aerosols, which are transported to the higher latitudes, and their effects on Arctic climate. RALI is designed to retrieve cloud properties and characterize its phase at scales smaller than a kilometer. Such details are important for cloud process analysis, and RALI capabilities were illustrated for Arctic cloud data collected on Apr. 1, 2008, north of Kiruna, Sweden, over northern Norway and the Arctic Ocean (Fig. 2). RALI combines two nadir-pointing instruments: (a) the 95-GHz Doppler radar System Airborne (RASTA) and (b) the Leandre New Generation (LNG) lidar. LNG was set up in its backscatter configuration, operating at three wavelengths (355, 532, and 1064 nm), including depolarization at 355 nm (for details, see Delanoë et al., 2013).

Taking simultaneous radar and lidar measurements and exploiting the advantages of their synergy for cloud studies originated in the early studies of Arctic clouds (Intrieri et al., 2002; Shupe et al., 2006). Since radar and lidar operate at different wavelengths, they show different sensitivities to the same set of hydrometeors. The lidar signal is dominated by the concentration of the cloud particles. In contrast, the radar reflectivity is very sensitive to the particle size. The difference in sensitivity can also be utilized. The lidar is more sensitive to small liquid drops or small crystals and can detect thin clouds containing both ice crystals and liquid droplets that are not detectable by the radar. The radar has more sensitivity to large crystals, and can deeply penetrate ice clouds when the lidar signal is too weak (Delanoë et al., 2013).

2.7 The Indirect and Semi-Direct Aerosol Campaign (ISDAC)

The Indirect and Semi-Direct Aerosol Campaign (ISDAC) was conducted over Alaska during Apr. 2008 of the International Polar Year. The primary goal of ISDAC was to understand how changes in aerosols influence cloud properties and the related radiative forcing. ISDAC in Apr. 2008 built upon the success of M-PACE, operated in Oct. 2004, by allowing Arctic aerosol and cloud properties to be contrasted between the somewhat pristine fall and more polluted spring seasons (McFarquhar et al., 2011). The measurements gathered by 41 instruments on the NRC Convair-580 during ISDAC provide a complete airborne dataset on aerosol microphysical and radiative properties on Arctic BL and clouds over the North Slope of Alaska. During Apr. 2008, the Convair-580 aircraft collected data inside, above, and below a single-layer Arctic stratocumulus cloud. The extensive observations provided data needed to test models at various scales and validate ground-based remote sensing instruments at NSA.

One highlight of this experiment is represented by the comprehensive measurements of aerosol, cloud microphysics, radiative fluxes, and meteorological data required to study the longevity of single layer MPC and its dependence on aerosol properties. In this context, one application was to use data and test three aerosol indirect effects hypothesized to act in a MPC: (a) the riming indirect effect; (b) the glaciation indirect effect, and (c) the thermodynamic indirect effect. The study provided evidence for cases of glaciation and thermodynamic indirect effects (Jackson et al., 2012).

2.8 Aircraft Icing Experiments

Supercooled liquid water (SLW) droplets in the atmosphere can cause aircraft icing. Predicting SLW with numerical weather prediction (NWP) models is of great importance for aircraft safety. New cloud parameterizations in forecast models benefit greatly from in situ measurements of aircraft icing conditions. In this context, observations of aircraft icing environments that included SLW droplets were made during a series of field projects conducted by Environment Canada (EC) and the NASA Glenn Icing Technology Branch, during the period from 1995 through 2000 (Cober and Isaac, 2012). These field projects included the following:

The Canadian Freezing Drizzle Experiment (CFDE) had three phases. The CFDE I project was conducted in Mar. 1995, with numerous flights over Newfoundland and the Atlantic Ocean. The CFDE II, conducted during Dec. 1996 and Jan. 1997, was mainly a test program, in preparation for the next missions. The CFDE III started in Dec. 1997 and ended in Feb. 1998; during CFDE III, the aircraft flew over Southern Ontario, Quebec, Lake Ontario, and Lake Erie (Cober et al., 2001; Isaac et al., 2001).

The FIRE-ACE (reviewed above) was conducted by EC in Apr. 1998 (Curry et al., 2000), and the NRC Convair-580 research aircraft, based out of Inuvik in the Northwest Territories, Canada, sampled boundary layer and midlevel Arctic clouds during 18 flights. This experiment is mentioned here in the context of EC long-term investigations of aircraft icing.

The First Alliance Icing Research Study (AIRS) (Cober and Isaac, 2012) occurred during Dec. 1999 to Feb. 2000. Five research aircraft were involved in the field project. These aircraft flew special flight operations over a network of ground in situ and remote-sensing meteorological measurement systems, located at Mirabel, Quebec. The field phase of the Second Alliance Icing Research Study (AIRS II) was conducted during the winter of 2003/2004 in the same area. AIRS II was a project endorsed by the Aircraft Icing Research Alliance (AIRA), which consists of government organizations interested in aircraft icing. It was also supported by the World Meteorological Organization (WMO) World Weather Research Program (WWRP) project on Aircraft In-Flight Icing (Fig. 3).

AIRS II was planned based on the results of AIRS I, and had operational objectives to improve the characterization of the aircraft icing process and its effects on aircraft operations. To support the operational goals of the program, a series of objectives were addressed to investigate the conditions associated with supercooled large drop formation,

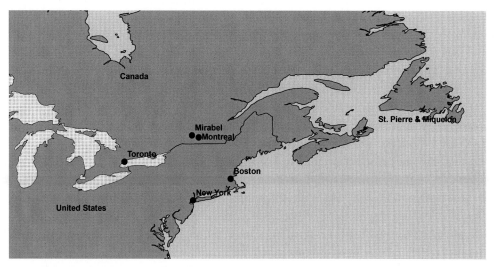

Fig. 3 Map of the CFDE and AIRS missions conducted mainly in eastern Canada. The CFDE I was conducted during Mar. 1995 and it was based from St. John's, Newfoundland, Canada. The CFDE III was conducted during the period of Dec. 1997 to Feb. 1998 and the NRC Convair-580 research aircraft conducted flights over southern Ontario, southern Quebec (Canada), Lake Ontario, and Lake Erie. AIRS I was conducted during Dec. 1999 and Feb. 2000, it was based from Ottawa and the majority of the research flights were conducted in the vicinity of Mirabel, Quebec. AIRS II was conducted between Nov. 2003 and Feb. 2004, with several research aircraft operated out of Ottawa, Ontario, Cleveland, Ohio, and Bangor, Maine (see Cober and Isaac, 2012).

cloud glaciation, to determine the spatial distribution of ice crystals and supercooled water, and verify the response of remote sensors to various cloud particles (Isaac et al., 2005). Such field measurements were complemented by wind tunnel experiments and numerical simulations of aircraft icing.

3. GROUND BASED PROGRAMS

Before the creation of permanent ground-based observation systems, the cloud research was based to a large extent on short-term field experiments. These campaigns, lasting from a few weeks to several months, produced excellent results but were not sufficient to generate climatically representative datasets. Nevertheless, they were critical in the development of permanent ground-based comprehensive monitoring systems such as ARM and Cloudnet, described in this section. We note that while ARM is presented here as an example for its ground-based network, the program is more comprehensive and includes both mobile and airborne observing facilities.

3.1 Atmospheric Radiation Measurement (ARM)

The US DOE ARM Program is summarized here from the perspective of contribution to MPC characterization. The primary goal of the ARM Program was to improve cloud parameterizations in global climate models (GCMs) through a better understanding of cloud and radiation processes obtained from a combination of modeling and data analysis (Mather et al., 2016). Since the early 1990s, the ARM Program developed field measurements and modeling studies to improve the representation of clouds in the Earth's climate. The ARM Program established several ground stations for studying aerosol, cloud, and radiative transfer in the atmosphere. These highly instrumented stations also offer enhanced sites for periodic airborne or intensive observational periods (IOP) and remote-sensing studies and complement satellite atmospheric observations. The ARM Program evolved to provide a continuous data record on climatologically time scales and address the observational and modeling complexities of clouds (Stokes and Schwartz, 1994; Stokes, 2016). Concerning the MPC characterization, the ARM program recognized, from its early stages, the importance of the cloud phase identification, particularly in the application of cloud retrieval methods. The success in the detection and characterization of MPCs is possible by the development of measurements that have complementary phase-specific signatures, which can constrain cloud phase. The classification of cloud according to meteorological type and phase type is constructed on multisensory, threshold-based techniques. Over more than two decades, the ARM program developed and refined a set of instruments capable of providing very detailed physical description of MPCs (Shupe et al., 2016).

The ARM fixed research sites were selected to represent three different climatic regimes: (1) Southern Great Plains (SGP) to address variable mid-latitude climate

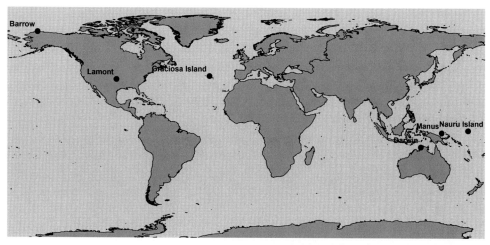

Fig. 4 Locations of the ARM program fixed sites. The Western Pacific sites are no longer in operation.

conditions, (2) North Slope of Alaska (NSA) to address land and land–sea-ice arctic climate, and (3) Eastern North Atlantic (ENA) to address marine stratocumulus. Also, two ARM Mobile Facilities (AMF1 and AMF2) are deployed for short-term field campaigns (approximately 1 year) at sites around the world. AMF2 is designed for deployments in a marine environment. A third mobile facility (AMF3) is currently deployed for an extended period at Oliktok, Alaska. Between 1996 and 2014, ARM operated instruments at three sites in the Tropical Western Pacific (TWP). While these sites are no longer operating, the data from these locations are available at the ARM data archive. The positions of the ARM fixed research sites are shown in Fig. 4.

3.1.1 Southern Great Plains (SGP)

The SGP site consists of both in situ and remote-sensing instruments placed across north-central Oklahoma and south-central Kansas. The site choice was greatly influenced by the potential synergistic opportunities of collaboration with other state and federal observational networks and research programs. It started data collection in 1992 and provided many valuable lessons to be applied later in developing other ARM sites. The ARM SGP site includes a Central Facility (CF) (36°36′N, 97°29′W) with routine operations of Raman lidars, millimeter-wavelength cloud radar, micropulse lidar, microwave radiometer, and radar wind profilers. These remote sensors are augmented by surface radiation measurements, balloon-borne atmospheric profiling, and surface latent and sensible heat flux measurements. The site is expanded to boundary-layer profiling sites, a soil moisture network, in situ precipitation instruments, and upgraded soil moisture measurements, to support the development of continuous model forcing data sets. Four boundary facilities (BF) placed approximately 200 km away from the CF on the sides of SGP domain provided data on large-scale motion of the atmosphere, passing into and out of the domain. The data set will be used to run and validate a high-resolution model on a routine basis. With the expanded capabilities, the SGP site is being designated a "megasite" and continues to evolve and provide comprehensive atmospheric observations (Sisterson et al., 2016).

3.1.2 North Slope of Alaska (NSA)

The NSA site provides continuous observations of cloud and radiative processes at high latitudes. The site consists of a facility at Barrow, Alaska (71°19'N, 156°37'W), which includes a subset of the instruments available at the SGP Central Facility. NSA operating instruments are: a millimeter-wavelength cloud radar, a micropulse lidar, radiometers, and instruments for atmospheric profiling and measurements of surface meteorology. NSA started data collection in 1997. Data from these instruments are being used to understand cloud processes in the Arctic region and to refine models and parameterizations related to Arctic climate. The NSA site provides a test bed for studies of climate change at high latitudes. One major scientific contribution of NSA was the characterization in detail of the Arctic MPCs. Observations collected at NSA have been used to evaluate satellite cloud detection. The campaigns M-PACE and ISDAC contributed greatly to this effort and allowed a closer look at the impacts of aerosols on Arctic MPCs. Like the SGP, the NSA, in combination with the AMF3 site at Oliktok, is being developed as a second megasite. Much of the megasite development is being carried out at Oliktok. Like the SGP, the goal is to expand observational capabilities at the NSA to better support model simulations and process study research (Verlinde et al., 2016). Table 1 gives a subset of instruments at NSA, capable of characterizing MPCs.

3.1.3 Eastern North Atlantic (ENA)

The ENA site is located on Graciosa Island (39°5'N, 28°1'W) in the Azores archipelago, in the northeastern Atlantic Ocean. The ARM measurements started in 2013. The region is characterized by marine stratocumulus clouds. The response of these clouds to changes in atmospheric gases and aerosols is a major source of uncertainty in global climate models (Wood, 2012).

Other ARM Facilities include Mobile Facilities (AMFs) and Aerial Facility (AAF). The AMFs were designed to be deployed to locations around the world for campaigns lasting 6–24 months. They can operate in any environment, from poles to tropics. The AAF, including manned aircraft and fixed-wing unmanned aerial systems (UAS), provides airborne measurements required to answer science questions within the ARM program.

The ARM Facility is evolving toward the application of ARM observations and data processing for the understanding of important atmospheric processes and their representation in global climate models. This aim will be achieved by a series of developments:

Table 1 Active remote sensing instruments at Barrow Alaska (the ARM NSA site) used to monitor mixed-phase cloud properties

Name	Full name	Measured parameters	Notes
TSI	Total Sky Imager	Fraction of the sky view covered by clouds	[a]
MPL	Micropulse Lidar	Altitude of clouds	[b]
HSRL	High Spectral Resolution Lidar	Vertical profiles of optical depth, backscatter cross-section, depolarization, and backscatter phase function	[c]
MWR	Microwave Radiometer	Column-integrated amounts of water vapor and liquid water	[d]
KAZR	Ka ARM Zenith Radar	Zenith-pointing Doppler radar that operates at a frequency of ~35 GHz	[e]
CEIL	Ceilometer	Cloud height, vertical visibility, and potential backscatter signals by aerosols	[f]

[a] The total sky imager (TSI) measures the fraction of the sky view covered by clouds.

[b] Micropulse Lidar (MPL) is a system designed primarily to determine the altitude of clouds; however, it is also used for detection of atmospheric aerosols. Similar to a radar, pulses of energy are transmitted into the atmosphere. The energy scattered back to the transceiver is collected and measured as a time-resolved signal, detecting clouds and aerosols in real time.

[c] The High Spectral Resolution Lidar (HSRL) system collects data about clouds and aerosols. It provides vertical profiles of optical depth, backscatter cross-section, depolarization, and backscatter phase function. All HSRL measurements are absolutely calibrated by reference to molecular scattering, which is measured at each point in the lidar profile. The HSRL can measure backscatter cross-sections and optical depths without prior assumptions about the scattering properties of the atmosphere. The depolarization observations also allow robust discrimination between ice and water clouds.

[d] The Microwave Radiometer (MWR) provides time-series measurements of column-integrated amounts of water vapor and liquid water. The instrument detects the microwave emissions of the vapor and liquid water molecules in the atmosphere at two frequencies: 23.8 and 31.4 GHz. Integrated water vapor and liquid water path are derived from radiance measurements with a statistical retrieval algorithm that uses monthly derived and location-dependent linear regression coefficients.

[e] The Ka-band ARM zenith radar (KAZR) remotely probes the extent and composition of clouds at millimeter wavelengths. The KAZR is a zenith-pointing Doppler radar that operates at a frequency of about 35 GHz. The main purpose of this radar is to determine the first three Doppler moments (reflectivity, vertical velocity, and spectral width) at a range resolution of approximately 30 m from near-ground to nearly 20 km in altitude. The KAZR replaces the millimeter-wavelength cloud radar (MMCR) and uses a new digital receiver that provides higher spatial and temporal resolution than the MMCR.

[f] The ceilometer (CEIL) measures cloud height, vertical visibility, and potential backscatter signals by aerosols. It detects up to three cloud layers simultaneously. Operating through a maximum vertical range of 7700 m, the CEIL transmits near-infrared pulses of light and the receiver detects the light scattered back by clouds and precipitation.

This is a subset of available instruments at the NSA site.

(1) Enhancing ARM observations to enable the routine operation and evaluation of high-resolution models. Particularly the addition of cloud scanning radars and precipitation radars to provide a 3D view of cloud systems (K- and W- or K- and X-band systems) and precipitation (C- and X-band systems) will provide more complete data in the area. These additions will permit ARM to have a major role in the validation of satellite-based cloud and precipitation retrievals.

(2) Regular operation of high-resolution models at ARM sites. This approach, similar to that demonstrated by the Cloudnet program described in the next section, will allow

direct comparison of high-resolution models with detailed cloud observations. This effort will be complemented by the development of data products and analysis tools for the evaluation of models using ARM data.

3.2 Cloudnet

The Cloudnet program was conducted during 2001–05 and its activities continue under the European Project ACTRIS (Aerosols, Clouds, and Trace gases Research InfraStructure Network). ACTRIS is aiming at integrating European ground-based stations equipped with advanced atmospheric probing instrumentation for aerosols, clouds, and short-lived gas-phase species. The goal of Cloudnet is to give a systematic evaluation of clouds in NWP models. In essence, cloud fraction, liquid and ice water contents—derived from long-term radar, lidar, and microwave radiometer data—are carefully compared to models to quantify and improve their performance (Illingworth et al., 2007).

The difficulty of making accurate observations delayed the effort to improve the representation of clouds in forecast models. While in situ aircraft measurements reveal the macroscopic structure and typical water contents of clouds and the habits of cloud ice crystals, they suffer from limitations in sampling. The solution was presented and demonstrated by the ARM Program with a network of several ground stations, which continuously monitor cloud-related variables over multiyear time periods. This approach showed how to fill the gap between the ground-based case studies and satellite remote sensing.

To address the problem of continuous high-resolution measurements of cloud properties and their representation in NWP models, the Cloudnet program: (1) established a ground-based network with active remote sensing instruments (Fig. 5); and (2) involved several European operational forecast centers to evaluate their models and improve skill in cloud predictions. For continuous long-term vertical profiles of cloud fraction, liquid water content, and ice water content suitable for evaluating models, the minimum three instruments are needed: (a) a Doppler cloud radar; (b) a ceilometer to detect the cloud base of liquid water clouds; (c) a dual-frequency microwave radiometer to derive accurate liquid water path. Also, a rain gauge is present at all stations. Derived meteorological products averaged to the grid of each model, together with the model value for comparison are cloud fraction in each model grid box, grid-box mean liquid water content, and grid-box mean ice water content. The operational forecast centers involved are European Centre for Medium-Range Weather Forecasts (ECMWF), Met Office, Météo France, Royal Netherlands Meteorological Institute (KNMI), Swedish Meteorological and Hydrological Institute (SMHI), and Deutscher Wetterdienst (DWD), which used models ranging from mesoscale to global scale (for details, see Illingworth et al., 2007). A detailed analysis of using Cloudnet data for evaluation several NWP models is presented by Bouniol et al. (2010).

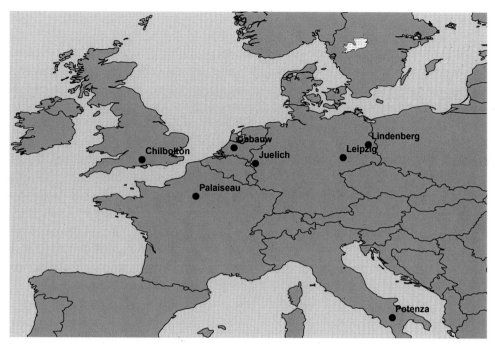

Fig. 5 The Cloudnet observing stations. The original stations are: Chilbolton (51°08′N, 1°26′W), United Kingdom; Palaiseau (48°43′N, 2°13′E), France; Cabauw (51°58′N, 4°55′E), Netherlands (see Illingworth et al., 2007).

It is important to note that Cloudnet is continuing to expand and include more ground observation sites and modeling groups. While there is some variability in the available instrumentation at various sites, data from the new sites are analyzed using the same algorithms. An example of addition is the Leipzig Aerosol and Cloud Remote Observations System (LACROS) (51°20′N, 12°23′E), which joined the Cloudnet consortium in 2011 (Bühl et al., 2016). The analysis was extended to Lindenberg (52°21′N, 14°12′E), Germany, and the ARM sites in Oklahoma, the North Slope of Alaska, and the two stations in the western Pacific at Nauru and Manus. The Cloudnet analysis can be extended to additional observing sites in different geographical locations and other forecasting models. More details about the current status of Cloudnet and ACTRIS, as well as ongoing collaborations with the ARM program, are given by Haeffelin et al. (2016).

4. CONCLUDING REMARKS

Over the last decades, the characterization of MPCs was achieved by aircraft in situ measurements, ground-based monitoring sites, satellite remote sensing, and numerical modeling. This chapter reviewed the contributions from several outstanding field campaigns and ground-based networks.

Earlier field campaigns focused on measurements of the microphysical characteristics and dynamical conditions of MPC formation and evolution. These studies contributed to solving problems such as aircraft icing and cloud seeding for weather modification. In situ aircraft measurements documented the presence of MPCs with a layer of supercooled liquid water on the top of an ice cloud. As measurements accumulated from observations in various parts of the world, it became evident that MPCs can be found in low-level, mid-level, and convective clouds. The frequency of these clouds increases with latitude and observations showed that Arctic MPCs can be persistent, with significant implications for the climate. Gradually, the experimental data were used to validate and refine numerical models of increasing complexity, from conceptual models to LES, and detailed regional simulations. It has been demonstrated that the persistence of a single layer MPC is the result of competition between supercooled cloud droplets formation and removal of ice crystals as they grow and fall out of cloud. A MPC is often the result of a delicate balance between microphysical and dynamical processes frequent in Arctic regions.

With the start of the US DOE ARM program and a focus on the role of clouds in the climate system, many field missions were directed to observations in Arctic regions, especially in the North Slope of Alaska (NSA), which became a permanent observational station in 1996. Advances in ground-based remote sensing capabilities developed by the ARM program, aided by field campaigns, resulted in precise methods to observe atmospheric processes related to water vapor, aerosol, clouds, and radiation. The ability to detect and characterize MPCs at ARM sites created the starting point for the development of additional observation stations in different parts of the world.

The Cloudnet program provided the basis for comparisons between continuous high-resolution cloud data at several European sites and a set of operational numerical weather prediction (NWP) models. The effort continues under the ACTRIS program, and is expanding to include more locations and participating NWP centers. One significant result of the Cloudnet program was the establishment of a standard set of ground-based remote sensing capabilities to provide cloud parameters that can be compared with current operational NWP models. These instruments are MMCR, lidar, MWR, and ceilometer, complemented by other equipment. Moreover, the Cloudnet program produced a set of algorithms able to process data from participating sites. These algorithms provide a standardized procedure to compare cloud observations with NWP model results. Similarly, the ARM program established several ground-based stations, mobile and aerial facilities to obtain cloud characterization for climate modeling. An ongoing effort is to use the high-resolution data and run high-resolution models for comparison and improvement of cloud parameterizations.

While this chapter did not deal with the satellite remote sensing, the global monitoring by CloudSat and CALIPSO opened new possibilities for global monitoring of cloud properties. At the same time, the array of available numerical models from LES to GCMs are instrumental in advancing MPC representation and improving weather forcasting and climate projections.

ACKNOWLEDGMENTS

I would like to thank Elsevier for the opportunity to write this chapter, as well as two anonymous reviewers for many helpful comments.

REFERENCES

Bouniol, D., et al., 2010. Using continuous ground-based radar and lidar measurements for evaluating the representation of clouds in four operational models. J. Appl. Meteorol. Climatol. 49, 1971–1991. https://doi.org/10.1175/2010JAMC2333.1.

Bühl, J., Ansmann, A., Seifert, P., Baars, H., Engelmann, R., 2013. Towards a quantitative characterization of heterogeneous ice formation with lidar/radar: comparison of CALIPSO/CloudSat with ground-based observations. Geophys. Res. Lett. 40, 4404–4408. https://doi.org/10.1002/grl.50792.

Bühl, J., Seifert, P., Myagkov, A., Ansmann, A., 2016. Measuring ice- and liquid-water properties in mixed-phase cloud layers at the Leipzig Cloudnet Station. Atmos. Chem. Phys. 16, 10609–10620. https://doi.org/10.5194/acp-16-10609-2016.

Campos, E., Ware, R., Joe, P., Hudak, D., 2014. Monitoring water phase dynamics in winter clouds. Atmos. Res. 147–148, 86–100.

Choi, Y.S., Ho, C.H., Park, C.E., Storelvmo, T., Tan, I., 2014. Influence of cloud phase composition on climate feedbacks. J. Geophys. Res. Atmos. 119, 3687–3700. https://doi.org/10.1002/2013JD020582.

Cober, S.G., Isaac, G.A., 2012. Characterization of aircraft icing environments with supercooled large drops for application to commercial aircraft certification. J. Appl. Meteorol. Climatol. 51, 265–284.

Cober, S.G., Isaac, G.A., Strapp, J.W., 2001. Characterizations of aircraft icing environments that include supercooled large drops. J. Appl. Meteorol. 40, 1984–2002.

Cooper, W.A., Vali, G., 1981. The origin of ice in mountain cap clouds. J. Atmos. Sci. 38, 1244–1259.

Cotton, W.R., Pielke, R.A., 1995. Human impacts on weather and climate. Cambridge University Press; Cambridge, United Kingdom. 288 pp.

Curry, J.A., Rossow, W.B., Randall, D., Schramm, J.L., 1996. Overview of arctic cloud and radiation properties. J. Clim. 9, 1731–1764.

Curry, J.A., Pinto, J.O., Benner, T., Tschudi, M., 1997. Evolution of the cloudy boundary layer during the autumnal freezing of the Beaufort Sea. J. Geophys. Res. 102, 13851. https://doi.org/10.1029/96JD03089.

Curry, J.A., Hobbs, P.V., King, M.D., Randall, D.A., Minnis, P., Isaac, G.A., et al., 2000. FIRE arctic clouds experiment. Bull. Am. Meteorol. Soc. 81, 5–29. https://doi.org/10.1175/1520-0477(2000) 081<0005:FACE>2.3.CO;2.

Delanoë, J., Protat, A., Jourdan, O., Pelon, J., Papazzoni, M., Dupuy, R., et al., 2013. Comparison of airborne in situ, airborne radar–lidar, and spaceborne radar–lidar retrievals of polar ice cloud properties sampled during the POLARCAT campaign. J. Atmos. Ocean. Technol. 30, 57–73. https://doi.org/10.1175/JTECH-D-11-00200.1.

Ehrlich, A., Wendisch, M., Bierwirth, E., Herber, A., Schwarzenböck, A., 2008a. Ice crystal shape effects on solar radiative properties of arctic mixed-phase clouds—dependence on microphysical properties. Atmos. Res. 88, 266–276.

Ehrlich, A., Bierwirth, E., Wendisch, M., Gayet, J.-F., Mioche, G., Lampert, A., Heintzenberg, J., 2008b. Cloud phase identification of arctic boundary-layer clouds from airborne spectral reflection measurements: test of three approaches. Atmos. Chem. Phys. 8, 7493–7505. https://doi.org/10.5194/acp-8-7493-2008.

Ehrlich, A., Wendisch, M., Bierwirth, E., Gayet, J.-F., Mioche, G., Lampert, A., Mayer, B., 2009. Evidence of ice crystals at cloud top of arctic boundary-layer mixed-phase clouds derived from airborne remote sensing. Atmos. Chem. Phys. 9, 9401–9416. https://doi.org/10.5194/acp-9-9401-2009.

Field, P.R., Hogan, R.J., Brown, P.R.A., Illingworth, A.J., Choularton, R.W., Kaye, P.H., Hirst, E., Greenaway, R., 2004. Simultaneous radar and aircraft observations of mixed-phase cloud at the 100 m scale. Q. J. R. Meteorol. Soc. 130, 1877–1904.

Field, P.R., Hill, A., Furtado, K., Korolev, A., 2014. Mixed phase clouds in a turbulent environment. Part 2: analytic treatment. Q. J. R. Meteorol. Soc. 21, 2651–2663. https://doi.org/10.1002/qj.2175.

Findeisen, W., 1938. Die kolloidmeteorologischen Vorgänge bei der Niederschlagsbildung (Colloidal meteorological processes in the formation of precipitation). Meteor. Z. 55, 121.

Fleishauer, R.P., Larson, V.E., Vonder Haar, T.H., 2002. Observed microphysical structure of midlevel, mixed-phase clouds. J. Atmos. Sci. 59, 1779–1804.

Furtado, K., Field, P.R., Boutle, I.A., et al., 2016. A physically-based, subgrid parametrization for the production and maintenance of mixed-phase clouds in a general circulation model. J. Atmos. Sci. 73 (1), 279–291. ISSN 0022-4928.

Gayet, J.-F., Asano, S., Yamazaki, A., Uchiyama, A., Sinyuk, A., Jourdan, O., et al., 2002. Two case studies of winter continental-type water and mixed phase stratocumuli over the sea 1. Microphysical and optical properties. J. Geophys. Res. 107, 4569. https://doi.org/10.1029/2001JD001106.

Gayet, J.-F., Treffeisen, R., Helbig, A., Bareiss, J., Matsuki, A., Herber, A., et al., 2009a. On the onset of the ice phase in boundary layer Arctic clouds. J. Geophys. Res. 114, D19201. https://doi.org/10.1029/2008JD011348.

Gayet, J.-F., Mioche, G., Dörnbrack, A., Ehrlich, A., Lampert, A., Wendisch, M., 2009b. Microphysical and optical properties of Arctic mixed-phase clouds. The April 9, 2007 case study. Atmos. Chem. Phys. 9 (2009), 6581–6595.

Haeffelin, M., et al., 2016. Parallel developments and formal collaboration between European atmospheric profiling observatories and the U.S. ARM research program. The atmospheric radiation measurement (ARM) program: the first 20 years, Meteor. Monogr., No. 57. Am. Meteorol. Soc. https://doi.org/10.1175/AMSMONOGRAPHS-D-15-0045.1.

Heymsfield, A.J., Miloshevich, L.M., 1993. Homogeneous ice nucleation and supercooled liquid water in orographic wave clouds. J. Atmos. Sci. 50, 2235–2353.

Heymsfield, A.J., Miloshevich, L.M., Slingo, A., Sassen, K., Starr, D.O'.C., 1991. An observational and theoretical study of highly supercooled altocumulus. J. Atmos. Sci. 48, 923–945.

Hill, A.A., Field, P.R., Furtado, K., Korolev, A., Shipway, B.J., 2014. Mixed-phase clouds in a turbulent environment. Part 1. Large-eddy simulation experiments. Q. J. R. Meteorol. Soc. 140 (680), 855–869.

Hobbs, P.V., 1975. The nature of winter clouds and precipitation in the Cascade Mountains and their modification by artificial seeding. Part III: Case studies of the effects of seeding. J. Appl. Meteorol. 14, 819–858.

Hobbs, P.V., Rangno, A.L., 1985. Ice particle concentrations in clouds. J. Atmos. Sci. 23, 2523–2549.

Hogan, R.J., Jakob, C., Illingworth, A.J., 2001. Comparison of ECMWF winter-season cloud fraction with radar derived values. J. Appl. Meteorol. 40, 513–525.

Hogan, R.J., Field, P.R., Illingworth, A.J., Cotton, R.J., Choularton, T.W., 2002. Properties of embedded convection in warm-frontal mixed-phase cloud from aircraft and polarimetric radar. Q. J. R. Meteorol. Soc. 128, 451–476.

Hogan, R.J., Illingworth, A.J., O'Connor, E.J., Poiares Baptista, J.P.V., 2003. Characteristics of mixed-phase clouds. II: A climatology from ground-based lidar. Q. J. R. Meteorol. Soc. 129, 2117–2134.

Illingworth, A.J., et al., 2007. CloudNet: continuous evaluations of cloud profiles in seven operational models using ground-based observations. Bull. Am. Meteorol. Soc. 88, 883–898.

Intrieri, J.M., Shupe, M.D., Uttal, T., McCarty, B.J., 2002. An annual cycle of Arctic cloud characteristics observed by radar and lidar at SHEBA. J. Geophys. Res. 107, 8030. https://doi.org/10.1029/2000JC000423.

Isaac, G.A., Cober, S.G., Strapp, J.W., Korolev, A.V., Tremblay, A., Marcotte, D.L., 2001. Recent Canadian research on aircraft in-flight icing. Can. Aeronaut. Space J. 47, 213–221.

Isaac, G.A., et al., 2005. First results from the Alliance Icing Research Study II. Preprints, In: AIAA 43d Aerospace Science Meeting and Exhibit, AIAA 2005-0252, Reno, NV. American Institute of Aeronautics and Astronautics, pp. 1–18.

Jackson, R.C., McFarquhar, G.M., Korolev, A.V., Earle, M.E., Liu, P.S.K., Lawson, R.P., et al., 2012. The dependence of ice microphysics on aerosol concentration in arctic mixed-phase stratus clouds during ISDAC and M-PACE. J. Geophys. Res. 117, D15207. https://doi.org/10.1029/2012JD017668.

Jackson, R., French, J., Finlon, J., 2017. Microphysical properties of convectively forced mixed phase clouds. In: Mixed-Phase Clouds: Observations and Modeling. Elsevier.

Klein, S.A., et al., 2009. Intercomparison of model simulations of mixed-phase clouds observed during the ARM Mixed-Phase Arctic Cloud Experiment. I: single-layer cloud. Q. J. R. Meteorol. Soc. 135 (641), 979–1002.

Köhler, F., Görsdorf, U., 2014. Towards 3D prediction of supercooled liquid water for aircraft icing: modifications of the microphysics in COSMO-EU. Meteorol. Z. 23 (3), 253–262.

Komurcu, M., Storelvmo, T., Tan, I., Lohmann, U., Yun, Y.X., Penner, J.E., et al., 2014. Intercomparison of the cloud water phase among global climate models. J. Geophys. Res. Atmos. 119, 3372–3400. https://doi.org/10.1002/2013JD021119.

Korolev, A., Field, P.R., 2008. The effect of dynamics on mixed-phase clouds: theoretical considerations. J. Atmos. Sci. 65, 66–86. https://doi.org/10.1175/2007JAS2355.1.

Lampert, A., Ritter, C., Hoffmann, A., Gayet, J.-F., Mioche, G., Ehrlich, A., et al., 2010. Lidar characterization of the Arctic atmosphere during ASTAR 2007: four cases studies of boundary layer, mixed-phase and multi-layer clouds. Atmos. Chem. Phys. 10, 2847–2866. https://doi.org/10.5194/acp-10-2847-2010.

Lohmann, U., Hoose, C., 2009. Sensitivity studies of different aerosol indirect effects in mixed-phase clouds. Atmos. Chem. Phys. 9, 8917–8934. www.atmos-chem-phys.net/9/8917/2009/.

Lubin, D., 2004. Thermodynamic phase of maritime Antarctic clouds from FTIR and supplementary radiometric data. J. Geophys. Res. 109, D04204. https://doi.org/10.1029/2003JD003979.

Marsham, J.H., Dobbie, S., Hogan, R.J., 2006. Evaluation of a large-eddy model simulation of a mixed-phase altocumulus cloud using microwave radiometer, lidar and Doppler radar data. Q. J. R. Meteorol. Soc. 132, 1693–1715.

Mather, J.H., Turner, D.D., Ackerman, T.P., 2016. Scientific maturation of the ARM Program. The Atmospheric Radiation Measurement (ARM) Program: the First 20 Years, Meteor. Monogr. Am. Meteor. Soc. 57. https://doi.org/10.1175/AMSMONOGRAPHS-D-15-0053.1.

McCoy, D., Tan, I., Hartmann, D., Zelinka, M., Storelvmo, T., 2016. On the relationships among cloud cover, mixed-phase partitioning, and planetary albedo in GCMs. J. Adv. Model. Earth Syst. https://doi.org/10.1002/2015MS000589.

McFarquhar, G., Zhang, G., Poellot, M., Kok, G., McCoy, R., Tooman, T., Fridlind, A., Heymsfield, A., 2007. Ice properties of single-layer stratocumulus during the Mixed-Phase Arctic Cloud Experiment: 1. Observations. J. Geophys. Res. 112.

McFarquhar, G.M., Ghan, S., Verlinde, J., Korolev, A., Strapp, J.W., Schmid, B., et al., 2011. Indirect and semi-direct aerosol campaign: the impact of arctic aerosols on clouds. Bull. Am. Meteorol. Soc. 92, 183–201. https://doi.org/10.1175/2010BAMS2935.1.

Mioche, G., Jourdan, O., Ceccaldi, M., Delanoë, J., 2015. Variability of mixed-phase clouds in the Arctic with a focus on the Svalbard region: a study based on spaceborne active remote sensing. Atmos. Chem. Phys. 15, 2445–2461. https://doi.org/10.5194/acp-15-2445-2015.

Morrison, H., Pinto, J., 2005. Mesoscale modeling of springtime Arctic mixed-phase stratiform clouds using a new two-moment bulk microphysics scheme. J. Atmos. Sci. 62, 3683–3704.

Morrison, H., et al., 2009. Intercomparison of model simulations of mixed-phase clouds observed during the ARM Mixed-Phase Arctic Cloud Experiment, Part II: multi-layered cloud. Q. J. R. Meteorol. Soc. 135, 1003–1019. https://doi.org/10.1002/qj.415.

Morrison, H., De Boer, G., Feingold, G., Harrington, J., Shupe, M.D., Sulia, K., 2012. Resilience of persistent Arctic mixed-phase clouds. Nat. Geosci. 5, 11–17. https://doi.org/10.1038/ngeo1332.

Noh, Y.-J., Seaman, C.J., Vonder Haar, T.H., Liu, G., 2013. In situ aircraft measurements of the vertical distribution of liquid and Ice water content in midlatitude mixed-phase clouds. J. Appl. Meteorol. Climatol. 52, 269–279. https://doi.org/10.1175/JAMC-D-11-0202.1.

Phillips, V.T.J., DeMott, P.J., Andronache, C., 2008. An empirical parameterization of heterogeneous ice nucleation for multiple chemical species of aerosol. J. Atmos. Sci. 65 (9), 2757–2783.

Platt, C.M.R., 1977. Lidar observations of a mixed-phase altostratus cloud. J. Appl. Meteorol. 16, 339–345.

Politovich, M.K., Vali, G., 1983. Observations of liquid water in orographic clouds over Elk Mountain. J. Atmos. Sci. 40, 1300–1312.

Rauber, R.M., 1987. Characteristics of cloud ice and precipitation during wintertime storms over the mountains of northern Colorado. J. Clim. Appl. Meteorol. 26, 488–524.

Rauber, R.M., Grant, L.O., 1986. The characteristics and distribution of cloud water over the mountains of northern Colorado during wintertime storms. Part II: spatial distribution and microphysical characteristics. J. Clim. Appl. Meteorol. 25, 489–504.

Rauber, R.M., Tokay, A., 1991. An explanation for the existence of supercooled water at the top of cold clouds. J. Atmos. Sci. 48, 1005–1023.

Sassen, K., Khvorostyanov, V.I., 2007. Microphysical and radiative properties of mixed-phase altocumulus: a model evaluation of glaciation effects. Atmos. Res. 84, 390–398. https://doi.org/10.1016/j.atmosres.2005.08.017.

Schmidt, J., Ansmann, A., Bühl, J., Wandinger, U., 2015. Strong aerosol–cloud interaction in altocumulus during updraft periods: lidar observations over central Europe. Atmos. Chem. Phys. 15, 10687–10700. https://doi.org/10.5194/acp-15-10687-2015.

Shupe, M., 2007. A ground-based multisensor cloud phase classifier. Geophys. Res. Lett. 34 (L22), 809.

Shupe, M., Uttal, T., Matrosov, S., 2005. Arctic cloud microphysics retrievals from surface-based remote sensors at SHEBAs. J. Appl. Meteorol. 44, 1544–1562.

Shupe, M., Matrosov, S., Uttal, T., 2006. Arctic mixed-phase cloud properties derived from surface-based sensors at SHEBA. J. Atmos. Sci. 63, 697–711.

Shupe, M., Kollias, P., Persson, P., McFarquhar, G., 2008a. Vertical motions in Arctic mixed-phase stratiform clouds. J. Atmos. Sci. 65, 1304–1322.

Shupe, M., et al., 2008b. A focus on mixed-phase clouds: the status of ground-based observational methods. Bull. Am. Meteorol. Soc. 87, 1549–1562.

Shupe, M.D., Comstock, J.M., Turner, D.D., Mace, G.G., 2016. Cloud property retrievals in the ARM Program. The Atmospheric Radiation Measurement (ARM) Program: the First 20 Years, Meteor. Monogr. Amer. Meteorol. Soc. 57. https://doi.org/10.1175/AMSMONOGRAPHS-D-15-0030.1.

Sisterson, D., Peppler, R., Cress, T.S., Lamb, P., Turner, D.D., 2016. The ARM Southern Great Plains (SGP) site. The atmospheric radiation measurement (ARM) program: the first 20 years, Meteor. Monogr. Am. Meteorol. Soc. 57. https://doi.org/10.1175/AMSMONOGRAPHS-D-16-0004.1.

Stephens, G.L., et al., 2002. The CloudSat mission and the A-Train. Bull. Am. Meteorol. Soc. 83, 1771–1790. https://doi.org/10.1175/BAMS 83 12 1771.

Stephens, G.L., et al., 2008. The CloudSat mission: performance and early science after the first year of operation. J. Geophys. Res. 113. D00A18, https://doi.org/10.1029/2008JD009982.

Stokes, G.M., 2016. Original ARM Concept and Launch. The atmospheric radiation measurement (ARM) program: the first 20 years, Meteor. Monogr. Am. Meteorol. Soc. 57. https://doi.org/10.1175/AMS-MONOGRAPHS-D-15-0054.1.

Stokes, G.M., Schwartz, S.E., 1994. The Atmospheric Radiation Measurement (ARM) Program: programmatic background and design of the cloud and radiation test bed. Bull. Am. Meteorol. Soc. 75 (7), 1201–1221. https://doi.org/10.1175/1520-0477(1994)075<1201:RARMPP>2.0.CO;2.

Tan, I., Storelvmo, T., Zelinka, M., 2016. Observational constraints on mixed-phase clouds imply higher climate sensitivity. Science 352. https://doi.org/10.1126/science/aad530.

Verlinde, J., et al., 2007. The mixed-phase arctic cloud experiment. Bull. Am. Meteorol. Soc. 88, 205–221.

Verlinde, J., Zak, B., Shupe, M.D., Ivey, M., Stamnes, K., 2016. The ARM North Slope of Alaska (NSA) sites. The atmospheric radiation measurement (ARM) program: the first 20 years, Meteor. Monogr. Am. Meteorol. Soc. 57. https://doi.org/10.1175/AMSMONOGRAPHS-D-15-0023.1.

Wood, R., 2012. Stratocumulus clouds. Mon. Weather Rev. 140, 2373–2423.

Zhang, D., Wang, Z., Liu, D., 2010. A global view of midlevel liquid layer topped stratiform cloud distribution and phase partition from CALIPSO and CloudSat measurements. J. Geophys. Res. Atmos. 115. D00H13, https://doi.org/10.1029/2009JD012143.

CHAPTER 6

Spaceborne Remote Sensing and Airborne In Situ Observations of Arctic Mixed-Phase Clouds

Guillaume Mioche*,†, Olivier Jourdan*,†
*Université Clermont Auvergne, Clermont-Ferrand, France
†CNRS, Aubière, France

Contents

1. INTRODUCTION

The Arctic is a sentinel for global climate change as it is warming at more than twice the global mean rate showing acute visible signs such as the fast retreat of the sea ice (Deser and Teng, 2008; IPCC, 2013; Serreze et al., 2007; Stroeve et al., 2007). This region also experiences a significant increase of aerosol loading as a receptor for pollution transported from distant source regions at midlatitudes (Law et al., 2014; Stohl, 2006) combined with an enhancement of local emission sources. The dramatic changes observed in the Arctic climate also have strong consequences on the global climate. However, the observed rate of climate change in the Arctic is not accurately reproduced in climate models (Eckhardt et al., 2013). The predictive capability of Arctic climate change is severely hampered by a lack of understanding about key processes influencing the atmosphere–ice–terrestrial-ocean system and feedbacks. Within this system, cloud-related processes play a crucial role in Arctic climate system and its evolution, impacting and determining the local radiation budget (Curry, 1995; Curry et al., 1996; Kay et al., 2012; Kay and Gettelman, 2009; Morrison et al., 2012). Clouds interact with shortwave and longwave radiations, cooling or warming the surface and the atmosphere depending on their macrophysical,

Mixed-Phase Clouds
https://doi.org/10.1016/B978-0-12-810549-8.00006-4

microphysical, and optical properties. The low sun elevation in summer and the lack of solar radiation during the winter polar night are responsible for the predominant long-wave radiative effect in the Arctic (Lubin and Vogelmann, 2006), tending to a regional net warming effect (Solomon et al., 2007; Stephens, 2005). However, major uncertainties surround our knowledge of the complex and numerous interactions and feedbacks between the physical processes involved in the cloud life cycle (Liu et al., 2012a). This complexity reflects in the large discrepancies among cloud-related processes represented in models at all scales. In the last IPCC report Boucher et al. (2013) showed that, in particular, ice cloud microphysics must be significantly improved in order to reduce the low confidence attributed to model estimates of aerosol-cloud feedbacks. For instance our knowledge of the local and large-scale processes responsible for the persistence of Arctic mixed-phase clouds (MPCs) is still very basic.

MPCs are characterized by a microphysically unstable mixture of liquid droplets and ice crystals. However, observations show that these clouds are ubiquitous in the Arctic and persist for several days under a variety of meteorological conditions (Mioche et al., 2015; Morrison et al., 2012; Shupe et al., 2011; Shupe and Intrieri, 2004). They occur as single or multiple stratiform layers of supercooled droplets near the cloud top from which ice crystals form and precipitate (Gayet et al., 2009a; McFarquhar et al., 2007). The strong impact of MPC on the energy budget stems from their persistence and microphysical properties. The longevity of MPC results from interactions between fast local dynamical, radiative, surface, and microphysical processes and larger scale meteorological or environmental conditions that greatly complicate their understanding and modeling. In particular, the ice microphysical processes that impact cloud top radiative cooling trigger the turbulence responsible for the maintenance of MPC. Aerosol-cloud interaction processes also play a key role as atmospheric aerosols can influence the persistence of MPC by changing their microphysical properties. In MPC, ice crystals formation is expected to be initiated by the presence of the liquid phase through different heterogeneous nucleation mechanisms depending on the temperature, supersaturation, and aerosol properties. The balance between cloud condensation nuclei (CCN) and ice nuclei (IN) concentration is critical as it directly impacts the ice production. On one hand, modeling studies have shown that even a slight increase of IN concentration can lead to a rapid conversion of MPC to pure ice cloud (Harrington et al., 1999; Jiang et al., 2000; Morrison et al., 2011; Pinto, 1998; Prenni et al., 2007). Under weak updraft velocities or high IN concentration, ice crystals can grow rapidly by vapor deposition at the expense of liquid by the Wegener-Bergeron-Findeisen process if the liquid saturation is not reached (Ervens et al., 2011). On the other hand, the IN depletion through sedimentation of ice forming particles limits the continuous ice production and further maintenance of the MPC system (Westbrook and Illingworth, 2013). Moreover, the number and the shape of ice crystals influence their growth and sedimentation contributing to the prevention of cloud glaciation (Ervens et al., 2011).

Tan et al. (2016) showed that as ice forms, the liquid water is depleted restricting further ice formation through the competition of water vapor and moderating the loss of supercooled water, thus forming a self-regulating process. This ice-liquid phase partitioning is also governed by the feedbacks between dynamical and microphysical processes (Avramov et al., 2011) as turbulent mixing favors the entrainment of IN particles into the cloud-driven mixed layer. However, Fridlind et al. (2012) showed that this process alone does not lead to ice concentrations in accordance with observations. Indeed, despite substantial progress, most models require an unrealistic concentration of IN to maintain ice formation (Savre and Ekman, 2015) or fail to reproduce the physical processes involved in the life cycle of MPC. The formation of ice crystals in the upper part of the cloud is a prerequisite to maintaining the mixed phase conditions. Large-eddy simulations coupled with heterogeneous ice nucleation parameterizations showed that the persistence of ice production is determined by the competition between radiative cooling, cloud top entrainment, and nucleation scavenging of the IN (Savre and Ekman, 2015). They emphasized that the accurate representation of processes involved in MPC requires modeling the time-evolving ability of the IN population to form ice crystals. Solomon et al. (2015) suggested that the recycling of IN particles through subcloud sublimation could maintain the ice production, slow the rate of ice loss from the mixed layer and regulate the liquid production.

Although these case studies provide a framework to understand the mechanisms responsible for the persistence of MPC, measurements at different scales are still needed to characterize the microphysical properties and the phase partitioning within such clouds. Crucial properties such as cloud top liquid water profile and the vertical particle size and shape distribution of ice crystals have to be accurately assessed as they influence the local radiative and microphysical processes maintaining the phase partitioning in the MPC. Airborne campaigns can contribute to a process-level understanding of the mechanisms involved in the MPC life cycle even though large uncertainties in the counting and sizing of ice particles still exist.

Remote sensing observations from space or ground based stations are also needed to generate statistical datasets and allow reliable studies of MPC properties' variability at a regional scale (Dong et al., 2010; Kay and Gettelman, 2009; Liu et al., 2012b; Shupe et al., 2011). These types of measurements coupled with a more integrated strategy may increase our level of understanding on the relationships between large-scale arctic environmental properties and MPC characteristics (Morrison et al., 2012).

This chapter is a modest contribution to the field of microphysical properties and processes occurring within arctic MPCs. We start with a summary of the main results from a regional study of the cloud phase partitioning and distribution undertaken with the CALISPO/CloudSat satellite remote sensing observations (Section **2**). Then, a quick review of the major projects devoted to the study of Arctic MPC from in situ aircraft observations is made in Section **3**. Section **4** presents the main findings obtained from aircraft measurements concerning the microphysical and optical properties of Arctic MPC. Finally, Section 5 is more specially focused on an example of possible synergies between in situ airborne measurements and satellite observations to evaluate MPC retrieval products.

2. OBSERVATIONS AT REGIONAL SCALE FROM SATELLITE

Spaceborne, ground based, and airborne radar-lidar observations, when analyzed at regional and global scales, offer new perspectives to better understand the cloud life cycle (formation, maintenance, dissipation) and the variability of dominant processes over different climatic regions. The recent development of ground based stations [such as Barrow (71°N), Eureka (80°N), Summit (72°N), or Ny-Ålesund (78°N)], well equipped for cloud observations plays a key role in the characterization of the cloud phase and cloud microphysical variability at a regional scale (Chernokulsky and Mokhov, 2012; Dong et al., 2010; Kay and Gettelman, 2009; Liu et al., 2012b; Shupe et al., 2011). However, the ground-based instrumental payload may differ significantly from one site to another. Climatologies based on specific instruments can be biased by the sensitivity of the instruments and their capability to operate in extreme conditions. For instance, it is quite difficult to compare results derived from a lidar system with the ones obtained from radar as they are sensitive to different types of hydrometeors.

Since 2006, new observations of cloud properties from space are possible with active instruments like the Cloud-Aerosol Lidar with Orthogonal Polarization (CALIOP) lidar at 532 and 1064 nm onboard CALIPSO and the 95 GHz cloud profiling radar (CPR) on CloudSat satellites (Winker et al., 2003; Stephens et al., 2002) as part of the A-Train constellation. These radar-lidar observations constitute an unprecedented dataset documenting the cloud vertical structure with a high spatial resolution at a global scale. Compared to ground-based observations, they present the great advantage to involve a uniform measurement technique with a very large coverage.

Before the emergence of these satellites, polar cloud climatologies from space were performed based on passive sensors, mainly the:
- Operational Vertical Sounder-Polar Pathfinder (TIROS N TOVS-Path P; Schweiger et al., 1999)
- International Satellite Cloud Climatology Project (ISCCP; Rossow and Schiffer, 1999)
- Advanced Very High Resolution Radiometer-Polar Pathfinder (AVHRR; Wang and Key, 2005)
- Moderate Resolution Imaging Spectroradiometer (MODIS; Ackerman et al., 1998)

These instruments provided observations at regional scale in the Arctic allowing the first climatologies of Arctic clouds (Schweiger and Key, 1992). However, it is now well known that these measurement techniques suffer from large shortcomings which limit accuracy in the retrieval of cloud properties (cloud fraction, top and base altitudes, etc.), especially in the Polar regions. Passive remote sensing technique is based on the spectral signature difference between clouds and the surface in visible, near-IR and longwave IR bands to detect clouds. The special conditions encountered in the Arctic significantly hamper cloud detection. In particular, the lack of sunlight in winter makes the visible channels useless during this period of the year. The weak contrast between the clouds and the underlying ice-covered surfaces can also significantly impact the cloud detection close to the surface.

Therefore space-borne passive remote sensing measurements may lead to large uncertainties in Arctic cloud climatologies (Frey et al., 2008; Lubin and Morrow, 1998). For instance, it has been shown that the cloud amount retrieved from these observations is generally 5%–35% less than the one derived from ground based observations, with some regional differences up to 45% (Schweiger and Key, 1992). Chan and Comiso (2013) showed that AVHRR detects only 44% of clouds measured by CALIOP. They also showed that MODIS underestimates by 13%–30% the cloud amount, depending on the season, surface type, day/night conditions or cloud altitude and thickness, although it has more channels available than AVHRR (36 and 5, respectively).

Active remote sensing observations may in part overcome the shortcomings associated with passive measurements and improve the cloud detection since they are less impacted by the lack of sunlight or the ice-covered surfaces. Based on these measurements, studies on the total cloud fraction over the Arctic region have been recently performed (Chan and Comiso, 2013; Chernokulsky and Mokhov, 2012; Liu et al., 2012b; Zygmuntowska et al., 2012). These studies showed that cloud cover ranges between 73% and 86% over the whole Arctic region, but present a clear spatial and vertical variability and noticeable differences according to the seasons and the surface type. The maximum cloud cover is observed in autumn (86%), and the minimum in winter (73%), while spring and summer exhibit intermediate values (76%–79%). The frequency of cloud occurrence is higher over the open sea (around 80%) than over ice-covered surfaces (around 74%). The seasonality is more pronounced on the Arctic Pacific side than over the Atlantic side (Barents and Greenland Seas). The large and almost constant cloud cover (85%) observed all year long on the Atlantic side has been associated with the frequent synoptic activities and high atmospheric humidity and temperature over open water (Serreze and Barry, 2005). The minimum cloud fraction observed during winter, especially in the Pacific side and in the Central Arctic Ocean, has been linked to the lack of moisture and the strong high pressure systems at this season.

Arctic clouds also present a strong vertical variability as they are mainly distributed below an altitude of 2 km and between 7 and 9 km. The low-level cloud seasonality is associated with low-level moisture advection, radiative cooling, and boundary layer turbulence (Curry et al., 1988), whereas the almost constant midlevel cloud fraction through the year is linked to the large-scale transport of moisture.

These works focused on the total cloud fraction. Studies addressing the variability of the MPCs at the Arctic regional scale are scarce. They mostly involved local remote sensing measurements from ground-based stations (Eureka, Barrow, Summit, Ny-Alesund, among others) or from icebreaker ships drifting with the ice pack (SHEBA and ASCOS experiments). These measurements remain local in space and will not be described in the present chapter since the focus is on satellite observations and in situ airborne measurements only.

MPC variability at the Arctic regional scale has been particularly investigated by Mioche et al. (2015) from CALIPSO and CloudSat observations between 2007 and 2010 using the DARDAR retrieval algorithm. The DARDAR algorithm (Ceccaldi et al., 2013; Delanoë and Hogan, 2010) retrieves the atmospheric properties, in particular the scene classification (cloud, aerosol, clear sky, etc.) and the thermodynamical phase from CALIOP lidar and CloudSat CPR measurements. The great advantage of DARDAR algorithm is that lidar and radar observations are merged on the same resolution grid (1700 m horizontal and 60 m vertical), which allows the retrieval of ice, liquid and MPC type as the radar is more sensitive to ice crystals and thick clouds; lidar to liquid droplets and thin ice clouds.

This study shows that MPC prevail in the Arctic all year long, as previously observed from local ground based stations (Curry et al., 1996; Intrieri et al., 2002). MPC represent between 35% (in winter) and 65% (in autumn and spring) of the clouds in average over the whole Arctic region. MPC are mainly located at low altitudes. Below 3 km, their occurrence range lies between 70% and 90%, especially in winter, spring and autumn. During summer, the MPC are more frequent in midlevel altitudes (3–6 km). Single layer MPCs represent between 55% and 70% of the MPC amount and exhibit similar spatial, seasonal, and vertical variability properties. MPCs are statistically more frequent above open sea than over land or sea ice surfaces.

Local differences are pointed out when comparing MPC properties over the entire Arctic region with other areas such as the Svalbard region or the Western Arctic as shown in Fig. 1. On the Atlantic side (around Svalbard archipelago, Greenland and Barents Seas), MPC occurrence is almost constant through the year (around 55% in average). On the Pacific side (Northern Alaska, Beaufort and Chukchi Seas) or in the Central Arctic Ocean, MPC occurrence exhibits a clear variability, with the largest occurrences during autumn, and the minimum during winter.

These local differences in MPC seasonality can be partially attributed to the transport of moist air and warm water from the North Atlantic Ocean through the Arctic Ocean during the year that is associated with a temperature range favorable to mixed-phase conditions. The North Atlantic Ocean supplies more moisture to the Svalbard region than the rest of the Arctic (Serreze and Barry, 2005), making easier the vertical transfer of humidity. This amplifies the cloud formation because the supply of humidity is favorable for the initiation of liquid droplets. During the melting seasons

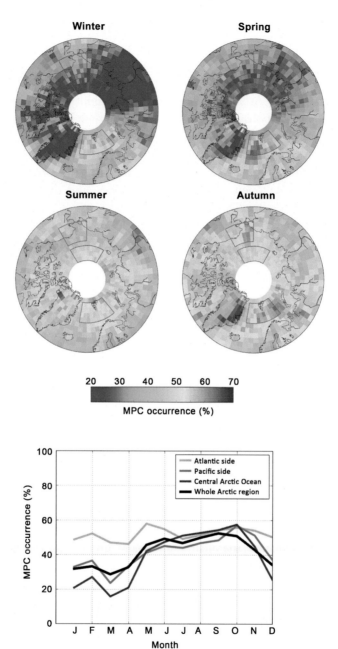

Fig. 1 Top panel: stereographic projections of the seasonal occurrence of MPC (referring to clouds). Occurrences are computed taking into account the 500–12,000 m altitude range, and bottom panel: monthly total MPC occurrence over the whole Arctic region *(black)*, the Atlantic side *(blue)*, the Pacific side *(orange)* and the Central Arctic Ocean *(red)*, as determined in Mioche et al. (2015). The location of the three regions are indicated by the purple boxes in the stereographic projections on top panel.

(late spring to autumn), the fraction of open seas increases and more warm water can be transported through the Arctic region resulting in warm and moist air advection in the western Arctic. Therefore, the cloud formation and initiation of liquid phase may be amplified in this region, leading to the increase of MPC occurrence observed during this period on the Arctic Pacific side. The variability of sea ice concentration, temperature, and humidity has been analyzed to strengthen this assumption, and a link with the MPC and cloud variability has been established. The results highlighted a strong negative correlation between sea ice concentration and MPC occurrence (see Figure 10 from Mioche et al., 2015), confirming the previous findings that MPC prevail more over open seas than iced surfaces, and the role of the melting of sea ice in the MPC seasonality. In addition, stable atmospheric conditions encountered during winter and transitions seasons in the Arctic region (Orbaek et al., 1999), as well as strong temperature and humidity inversions at cloud top (Nygård et al., 2014) contribute to limit the vertical extension of Arctic clouds and MPCs and thus maintain them at low altitude levels.

Conversely, during summer, the air temperature increases and atmospheric conditions are less stable than during the rest of the year, allowing a larger vertical extension of MPC.

Obviously, MPC variability and life cycle are also dependent on numerous other regional or local scale processes. The study of the role of aerosol particles acting as CCN or IN, their concentrations and composition, as well as the contributions of local sources versus long-range aerosol transport in the Arctic is necessary to investigate the cloud-aerosol interactions in the Arctic and improve the cloud representation in models.

This cloud phase distribution analysis at the regional scale can also be used to assess how local airborne experiments are representative of the variety of clouds encountered in the Arctic.

Additionally, active remote sensing measurements have inherent and well-known shortcomings near the ground level that may impact the determination of cloud and MPC amount. It is mainly the lidar laser beam attenuation by liquid layers for CALIOP observations and the contamination by radar ground echoes for CloudSat observations. Comparisons with ground-based observations (Blanchard et al., 2014; Mioche et al., 2015) estimated that the uncertainties in the cloud and MPC occurrence determined from the CALIOP and CloudSat observations processed with the DARDAR retrieval algorithm can reach 20% between 500 and 2 km, and 25% below 500 m.

This strengthens the fact that an analysis of in situ and ground-based observations of the cloud properties at low altitude is still necessary to characterize the MPC phase distribution. For instance, microphysical and optical in situ measurements from airborne campaigns provide a more accurate description of MPC at small scale.

Therefore, even though these measurements are localized in time and space, they can contribute to improving our understanding of the microphysical processes involved in MPCs.

3. AIRBORNE EXPERIMENTS IN THE ARCTIC REGION

Before the 1990s, airborne experiments in the Arctic region remained sparse due to the limited access to the region. The earliest in situ airborne measurements of cloud microphysical properties in the Arctic concern only liquid water clouds during summer (Curry, 1986; Dergach et al., 1960; Koptev and Voskresenskii, 1962; Tsay and Jayaweera, 1984). The first measurements of the ice phase in MPC (Curry et al., 1990; Jayaweera and Ohtake, 1973; Witte, 1968; among others) showed that ice occur for a wide range of temperatures, from $-8°C$ to $-20°C$. These studies are described in the review paper of Curry et al. (1996). However, regarding the large uncertainties and shortcomings associated with the measurement techniques, accurate quantitative estimates of ice crystal properties such as size, number, and shape was not possible at that time.

The significant climate sensitivity of the Arctic coupled with the large impact of clouds on the surface energy budget led the scientific community to draw special attention to Arctic cloud observations. In the last 20 years, substantial progress has been made on in situ measurement techniques, in particular to characterize the ice phase with promising aircraft experiments in the Arctic. Several major international projects emerged and still continue to be planned today. These main projects are summarized in Table 1. Since 1994, 14 major airborne campaigns were carried out, equally shared between two main locations: 7 took place in the Arctic Pacific side (also called Western Arctic) around regions of Northern Alaska and the Beaufort Sea, and 7 in the North Atlantic side, around the Svalbard archipelago and the Greenland and Barents Seas. Obviously, these locations are strongly constrained by the infrastructure facilities needed to perform airborne campaigns (mainly Barrow, Fairbanks, or Inuvik for the Western Arctic; and Longyearbyen or Kiruna on the Atlantic side). Most of the campaigns took place in spring and few of them in summer and autumn.

During all these campaigns, a wide variety of aircraft and in situ instrumentation were used. Cloud properties were assessed based on different measurement techniques (light scattering spectrometers, hot-wire probes, imaging probes, etc., see Baumgardner et al. (2012) for details about in situ instrumentation). Most of the time, the flight patterns consisted of ascent and descent sequences above, into, and below MPC, or horizontal flight legs at different altitude levels. Accordingly, the horizontal and vertical variability of relevant cloud properties (liquid droplet and ice crystal number and size, ice and liquid water contents, ice crystal morphology, asymmetry parameter, optical depth) was investigated to better understand the microphysical processes (formation, growth, and dissipation) of ice crystals and supercooled liquid droplets at small scale.

Table 1 Main field experiments involving airborne in situ measurements of Arctic mixed-phase clouds

Year	Period	Experiment		Location	Reference
1994	Sep.–Oct.	BASE	Beaufort and Arctic Storms Experiment	Beaufort Sea	Curry et al. (1997)
1998	Apr.–Jul.	FIRE-ACE	First International Satellite Cloud Climatology Project (ISCCP) Regional Experiment: Arctic Clouds Experiment	Beaufort Sea, Northern Alaska	Curry et al. (2000)
2004	Sep.–Oct.	M-PACE	Mixed-Phase Arctic Cloud Experiment	Northern Alaska	Verlinde et al. (2007)
2004	May–Jun.	ASTAR	Arctic Study of Tropospheric cloud, Aerosol and Radiation	Greenland/Barents Seas and Svalbard	Jourdan et al. (2010)
2007	Apr.	ASTAR	Arctic Study of Tropospheric cloud, Aerosol and Radiation	Greenland/Barents Seas and Svalbard	Gayet et al. (2009a)
2008	Mar.–Apr.	POLARCAT	Polar Study using Aircraft, Remote Sensing Surface Measurements and Models of Climate, Chemistry, Aerosols and Transport	Greenland/Barents Seas and Svalbard	Delanoë et al. (2013)
2008	Aug.	AMISA	Arctic Mechanisms for the Interaction of the Surface and Atmosphere	Arctic Ocean, North of Svalbard	Persson (2010)
2008	Apr.	ISDAC	Indirect and Semi-Direct Aerosol Campaign	Beaufort Sea, Northern Alaska	McFarquhar et al. (2011)
2008	Apr.	ARCPAC	Aerosol, Radiation and Cloud Processes affecting the Arctic Climate	Northern Alaska	Lance et al. (2011)
2010	May	SORPIC	Solar Radiation and Phase Discrimination of Arctic Clouds experiment	Greenland/Barents Seas and Svalbard	Bierwirth et al. (2013)
2012	May	VERDI	Study on the Vertical Distribution of Ice in Arctic clouds	Beaufort Sea	Klingebiel et al. (2014)
2013	Mar.–Apr.–Jul.	ACCACIA	Aerosol–Cloud Coupling And Climate Interactions in the Arctic	Greenland Sea	Lloyd et al. (2015)
2014	May	RACEPAC	Radiation–Aerosol–Cloud Experiment in the Arctic Circle	Beaufort Sea	
2017	Jun.	ACLOUD	Arctic Cloud Observations Using airborne measurements in polar Day conditions	Arctic Ocean, north of Svalbard	

4. IN SITU CHARACTERIZATION OF MPC PROPERTIES

This section summarizes the main findings related to the characterization of MPC properties inferred from the airborne campaigns listed in Table 1. In particular, Section 4.1 focuses on the discrimination of the thermodynamical phase, the interaction between the liquid and ice phases, the quantitative assessment of liquid droplets and ice crystal properties (N, size, water content, shape, etc.) as well as the main microphysical processes involved in the formation and growth of cloud particles. Then Section 4.2 describes a case study on the link between optical and microphysical properties of MPC.

One can note that the previous studies dedicated to the assessment of the microphysical properties of Arctic clouds based on in situ measurements (Avramov et al., 2011; Gayet et al., 2009a, 2009b; Rangno and Hobbs, 2001; Verlinde et al., 2007; among others) focused mainly on case studies. A few studies aimed to merge several in situ datasets to provide a statistical analysis and representative description of MPC properties (McFarquhar et al., 2007; Mioche et al., 2017).

4.1 Main Properties From Previous Experiments

Arctic MPCs occur as single or multiple stratiform layers of supercooled droplets near the cloud top from which ice crystals form and precipitate and can persist for several days. This peculiar structure has been widely observed in the airborne experiments listed in Table 1. The ice and liquid properties have been characterized from the in situ measurements performed during these campaigns. However, the formation, evolution, and persistence of these MPC result from numerous and complex interactions between aerosols, liquid droplets and ice crystals at both local and large scales. We propose here a summary of the main mechanisms, which have been highlighted from in situ airborne observations carried out on the Arctic Atlantic or Pacific side.

The formation of a MPC is initiated by the nucleation of supercooled water droplets at the cloud base from aerosol particles (droplet concentration is correlated with the aerosol number below the cloud). Then, due to adiabatic cooling, they grow during their ascent to the cloud top, leading to LWC values increasing with altitude. At cloud top of low-level MPC (down to $-30°C$), typical LWC values are around 0.2 g m^{-3}. Droplets are generally small (around 20 μm), but they can reach larger sizes of several hundreds of μm (drizzle) by collision-coalescence processes. An example of the vertical profiles of liquid phase properties derived from the statistical analysis of four airborne campaigns by Mioche et al. (2017) is displayed in Fig. 2A–D. At cloud top, the temperature inversion and the associated entrainment of dry air lead to the evaporation of a fraction of liquid droplets, capping the vertical extension of MPC. Humidity inversion may also occur at cloud top (Nygård et al., 2014), supplying moisture to the cloud top by entrainment. This may prevent the evaporation of liquid droplets, counteract the effect of mixing of dry air due to temperature inversion, and contribute to the persistence of the MPC. Moreover,

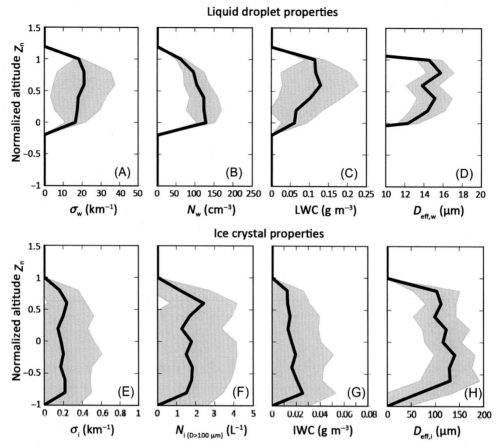

Fig. 2 Vertical profiles (expressed in normalized altitude) of liquid and ice phases from four airborne campaigns located in the Svalbard region (ASTAR 2004 and 2007, POLARCAT 2008, and SORPIC 2010 from Mioche et al., 2017). Liquid droplets properties are retrieved from FSSP measurements (3–45 μm size range): (A) extinction coefficient, (B) droplet concentration, (C) LWC, (D) droplet effective diameter, and ice crystal properties are determined from CPI measurements (15 μm–2.3 mm size range): (E) extinction coefficient, (F) ice crystal concentration, (G) IWC, (H) ice crystal effective diameter. The *black lines* are the average over all the campaigns, and the *gray shaded area* represents the standard deviations.

some studies have shown that supercooled droplets could exhibit a bimodal size distribution at cloud top. This feature may be caused by the entrainment and activation/condensation of new aerosol particles by the dry air layer above the cloud top (Lawson et al., 2001), or by the evaporation of the larger droplets due to turbulent entrainment of clear-air eddies (Klingebiel et al., 2014).

Understanding and characterizing processes responsible for the formation and growth of ice crystals from in situ measurements is more challenging, especially in the presence of liquid droplets. For example, the small ice particle detection (and thus the study of ice formation) is challenging, due to the shortcomings linked to shattering effect or minimum size detection thresholds. The presence of liquid water in the same sample volume also contributes to an increase of the uncertainties on the ice properties measurements (bulk measurements, discrimination of small spherical ice crystals from liquid droplets for example for imaging probes, ice crystal response for optical probes). Moreover, ice phase initiation depends on the IN concentration, which is very small compared to the CCN concentration. CCN concentration in the atmosphere ranges from 10 to 1000 cm^{-3} whereas IN concentration is 10^5–10^6 times smaller. Thus it is still very difficult to accurately measure the IN number in natural conditions, especially in the Arctic where the background aerosol concentration is lower than at lower latitudes. However, the numerous improvements made since the first airborne observations regarding the assessment of the ice phase properties allowed the observation and understanding of some of the main formation and evolution processes occurring in MPC.

Since the temperature range where MPC occur ($-40°C < T < 0°C$) is too warm for homogenous ice nucleation, the initiation of the ice phase is made through heterogeneous nucleation processes (deposition, immersion freezing, condensation freezing and contact freezing) induced by the presence of aerosol particles acting as IN particles. Ice crystals can grow through several mechanisms. Riming due to the collection of droplets by ice crystals is one of the main ice growth mechanisms, leading to complex-shapes and heavy particles that fall and precipitate. Rimed particles are observed everywhere in and below the cloud. Water vapor condensation is the second main mechanism responsible for the growth of ice crystals. If supersaturation with respect to ice is reached, the excess water vapor condensate upon ice particles, leading to regular shapes such as plates, stellars, or columns. If the humidity is saturated both with respect to liquid and ice, droplets and ice crystals can grow simultaneously. But conditions of supersaturation with respect to ice, associated to subsaturation with respect to liquid water, are most often encountered in MPC. Therefore, under these conditions, the Wegener-Bergeron-Findeisein (WBF) process occurs. Liquid droplets evaporate and the subsequent water vapor condensates over ice crystals. Through this mechanism, ice crystals grow to the expanse of liquid droplets, leading to the depletion of the liquid phase, and a potential rapid glaciation of the cloud. So other processes are needed to maintain the liquid phase and the cloud layer and to explain the large persistence observed. For example, the strong radiative cooling induced by the supercooled water may help to maintain the liquid phase. It leads to a decrease of the stability and produces turbulent updrafts that favor the condensational growth of droplets.

The role of dynamics and the updrafts and downdrafts play a crucial role by transporting the ice particles at all levels into the cloud. It ensures aggregation of ice particles leading to irregular shapes. These growth mechanisms occur everywhere in and below the cloud. Consequently, the size and concentration of ice crystals and IWC do not exhibit a large vertical variability, as shown in Fig. 2E–H from Mioche et al. (2017) statistical study.

Fig. 3 shows an example of the distribution of ice crystal shapes within MPC (Mioche et al., 2017). CPI measurements from four airborne campaigns around the Svalbard region (ASTAR 2004, ASTAR 2007, POLARCAT 2008, and SORPIC 2010, corresponding to a total of 18 flights) have been merged to study the distribution of particle shapes within Arctic MPC. High resolution images of ice particles (2.3 μm pixel size) give an insight of the main microphysical growth processes occurring in MPC. For example, regular shapes such as plates and dendrites are well detected and confirm the condensational growth processes (including the WBF mechanism). Also, the large occurrence of

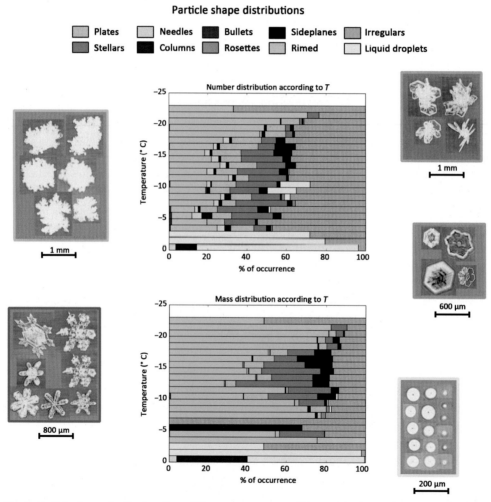

Fig. 3 Particle shape distribution (from CPI measurements and for particles larger than 100 μm) in MPC as a function of temperature (Mioche et al., 2017). Distributions are displayed according to particle number *(upper panel)* and mass *(bottom panel)*. Some images recorded by the CPI instrument are displayed.

rimed and irregular particles indicates a very efficient riming process and the role of the dynamics, respectively. From these results, it is thus clear that the assumption of spherical ice instead of nonspherical habits is unrealistic, may lead to an underestimation of ice growth, and should not be used in modeling.

Under specific mixed-phase conditions ($T > -8°C$, large droplets), ice crystals could also be produced by secondary production mechanisms caused by splintering of ice crystals during riming or shattering of isolated drops during freezing. Fragmentation of delicate ice crystals by collisions may also produce high ice concentration of small particles. Several studies (Bierwirth et al., 2013; Gayet et al., 2009b; Lawson et al., 2001; Lloyd et al., 2015; Rangno and Hobbs, 2001) reported these secondary ice production mechanisms. They observed a high number of small ice particles ($N_{ice} > 10$ L^{-1} according to Rangno and Hobbs, 2001; or $N_{ice} > 50$ L^{-1} according to Gayet et al., 2009b), which cannot be explained by the measured IN concentration or the one expected by Meyers et al. (1992) parameterization. Ice and liquid phase occur in pockets throughout the MPC. This distribution is linked to MPC aging since more patches are observed when the ice crystals have just blown up, according to Rangno and Hobbs (2001). These studies highlighted that the unexpected high concentration of small ice crystals depends on the largest cloud droplets formed into the cloud for "warm" MPC ($T > -8°C$), leading to large IWC and precipitating ice. On the contrary, in case of numerous small droplets, the concentration of precipitating ice particles is reduced.

All these findings show that the ice formation and growth processes and the subsequent cloud precipitation efficiency are closely linked to the liquid phase properties. Thus the aerosol concentration which drives the liquid phase properties also impacts significantly and indirectly the MPC properties and the ice phase through its impact on the liquid phase. Indeed, changes in CCN concentration influence the supercooled liquid droplets number and size, and thus the ice phase properties. Several works compared polluted and clean situations (Jackson et al., 2012; Lance et al., 2011) and showed that an increase of aerosol loading (from midlatitude long-range transport) increases the droplet concentration and reduce the ice crystal formation and growth efficiency. Hence the number of dense and precipitating rimed crystals is reduced, and thus the cloud precipitation efficiency too. Finally, the reduced precipitation may help with the maintenance of the liquid phase and the MPC longevity since droplets are not consumed by ice crystals. On the other hand, when ice precipitates, it is removed from the cloud. So, the depletion of liquid droplets due to ice growth by riming or WBF is reduced, leading to the maintenance of the liquid phase. Indeed, one of the particularities of Arctic MPC compared to the clouds at lower latitudes is that the liquid water fraction (LWF) increases with altitude. Examples are showed in works by Mioche et al. (2017) and McFarquhar et al. (2007). These results highlight that new LWF parameterizations are needed for Arctic single layer MPC as previous parameterizations established for MPC at different locations might not necessarily be representative in the Arctic.

Finally, in situ airborne cloud measurements performed in the Arctic from more than 20 years constitute a great step forward in the understanding of the small-scale processes involved in the life cycle of MPC. The formation, growth, and longevity of MPC depend on numerous interactions between aerosols, ice microphysics and liquid phase, controlling the liquid to ice conversion and precipitation rates. Obviously, these interactions and the associated feedbacks remain complex. Observations, coupled with modeling, are still needed for a better assessment of MPC lifetime processes and feedbacks.

In the following section, the focus is on the coupling of microphysical and optical properties from in situ measurements.

4.2 Case Study and Cloud-Radiation Interaction

Previous studies have shown that the presence of ice crystals in MPC is expected to enhance their warming effect (Ehrlich et al., 2009; Wendisch et al., 2013). The solar albedo effect can be significantly reduced for increasing ice water fraction in MPC. Therefore, coupling microphysical to optical properties is an important step to assess the radiative properties of such clouds.

Ice crystals in mixed-phase Arctic clouds are characterized by a wide variety of shape and size. The shape is determined by the growth process, which is related primarily to supersaturation and temperature regimes. Supersaturation is influenced by vertical velocity and dominates the internal structure and the degree of skeletal development (Keller and Hallett, 1982). Retrieval of cloud parameters from remote sensing requires, in particular, a precise knowledge of ice particle habits. However, most ice crystals in arctic clouds have non ideal shapes (irregulars) due to altering processes (alternating growth and sublimation), aggregation, and coagulation (Korolev et al., 1999). This wide variety of nonspherical shapes introduces significant challenges for computing reliable optical properties. The surface texture (the degree of surface roughness) is another important morphological parameter that can substantially modulate the single scattering properties of ice crystals (Yang et al., 2008). For instance, experimental laboratory studies performed by Ulanowski et al. (2006) and Schnaiter et al. (2016) implied that ice crystals with rough surfaces could reflect almost twice as much incident solar radiation back to space than their smooth counterparts. The effective size of cloud particles is also an important parameter to assess, as the size is a quantity playing a crucial role on the light scattering properties. Therefore, one of the main difficulties in predicting the radiative properties of arctic clouds is coupling the microphysical properties (mainly in the shape of a particle size distribution and liquid/ice partitioning) with a geometric model of the various ice crystal shapes such that its optical/radiative properties can be consistently assessed. Previous case studies have demonstrated the potential of the synergy between a cloud particle imager and a cloud nephelometer to link the microphysical and shape properties of arctic cloud particles to their single scattering characteristics (Gayet et al., 2009a; Jourdan et al., 2010; Lampert et al., 2009). These observations of arctic MPC properties have revealed that the ice crystal shapes and LWFs were connected to specific optical properties. Parameterizations of the cloud optical properties based on the synergetic use of statistical analysis of in situ measurements and light scattering modeling can be developed. The goal is to establish equivalent microphysical models, based on a limited set of free parameters (roughness, mixtures of idealized particle habits, and aspect ratio). These models are expected to reproduce the main optical features of cloud layers characterized by different LWFs and ice crystal habits.

Fig. 4 represents an example of this strategy implemented during the ASTAR campaign. A principal component analysis (PCA) methodology is applied to the polar nephelometer (PN) scattering phase function measurements. This analysis enables us to identify and extract specific cloud layers sharing similar optical properties. The optical patterns revealed by the PCA are related to specific microphysical properties (shape, PSD, IWC/LWC) measured by the microphysical probes. The behaviors of the average phase functions are consistent with the particle habit classification derived from the CPI measurements. Then an iterative inversion method (Jourdan et al., 2003; Oshchepkov et al., 2000) using light scattering modeling of irregular ice crystals (Yang and Liou, 1996) can be applied to the average phase function to retrieve equivalent microphysical models. The results show that the use of idealized geometric shape models representing ensembles of rough ice crystals combined with a population of water droplets is suitable to describe the scattering properties of arctic MPC. The optical properties of cloud layers dominated by the liquid water phase ($g > 0.83$) can be modeled by a mixture of water droplets ($>90\%$) and small droxtal shaped crystals. The optical properties of cloud layers where the ice phase prevails ($g < 0.80$) can be represented by a combination of water droplets (1535%) and rough plates or columns with a varying aspect ratio (65%–85%).

However, the microphysical and morphological properties derived from the CPI measurements are limited to particles with maximum dimension larger than 25–50 μm. Jourdan et al. (2010) showed, using light scattering modeling, that the optical contribution of small particles with sizes lower than 50 μm (droplets and ice crystals) was significant, always exceeding 50% of the total scattering signal. Thus the influence of small ice crystals on the optical properties needs to be more accurately quantified using in situ measurements. The combination of state-of-the-art instruments deployed during the RACEPAC or ACLOUD campaign should lead to a better estimate of the optical contribution of small particles as well as the impact of ice crystal habit on the scattering properties. This methodology applied to larger datasets should enable us to study more accurately the influence of small ice crystals, ice particle habits, and liquid-ice partitioning on the optical properties of MPC.

5. SATELLITE REMOTE SENSING EVALUATION

As shown in Section **2**, MPC can be observed on almost all the Arctic region using satellite measurements. However, these measurements are indirect and are based on retrieval algorithms involving hypothesis that need to be validated (Cesana et al., 2016; Mioche et al.,

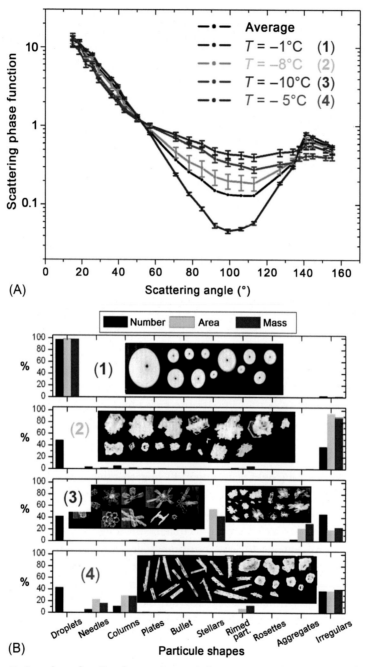

Fig. 4 (A) Scattering phase function from polar nephelometer measurements according to several cloud parts and (B) corresponding particle shape classification (in number, area, and mass) from CPI measurements.

2010). They also provide cloud properties typically averaged over 1 km, which may be insufficient to study cloud processes at a microphysical scale. Cesana et al. (2016) showed for example that cloud detection and phase retrieval product from CALIOP lidar measurements depend strongly on factors such as horizontal and vertical data averaging. Additionally, they suffer from inherent shortcomings at low altitude levels (Blanchard et al., 2014; Liu et al., 2017; Marchand et al., 2008). Moreover, the definition of the cloud thermodynamic phase strongly depends on the measurement technique and the observation scale.

In situ and ground-based remote sensing measurements may complement the satellite observations and partially overcome their limitations by providing a detailed characterization of cloud microphysical properties at low levels. In addition, in situ observations are based on direct measurement techniques and can provide data at a higher spatial resolution (generally < 100 m). In that sense, they can be used to assess the satellite remote sensing retrieval products, such as cloud detection and thermodynamical phase by providing accurate profiles of cloud properties at the very low altitude levels.

Thus it appears relevant to investigate the horizontal and vertical distribution of ice crystals and liquid water droplets, as well as the scale dependent liquid-ice partitioning for different observational techniques.

This section illustrates the potential of colocated in situ measurements to conduct satellite remote sensing validation exercises. The retrieval algorithm evaluated in the following is the DARDAR algorithm described in Ceccaldi et al. (2013) and Delanoë and Hogan (2008 and 2010). DARDAR algorithm merges CloudSat and CALIPSO observations on the same resolution grid (1700 m horizontal and 60 m vertical) and uses the combination of lidar and radar measurements to detect clouds and retrieve their phase and properties. Four flights sampling MPC during the ASTAR 2007 and POLARCAT 2008 experiments were successfully collocated with the A-Train track. The DARDAR algorithm was operated (from CALIPSO/CloudSat satellite data) for the cloud/no cloud detection and the retrieval of the MPC thermodynamical phase. Fig. 5 illustrates the

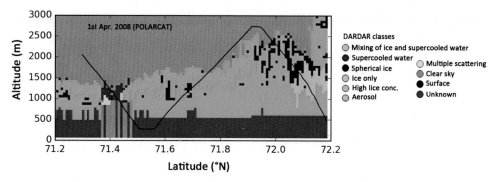

Fig. 5 Vertical profile of the cloud phase determined by the DARDAR retrieval algorithm for a satellite validation situation encountered during POLARCAT 2008 campaign. The *black line* shows the aircraft flight track.

vertical profile of DARDAR cloud phase product with the cloud type classification for one of the four situations. The flight track is superimposed in black line.

In order to evaluate the DARDAR cloud retrieval, the DARDAR cloud products along the flight track are compared to the asymmetry parameter (g) values determined from the PN in situ measurements. The method consists in oversampling the DARDAR products to match the PN resolution (around 80 m horizontal).

Cloud detection is first investigated by comparing the DARDAR cloud detection algorithm (i.e., all classes including a cloud type) along the flight tracks to the in situ PN measurements considered as the cloud/no cloud occurrence reference. The comparisons are summarized in Table 2 where the statistics of cooccurrences are displayed. A very good agreement is observed between DARDAR and in situ measurements both for cloud and clear sky cloud detection. 91% of the clear sky events and 86% of the cloudy pixels match with the PN measurements. The false detections can be explained by changes in the cloud structure (cloud top and base heights, dissipation) between the satellite overpass time and the aircraft measurements time (delay up to 85 min). Most of the undefined DARDAR class actually corresponds to clouds (60%). In particular, this occurs at low levels, where DARDAR retrievals are strongly impacted by the attenuation of the lidar laser beam by liquid layers, as well as the contamination by radar ground echoes. This assumption is strengthened by Fig. 5 showing that the undefined DARDAR pixels (brown) are mostly localized close to the surface.

The cloud phase retrieval is also evaluated by comparison with PN in situ measurements. g-Values of 0.80 and 0.83 are chosen to define the thresholds between the ice, mixed, and liquid phases (Jourdan et al., 2010). The quantitative and statistical approach is provided in Fig. 6 where the frequencies of occurrence of g-values are displayed for the three main phase classes derived from DARDAR cloud type classification. From these histograms, the validation scores are summarized in Table 3.

The g-values distribution corresponding to the mixing of ice and supercooled water DARDAR phase (green, hereafter called mixing class) is centered on 0.85. Nearly 90% of the DARDAR mixing class is associated with a liquid phase according to PN ($g > 0.83$). The remaining pixels are distributed more or less equally among the in situ ice and mixed phase (6% and 5%, respectively). The ice DARDAR class (blue) distribution exhibits two modes: the main one around 0.74, and the second around 0.84. The statistics scores show

Table 2 Statistics of the cloud detection validation

DARDAR class	PN (reference)	
	No cloud (%)	Cloud (%)
Clear sky	91	9
Cloud	14	86
Undefined	40	60

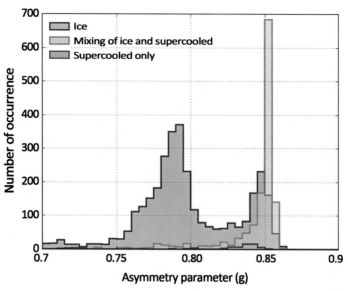

Fig. 6 Frequencies of occurrence of the asymmetry parameter from PN according to the DARDAR cloud phase retrieval in color.

Table 3 Statistics of cloud phase retrieval validation

DARDAR retrieval	PN (reference)		
	Ice phase $(0.75 < g < 0.80)$ (%)	Mixed phase $(0.80 < g < 0.83)$ (%)	Liquid phase $(g > 0.83)$ (%)
Ice	61	15	24
Mixing of ice and supercooled water	6	5	89
Supercooled water only	9	24	67

that 61% of the observations corresponding to ice DARDAR class are validated by PN data. The remaining DARDAR ice pixels are distributed among the in situ mixed (15%) and liquid (24%) phases. Finally, the distribution of the supercooled water class (red) shows that 67% of the DARDAR pixels detected as supercooled water only are validated by the in situ measurements while 24% correspond to an in situ mixed phase. The remaining 9% corresponds to an ice phase. However, the number of DARDAR pixels concerning this class is limited and thus might not be representative, making it difficult to draw conclusions.

The differences of time and space observation scales between in situ and satellite measurements may explain the main misclassifications of DARDAR pixels. Indeed, in situ measurements account for the small-scale inhomogeneities of the liquid and ice occurrences as they document with accuracy and at the high spatial resolution the cloud thermodynamic phase. On the other hand, satellite products are more representative of an averaged cloud phase since their spatial resolution is coarser (one order of magnitude lower than in situ measurements). Therefore a pixel classified as ice by DARDAR could correspond to a mixture of several small sequences or pockets of ice and supercooled liquid droplets, as shown by previous in situ measurement studies (cf. Section 4.1). On the contrary, mixed cloud layers optically dominated by supercooled water droplets could be considered as a liquid phase by the PN measurements while the radar could still detect the presence of a few ice crystals. These differences of detection techniques could be responsible for the disagreement between in situ and satellite mixed-phase class.

Additionally, the aircraft sampling time may differ from the actual satellite overpass time. During this delay, cloud top and base altitudes, and the cloud layer thickness, may vary and could contribute to the discrepancies observed between DARDAR and PN classification. For instance, this could explain the misclassification of 24% of the DARDAR pixels of the ice class that should belong to the liquid phase according to the PN measurements ($g > 0.83$). Physical assumptions considered in DARDAR algorithm, such as the supercooled liquid layer thickness, can also influence the retrievals of the cloud phase. The liquid layer is set to have a maximum thickness of 300 m (Delanoë and Hogan, 2010) whereas aircraft measurements have clearly shown that within MPC this liquid layer can be thicker than 300 m. This 300 m threshold corresponds to the full attenuation of the CALIOP laser beam meaning that DARDAR is not able to detect the liquid phase beyond this thickness. Since most of the misclassified pixels occur in the lower part of the liquid layer, this could be a plausible explanation for the observed disagreement.

The results presented here have been made by oversampling the satellite observations to match the in situ measurements resolution. Similar work has been performed averaging the PN data on the same resolution grid as DARDAR products (pixel size of 1700 and 60 m horizontal and vertical, respectively, corresponding to approximately 17 in situ data points) and it is quite remarkable that the results are very similar for the two methods, both in terms of cloud/no cloud detection and for the cloud phase product.

These comparisons illustrate the impact of sampling resolution on the cloud detection and cloud phase retrievals. Subpixel representativity of the spaceborne observations can be assessed showing that the satellite detection of cloudy pixels is consistent with higher spatial resolution in situ cloud measurements.

Finally, these findings also highlight that the characterization of the horizontal and vertical variability of MPC thermodynamical phase at different observation scales is still needed.

6. CONCLUSION

The investigation of Arctic MPCs is of primary importance to improve their representation in numerical models and thus improve the accuracy of the Arctic climate modeling, as well as global climate predictions. In particular, the liquid-ice partitioning, ice crystals, and liquid droplets properties (number, size, shape, mass) formation and growth processes and interactions with aerosols and radiation, need to be assessed.

Observations from satellites can be used to study Arctic clouds at a regional scale. In particular, the new satellite active remote sensing measurements from CALIPSO and CloudSat have greatly improved the accuracy of observations from space compared to the previous passive remote sensing measurements. The assessment of the seasonal, spatial, and vertical variability of Arctic MPC became possible. Main results (Cesana et al., 2012; Mioche et al., 2015) highlighted that MPC are present all year long in the Arctic, with important spatial and seasonal variability. The Arctic Atlantic side (region around Barents and Greenland Seas) exhibits high and almost constant occurrence of MPC throughout the year with higher frequency than the Arctic average. On the contrary, regions of Central Arctic Ocean and Pacific side (region around Northern Alaska, Barents and Chukchi Seas) exhibit a clear seasonal variability, with low MPC occurrences during winter and spring, and larger occurrences in summer and autumn. The vertical distribution of MPC showed that most of the MPC are present below 3 km, except during summer where these clouds are frequently observed at midaltitudes (3–6 km). This seasonal and vertical variability of MPC at the regional scale are related to the influence of the North Atlantic Ocean, and with the melting of sea ice. The Atlantic Ocean provides heat and moisture necessary to the formation of MPC in the Greenland and Barents Seas regions all year long. The open water resulting from the melting of sea ice from spring to autumn allows the transport of heat and moisture from the North Atlantic Ocean to the Central Arctic Ocean and Pacific side, promoting the formation of MPC in this region during these seasons.

However, active remote sensing retrieval algorithms need to be evaluated and validated. Moreover, observations from space are known to be biased close to the surface (<2 km). Ground-based and airborne remote sensing can provide accurate cloud measurements at low-level altitude. High spatial and temporal resolution measurements are also needed to study microphysical properties and processes. Therefore, even though they remain very local in time and space, airborne in situ measurements offer the high resolution needed to accurately characterize cloud particles' properties and microphysical processes.

Since the 1990s, numerous major field experiments involving in situ cloud airborne measurements have been carried out. Based on these measurements, properties of liquid droplets and ice crystals within MPC (number, size, shape, mass) and main processes responsible for their formation, growth, and persistence can be investigated. These campaigns datasets combined with modeling studies give an insight of the main processes involved in the MPC life cycle.

In particular, the main findings highlighted from in situ airborne measurements concerning Arctic MPC can be summarized as follows:

- Liquid droplets result from the activation of CCN particles at cloud base. They grow during their ascent towards the cloud top. Temperature inversion at cloud top limits the vertical development of MPC by entrainment of dry air and the evaporation of a part of the cloud droplets. If humidity inversion occurs at cloud top, it can counteract this effect by providing moisture and prevent the evaporation of droplets.
- Ice crystals are formed by heterogeneous ice nucleation processes or, under specific thermodynamical conditions, by secondary ice formation processes. Ice crystals can grow through riming, condensational (including Wegener-Bergeron-Findeisen process), and aggregation processes, depending on the supercooled liquid amount available, humidity (supersaturation) conditions, and dynamical conditions. Consequently to these numerous processes, ice crystal properties do not exhibit specific vertical variability and are almost uniformly distributed into MPC.
- The formation and growth of ice crystals and the subsequent precipitation efficiency depend strongly on the liquid phase and aerosols properties. In cases of polluted situations (from midlatitude long-range transport), a higher concentration of aerosols (and thus CCN) results in a large number of small liquid droplets and reduce the formation of the ice phase, its growth efficiency, and thus its precipitation ability. Moreover, this reduction of the precipitation efficiency may help to maintain the liquid phase and the MPC longevity since droplets are less consumed by ice crystals, resulting in a positive cloud feedback.
- As a result of the interaction processes between ice and liquid phases, it has been shown that Arctic MPC exhibit a LWF increasing with decreasing temperature into the cloud, contrary to MPCs at lower latitudes. That means that specific parameterizations for Arctic MPC are needed to improve modeling.
- There is a strong optical signature of the LWF and shape of ice crystals within the MPC layers with possible impacts on remote sensing retrievals and radiative properties. This confirms the importance of deriving equivalent microphysical models based on a limited set of free parameters (roughness, mixture of idealized particle habits, and aspect ratio of ice crystals) that reproduce the main optical and features of the cloud layers.

At last, in situ measurements are very useful to validate remote sensing observations. We showed that, when they are collocated with satellite observations such as the CloudSat radar and CALIOP lidar measurements, they can be used to evaluate retrieval algorithms. Validation works in terms of cloud detection and thermodynamical phase highlighted a general good accuracy of the retrieval products. Nevertheless, some issues have been identified regarding the retrieval of the supercooled water phase, which is mostly confounded with the mixing of ice and supercooled water class. The retrieval of the ice phase also presents some misclassifications as some part of the retrieved ice pixels is actually liquid water. These issues may be mainly attributed to the difference in resolution between satellite and in situ observations, which have a better ability to detect the heterogeneities (sequences of liquid and ice) of mixed phase at a small scale compared to large scale (satellite resolution). Finally, these results also highlight the need to link large and small scale observations (and allow the investigating of the subpixel representativity of satellite observations).

In conclusion, the synergy of satellite active remote sensing observation and in situ airborne cloud measurements performed in the Arctic region constitute a great step forward in the characterization of the macrophysical, microphysical, and optical properties of low-level MPC, and the understanding of the large and small scale processes governing their life cycle, as well as their interactions with aerosols and radiation. This is important to mitigate biases of the modeling and parameterizations of cloud properties and radiative impact.

However, all the radiative-dynamical-microphysical process interactions and feedbacks controlling the liquid to ice conversion, the precipitation rates, and the MPC maintenance remain complex and need to be better assessed. Many challenges and unanswered questions remain if we are to understand the MPC life cycle and improve the accuracy of the retrieval algorithm and MPC representation in global and regional models. More multiscale modeling studies constrained by larger comprehensive datasets are needed to unravel the web of interactions between processes occurring in MPC. Measurements planned during the ACLOUD (Arctic Cloud Observations Using airborne measurements during polar Day, 2017) campaign or the international initiative MOSAiC (Multidisciplinary drifting Observatory for the Study of Arctic Climate, 2020) in the Central Arctic Basin ice pack should provide an accurate dataset to improve our process-level understanding of MPC clouds. Ship-based, airborne, ground-based measurements coupled with satellite observations over sea ice and open ocean will be used to estimate the role of Arctic MPC on the amplified climate change occurring in the Arctic region.

ACKNOWLEDGMENTS

This work is part of the French scientific community EECLAT project (Expecting EarthCare, Learning from A-Train) and is supported by the French Centre National des Etudes Spatiales (CNES) and the Centre National de la Recherche Scientifique—Institut National des Sciences de l'Univers (CNRS-INSU).

REFERENCES

Ackerman, S.A., Strabala, K.I., Menzel, W.P., Frey, R.A., Moeller, C.C., Gumley, L.E., 1998. Discriminating clear sky from clouds with MODIS. J. Geophys. Res. Atmos. 103, 32141–32157. https://doi.org/10.1029/1998JD200032.

Avramov, A., Ackerman, A.S., Fridlind, A.M., van Diedenhoven, B., Botta, G., Aydin, K., 2011. Toward ice formation closure in Arctic mixed-phase boundary layer clouds during ISDAC. J. Geophys. Res. 116. https://doi.org/10.1029/2011JD015910.

Baumgardner, D., Avallone, L., Bansemer, A., Borrmann, S., Brown, P., Bundke, U., 2012. In situ, airborne instrumentation: addressing and solving measurement problems in ice clouds. Bull. Am. Meteorol. Soc. 93, ES29–ES34. https://doi.org/10.1175/BAMS-D-11-00123.1.

Bierwirth, E., Ehrlich, A., Wendisch, M., Gayet, J.-F., Gourbeyre, C., Dupuy, R., 2013. Optical thickness and effective radius of Arctic boundary-layer clouds retrieved from airborne nadir and imaging spectrometry. Atmos. Meas. Tech. 6, 1189–1200. https://doi.org/10.5194/amt-6-1189-2013.

Blanchard, Y., Pelon, J., Eloranta, E.W., Moran, K.P., Delanoë, J., Sèze, G., 2014. A synergistic analysis of cloud cover and vertical distribution from A-Train and ground-based sensors over the high arctic station Eureka from 2006 to 2010. J. Appl. Meteorol. Climatol. 53, 2553–2570. https://doi.org/10.1175/JAMC-D-14-0021.1.

Boucher, O., Randall, D., Artaxo, P., Bretherton, C., Feingold, G., Forster, P., 2013. Clouds and aerosols. In: Stocker, T.F., Qin, D., Plattner, G.-K., Tignor, M., Allen, S.K., Boschung, J., Nauels, A., Xia, Y., Bex, V., Midgley, P.M. (Eds.), Climate Change 2013: The Physical Science Basis. Contribution of Working Group I to the Fifth Assessment Report of the Intergovernmental Panel on Climate Change. Cambridge University Press, Cambridge, United Kingdom/New York, NY, pp. 571–658. https://doi.org/10.1017/CBO9781107415324.016.

Ceccaldi, M., Delanoë, J., Hogan, R.J., Pounder, N.L., Protat, A., Pelon, J., 2013. From CloudSat-CALIPSO to EarthCare: evolution of the DARDAR cloud classification and its comparison to airborne radar-lidar observations. J. Geophys. Res. Atmos. 118, 1–20. https://doi.org/10.1002/jgrd.50579.

Cesana, G., Kay, J.E., Chepfer, H., English, J.M., de Boer, G., 2012. Ubiquitous low-level liquid-containing Arctic clouds: new observations and climate model constraints from CALIPSO-GOCCP. Geophys. Res. Lett. 39, 1–6. https://doi.org/10.1029/2012GL053385.

Cesana, G., Chepfer, H., Winker, D., Getzewich, B., Cai, X., Jourdan, O., 2016. Using in situ airborne measurements to evaluate three cloud phase products derived from CALIPSO: CALIPSO cloud phase validation. J. Geophys. Res. Atmos. 121, 5788–5808. https://doi.org/10.1002/2015JD024334.

Chan, M.A., Comiso, J.C., 2013. Arctic cloud characteristics as derived from MODIS, CALIPSO, and CloudSat. J. Clim. 26, 3285–3306. https://doi.org/10.1175/JCLI-D-12-00204.1.

Chernokulsky, A., Mokhov, I.I., 2012. Climatology of total cloudiness in the Arctic: an intercomparison of observations and reanalyses. Adv. Meteorol. 2012, 1–15. https://doi.org/10.1155/2012/542093.

Curry, J.A., 1986. Interactions among turbulence radiation and microphysics in arctic stratus clouds. J. Atmos. Sci. 43, 90–106. https://doi.org/10.1175/1520-0469(1986)043<0090.IATRAM>2.0.CO;2.

Curry, J.A., 1995. Interactions among aerosols, clouds, and climate of the Arctic Ocean. Sci. Total Environ. 160–161, 777–791. https://doi.org/10.1016/0048-9697(95)04411-S.

Curry, J., Ebert, E., Herman, G., 1988. Mean and turbulence structure of the summertime Arctic cloudy boundary layer. Q. J. Roy. Meteor. Soc. 114, 715–746.

Curry, J.A., Meyer, F.G., Radke, L.F., Brock, C.A., Ebert, E.E., 1990. Occurrence and characteristics of lower tropospheric ice crystals in the arctic. Int. J. Climatol. 10, 749–764. https://doi.org/10.1002/joc.3370100708.

Curry, J.A., Schramm, J.L., Rossow, W.B., Randall, D., 1996. Overview of Arctic cloud and radiation characteristics. J. Clim. 9, 1731–1764. https://doi.org/10.1175/1520-0442(1996)009<1731:OOACAR>2.0.CO;2.

Curry, J.A., Pinto, J.O., Benner, T., Tschudi, M., 1997. Evolution of the cloudy boundary layer during the autumnal freezing of the Beaufort Sea. J. Geophys. Res. 102, 13851. https://doi.org/10.1029/96JD03089.

Curry, J.A., Hobbs, P.V., King, M.D., Randall, D.A., Minnis, P., Isaac, G.A., 2000. FIRE Arctic clouds experiment. Bull. Am. Meteorol. Soc. 81, 5–29. https://doi.org/10.1175/1520-0477(2000)081<0005:FACE>2.3.CO;2.

Delanoë, J., Hogan, R.J., 2008. A variational scheme for retrieving ice cloud properties from combined radar, lidar, and infrared radiometer. J. Geophys. Res. 113. https://doi.org/10.1029/2007JD009000.

Delanoë, J., Hogan, R.J., 2010. Combined CloudSat-CALIPSO-MODIS retrievals of the properties of ice clouds. J. Geophys. Res. 115. https://doi.org/10.1029/2009JD012346.

Delanoë, J., Protat, A., Jourdan, O., Pelon, J., Papazzoni, M., Dupuy, R., 2013. Comparison of airborne in situ, airborne radar-lidar, and spaceborne radar-lidar retrievals of polar ice cloud properties sampled during the POLARCAT campaign. J. Atmos. Ocean. Technol. 30, 57–73. https://doi.org/10.1175/JTECH-D-11-00200.1.

Dergach, A.L., Zabrodsky, G.M., Morachevsky, V.G., 1960. The results of a complex investigation of the type st-sc clouds and fogs in the Arctic. Bull Acad. Sci. USSR Geophys. Ser. 1, 66–70.

Deser, C., Teng, H., 2008. Evolution of Arctic sea ice concentration trends and the role of atmospheric circulation forcing, 1979–2007. Geophys. Res. Lett. 35. https://doi.org/10.1029/2007GL032023.

Dong, X., Xi, B., Crosby, K., Long, C.N., Stone, R.S., Shupe, M.D., 2010. A 10 year climatology of Arctic cloud fraction and radiative forcing at Barrow, Alaska. J. Geophys. Res. 115. https://doi.org/10.1029/2009JD013489.

Eckhardt, S., Hermansen, O., Grythe, H., Fiebig, M., Stebel, K., Cassiani, M., 2013. The influence of cruise ship emissions on air pollution in Svalbard – a harbinger of a more polluted Arctic? Atmos. Chem. Phys. 13, 8401–8409. https://doi.org/10.5194/acp-13-8401-2013.

Ehrlich, A., Wendisch, M., Bierwirth, E., Gayet, J.-F., Mioche, G., Lampert, A., 2009. Evidence of ice crystals at cloud top of Arctic boundary-layer mixed-phase clouds derived from airborne remote sensing. Atmos. Chem. Phys. 9, 9401–9416. https://doi.org/10.5194/acp-9-9401-2009.

Ervens, B., Feingold, G., Sulia, K., Harrington, J., 2011. The impact of microphysical parameters, ice nucleation mode, and habit growth on the ice/liquid partitioning in mixed-phase Arctic clouds. J. Geophys. Res. 116. https://doi.org/10.1029/2011JD015729.

Frey, R.A., Ackerman, S.A., Liu, Y., Strabala, K.I., Zhang, H., Key, J.R., 2008. Cloud detection with MODIS. Part I: improvements in the MODIS cloud mask for collection 5. J. Atmos. Ocean Tech. 25, 1057–1072. https://doi.org/10.1175/2008JTECHA1052.1.

Fridlind, A.M., van Diedenhoven, B., Ackerman, A.S., Avramov, A., Mrowiec, A., Morrison, H., 2012. A FIRE-ACE/SHEBA case study of mixed-phase arctic boundary layer clouds: entrainment rate limitations on rapid primary Ice nucleation processes. J. Atmos. Sci. 69, 365–389. https://doi.org/10.1175/JAS-D-11-052.1.

Gayet, J.-F., Mioche, G., Dörnbrack, A., Ehrlich, A., Lampert, A., Wendisch, M., 2009a. Microphysical and optical properties of Arctic mixed-phase clouds. The 9 April 2007 case study. Atmos. Chem. Phys. 9, 6581–6595. https://doi.org/10.5194/acp-9-6581-2009.

Gayet, J.-F., Treffeisen, R., Helbig, A., Bareiss, J., Matsuki, A., Herber, A., 2009b. On the onset of the ice phase in boundary layer Arctic clouds. J. Geophys. Res. 114. https://doi.org/10.1029/2008JD011348.

Harrington, J.Y., Reisin, T., Cotton, W.R., Kreidenweis, S.M., 1999. Cloud resolving simulations of Arctic stratus. Atmos. Res. 51, 45–75. https://doi.org/10.1016/S0169-8095(98)00098-2.

Intrieri, J.M., Shupe, M.D., Uttal, T., McCarty, B.J., 2002. An annual cycle of Arctic cloud characteristics observed by radar and lidar at SHEBA. J. Geophys. Res. 107. https://doi.org/10.1029/2000JC000423.

IPCC, (Ed.), 2013. Climate Change 2013 – The Physical Science Basis: Working Group I Contribution to the Fifth Assessment Report of the Intergovernmental Panel on Climate Change. Cambridge University Press, Cambridge.

Jackson, R.C., McFarquhar, G.M., Korolev, A.V., Earle, M.E., Liu, P.S.K., Lawson, R.P., 2012. The dependence of ice microphysics on aerosol concentration in arctic mixed-phase stratus clouds during ISDAC and M-PACE. J. Geophys. Res. 117. https://doi.org/10.1029/2012JD017668.

Jayaweera, K.O., Ohtake, T., 1973. Concentration of ice crystals in Arctic stratus clouds. J. Rech. Atmosph. 7, 199–207.

Jiang, H., Cotton, W.R., Pinto, J.O., Curry, J.A., Weissbluth, M.J., 2000. Cloud resolving simulations of mixed-phase arctic stratus observed during BASE: sensitivity to concentration of Ice crystals and large-scale heat and moisture advection. J. Atmos. Sci. 57, 2105–2117. https://doi.org/10.1175/1520-0469(2000)057<2105:CRSOMP>2.0.CO;2.

Jourdan, O., Oshchepkov, S., Shcherbakov, V., Gayet, J.-F., Isaka, H., 2003. Assessment of cloud optical parameters in the solar region: retrievals from airborne measurements of scattering phase functions. J. Geophys. Res. 108. https://doi.org/10.1029/2003JD003493.

Jourdan, O., Mioche, G., Garrett, T.J., Schwarzenböck, A., Vidot, J., Xie, Y., et al., 2010. Coupling of the microphysical and optical properties of an Arctic nimbostratus cloud during the ASTAR 2004 experiment: implications for light-scattering modeling. J. Geophys. Res. 115. https://doi.org/10.1029/2010JD014016.

Kay, J.E., Gettelman, A., 2009. Cloud influence on and response to seasonal Arctic sea ice loss. J. Geophys. Res. 114. https://doi.org/10.1029/2009JD011773.

Kay, J.E., Holland, M.M., Bitz, C.M., Blanchard-Wrigglesworth, E., Gettelman, A., Conley, A., 2012. The influence of local feedbacks and northward heat transport on the equilibrium arctic climate response to increased greenhouse gas forcing. J. Clim. 25, 5433–5450. https://doi.org/10.1175/JCLI-D-11-00622.1.

Keller, V.W., Hallett, J., 1982. Influence of air velocity on the habit of ice crystal growth from the vapor. J. Cryst. Growth 60, 91–106. https://doi.org/10.1016/0022-0248(82)90176-2.

Klingebiel, M., de Lozar, A., Molleker, S., Weigel, R., Roth, A., Schmidt, L., et al., 2014. Arctic low-level boundary layer clouds: in-situ measurements and simulations of mono- and bimodal supercooled droplet size distributions at the cloud top layer. Atmos. Chem. Phys. Discuss. 14, 14599–14635. https://doi.org/10.5194/acpd-14-14599-2014.

Koptev, A.P., Voskresenskii, A.I., 1962. On the radiation properties of clouds. Proc. Arct. Antarct. Res. Inst. 239, 39–47.

Korolev, A.V., Isaac, G.A., Hallett, J., 1999. Ice particle habits in Arctic clouds. Geophys. Res. Lett. 26, 1299–1302.

Lampert, A., Ehrlich, A., Dörnbrack, A., Jourdan, O., Gayet, J.-F., Mioche, G., 2009. Microphysical and radiative characterization of a subvisible midlevel Arctic ice cloud by airborne observations – a case study. Atmos. Chem. Phys. 9, 2647–2661. https://doi.org/10.5194/acp-9-2647-2009.

Lance, S., Shupe, M.D., Feingold, G., Brock, C.A., Cozic, J., Holloway, J.S., 2011. Cloud condensation nuclei as a modulator of ice processes in Arctic mixed-phase clouds. Atmos. Chem. Phys. 11, 8003–8015. https://doi.org/10.5194/acp-11-8003-2011.

Law, K.S., Stohl, A., Quinn, P.K., Brock, C.A., Burkhart, J.F., Paris, J.-D., 2014. Arctic air pollution: new insights from POLARCAT-IPY. Bull. Am. Meteorol. Soc. 95, 1873–1895. https://doi.org/10.1175/BAMS-D-13-00017.1.

Lawson, R.P., Baker, B.A., Schmitt, C.G., Jensen, T.L., 2001. An overview of microphysical properties of Arctic clouds observed in May and July 1998 during FIRE ACE. J. Geophys. Res. 106, 14989. https://doi.org/10.1029/2000JD900789.

Liu, C., Allan, R.P., Huffman, G.J., 2012a. Co-variation of temperature and precipitation in CMIP5 models and satellite observations: precipitation variation in CMIP5 models. Geophys. Res. Lett. 39. https://doi.org/10.1029/2012GL052093.

Liu, Y., Key, J.R., Ackerman, S.A., Mace, G.G., Zhang, Q., 2012b. Arctic cloud macrophysical characteristics from CloudSat and CALIPSO. Remote Sens. Environ. 124, 159–173. https://doi.org/10.1016/j.rse.2012.05.006.

Liu, Y., Shupe, M.D., Wang, Z., Mace, G., 2017. Cloud vertical distribution from combined surface and space radar/lidar observations at two Arctic atmospheric observations. Atmos. Chem. Phys. Discuss. 1–28. https://doi.org/10.5194/acp-2016-1132.

Lloyd, G., Choularton, T.W., Bower, K.N., Crosier, J., Jones, H., Dorsey, J.R., 2015. Observations and comparisons of cloud microphysical properties in spring and summertime Arctic stratocumulus clouds during the ACCACIA campaign. Atmos. Chem. Phys. 15, 3719–3737. https://doi.org/10.5194/acp-15-3719-2015.

Lubin, D., Morrow, E., 1998. Evaluation of an AVHRR cloud detection and classification method over the central Arctic Ocean. J. Appl. Meteorol. 37, 166–183. https://doi.org/10.1175/1520-0450(1998)037<0166:EOAACD>2.0.CO;2.

Lubin, D., Vogelmann, A.M., 2006. A climatologically significant aerosol longwave indirect effect in the Arctic. Nature 439, 453–456. https://doi.org/10.1038/nature04449.

Marchand, R., Mace, G.G., Ackerman, T., Stephens, G., 2008. Hydrometeor detection using CloudSat—an earth-orbiting 94-GHz cloud radar. J. Atmos. Ocean. Technol. 25, 519–533. https://doi.org/10.1175/2007JTECHA1006.1.

McFarquhar, G.M., Zhang, G., Poellot, M.R., Kok, G.L., McCoy, R., Tooman, T., 2007. Ice properties of single-layer stratocumulus during the mixed-phase Arctic cloud experiment: 1. Observations. J. Geophys. Res. 112. https://doi.org/10.1029/2007JD008633.

McFarquhar, G.M., Ghan, S., Verlinde, J., Korolev, A., Strapp, J.W., Schmid, B., 2011. Indirect and semi-direct aerosol campaign: the impact of arctic aerosols on clouds. Bull. Am. Meteorol. Soc. 92, 183–201. https://doi.org/10.1175/2010BAMS2935.1.

Meyers, M.P., DeMott, P.J., Cotton, W.R., 1992. New primary ice-nucleation parameterizations in an explicit cloud model. J. Appl. Meteorol. 31, 708–721. https://doi.org/10.1175/1520-0450(1992) 031<0708:NPINPI>2.0.CO;2.

Mioche, G., Josset, D., Gayet, J.-F., Pelon, J., Garnier, A., Minikin, A., 2010. Validation of the CALIPSO-CALIOP extinction coefficients from in situ observations in midlatitude cirrus clouds during the CIRCLE-2 experiment. J. Geophys. Res. 115. https://doi.org/10.1029/2009JD012376.

Mioche, G., Jourdan, O., Ceccaldi, M., Delanoë, J., 2015. Variability of mixed-phase clouds in the Arctic with a focus on the Svalbard region: a study based on spaceborne active remote sensing. Atmos. Chem. Phys. 15, 2445–2461. https://doi.org/10.5194/acp-15-2445-2015.

Mioche, G., Jourdan, O., Delanoë, J., Gourbeyre, C., Febvre, G., Dupuy, R., 2017. Characterization of Arctic mixed-phase cloud properties at small scale and coupling with satellite remote sensing. Atmos. Chem. Phys. Discuss. 1–52. https://doi.org/10.5194/acp-2017-93.

Morrison, H., Zuidema, P., Ackerman, A.S., Avramov, A., de Boer, G., Fan, J., 2011. Intercomparison of cloud model simulations of Arctic mixed-phase boundary layer clouds observed during SHEBA/FIRE-ACE: intercomparison of cloud model simulations of arctic mixed-phase. J. Adv. Model. Earth Syst. 3. https://doi.org/10.1029/2011MS000066.

Morrison, H., de Boer, G., Feingold, G., Harrington, J., Shupe, M.D., Sulia, K., 2012. Resilience of persistent Arctic mixed-phase clouds. Nat. Geosci. 5, 11–17. https://doi.org/10.1038/ngeo1332.

Nygård, T., Valkonen, T., Vihma, T., 2014. Characteristics of Arctic low-tropospheric humidity inversions based on radio soundings. Atmos. Chem. Phys. 14, 1959–1971. https://doi.org/10.5194/acp-14-1959-2014.

Orbaek, J.B., Hisdal, V., Svaasand, L.E., 1999. Radiation climate variability in Svalbard: surface and satellite observations. Polar Res. 18, 127–134.

Oshchepkov, S., Isaka, H., Gayet, J.-F., Sinyuk, A., Auriol, F., Havemann, S., 2000. Microphysical properties of mixed-phase & ice clouds retrieved from in situ airborne "polar nephelometer" measurements. Geophys. Res. Lett. 27, 209–212. https://doi.org/10.1029/1999GL010784.

Persson, P.O.G., 2010. Summary of Meteorological Conditions During the Arctic Mechanisms for the Interaction of the Surface and Atmosphere (AMISA) Intensive Observation Periods. U.S. Dept. of Commerce, National Oceanic and Atmospheric Administration, Office of Oceanic and Atmospheric Research, Earth System Research Laboratory, Physical Sciences Division, Boulder, CO.

Pinto, J.O., 1998. Autumnal mixed-phase cloudy boundary layers in the Arctic. J. Atmos. Sci. 55, 2016–2038. https://doi.org/10.1175/1520-0469(1998)055<2016:AMPCBL>2.0.CO;2.

Prenni, A.J., DeMott, P.J., Kreidenweis, S.M., Harrington, J.Y., Avramov, A., Verlinde, J., 2007. Can Ice-nucleating aerosols affect Arctic seasonal climate? Bull. Am. Meteorol. Soc. 88, 541–550. https://doi. org/10.1175/BAMS-88-4-541.

Rangno, A.L., Hobbs, P.V., 2001. Ice particles in stratiform clouds in the Arctic and possible mechanisms for the production of high ice concentrations. J. Geophys. Res. 106, 15065. https://doi.org/ 10.1029/2000JD900286.

Rossow, W.B., Schiffer, R.A., 1999. Advances in understanding clouds from ISCCP. Bull. Am. Meteorol. Soc. 80, 2261–2287. https://doi.org/10.1175/1520-0477(1999)080<2261:AIUCFI>2.0.CO;2.

Savre, J., Ekman, A.M.L., 2015. A theory-based parameterization for heterogeneous ice nucleation and implications for the simulation of ice processes in atmospheric models: a CNT-based ice nucleation model. J. Geophys. Res. Atmos. 120, 4937–4961. https://doi.org/10.1002/2014JD023000.

Schnaiter, M., Järvinen, E., Vochezer, P., Abdelmonem, A., Wagner, R., Jourdan, O., 2016. Cloud chamber experiments on the origin of ice crystal complexity in cirrus clouds. Atmos. Chem. Phys. 16, 5091–5110. https://doi.org/10.5194/acp-16-5091-2016.

Schweiger, A.J., Key, J.R., 1992. Arctic cloudiness: comparison of ISCCP-C2 and Nimbus-7 satellite-derived cloud products with a surface-based cloud climatology. J. Clim. 5, 1514–1527. https://doi. org/10.1175/1520-0442(1992)005<1514:ACCOIC>2.0.CO;2.

Schweiger, A.J., Lindsay, R.W., Key, J.R., Francis, J.A., 1999. Arctic clouds in multiyear satellite data sets. Geophys. Res. Lett. 26, 1845–1848. https://doi.org/10.1029/1999GL900479.

Serreze, M.C., Barry, R.G., 2005. The Arctic Climate System. Cambridge University Press, Cambridge, UK.

Serreze, M.C., Holland, M.M., Stroeve, J., 2007. Perspectives on the Arctic's shrinking sea-ice cover. Science 315, 1533–1536. https://doi.org/10.1126/science.1139426.

Shupe, M.D., Intrieri, J.M., 2004. Cloud radiative forcing of the Arctic surface: the influence of cloud properties, surface albedo, and solar zenith angle. J. Clim. 17, 616–628. https://doi.org/10.1175/1520-0442 (2004)017<0616:CRFOTA>2.0.CO;2.

Shupe, M.D., Walden, V.P., Eloranta, E., Uttal, T., Campbell, J.R., Starkweather, S.M., 2011. Clouds at Arctic atmospheric observatories. Part I: occurrence and macrophysical properties. J. Appl. Meteorol. Climatol. 50, 626–644. https://doi.org/10.1175/2010JAMC2467.1.

Solomon, S., Qin, D., Manning, M., Chen, Z., Marquis, M., Averyt, K.B., 2007. Climate Change 2007: The Physical Scence Basis. Cambridge University Press, Cambridge, UK.

Solomon, A., Feingold, G., Shupe, M.D., 2015. The role of ice nuclei recycling in the maintenance of cloud ice in Arctic mixed-phase stratocumulus. Atmos. Chem. Phys. 15, 10631–10643. https://doi.org/10.5194/acp-15-10631-2015.

Stephens, G.L., 2005. Cloud feedbacks in the climate system: a critical review. J. Clim. 18, 237–273.

Stephens, G.L., Vane, D.G., Boain, R.J., Mace, G.G., Sassen, K., Wang, Z., 2002. The CloudSat mission and the A-Train: a new dimension of space-based observations of clouds and precipitation. Bull. Am. Meteorol. Soc. 83, 1771–1790. https://doi.org/10.1175/BAMS-83-12-1771.

Stohl, A., 2006. Characteristics of atmospheric transport into the Arctic troposphere. J. Geophys. Res. 111. https://doi.org/10.1029/2005JD006888.

Stroeve, J., Holland, M.M., Meier, W., Scambos, T., Serreze, M., 2007. Arctic sea ice decline: faster than forecast: Arctic ice loss-faster than forecast. Geophys. Res. Lett. 34. https://doi.org/10.1029/2007GL029703.

Tan, I., Storelvmo, T., Zelinka, M.D., 2016. Observational constraints on mixed-phase clouds imply higher climate sensitivity. Science 352, 224–227. https://doi.org/10.1126/science.aad5300.

Tsay, S.-C., Jayaweera, K., 1984. Physical characteristics of Arctic stratus clouds. J. Clim. Appl. Meteorol. 23, 584–596. https://doi.org/10.1175/1520-0450(1984)023<0584:PCOASC>2.0.CO;2.

Ulanowski, Z., Hesse, E., Kaye, P.H., Baran, A.J., 2006. Light scattering by complex ice-analogue crystals. J. Quant. Spectrosc. Radiat. Transf. 100, 382–392. https://doi.org/10.1016/j.jqsrt.2005.11.052.

Verlinde, J., Harrington, J.Y., Yannuzzi, V.T., Avramov, A., Greenberg, S., Richardson, S.J., 2007. The mixed-phase Arctic cloud experiment. Bull. Am. Meteorol. Soc. 88, 205–221. https://doi.org/10.1175/BAMS-88-2-205.

Wang, X., Key, J.R., 2005. Arctic surface, cloud, and radiation properties based on the AVHRR polar pathfinder dataset. Part I: spatial and temporal characteristics. J. Clim. 18, 2558–2574. https://doi.org/10.1175/JCLI3438.1.

Wendisch, M., Yang, P., Ehrlich, A., 2013. Amplified Climate Changes in the Arctic: Role of Clouds and Atmospheric Radiation (Session Report). Sächsische Akademie der Wissenschaften zu Leipzig, Leipzig.

Westbrook, C.D., Illingworth, A.J., 2013. The formation of ice in a long-lived supercooled layer cloud: ice formation in altocumulus. Q. J. R. Meteorol. Soc. 139, 2209–2221. https://doi.org/10.1002/qj.2096.

Winker, D.M., Pelon, J.R., McCormick, M.P., 2003. The CALIPSO mission: spaceborne lidar for observation of aerosols and clouds. In: Proceedings of SPIE, Hangzhou, Chinavol. 4893. pp. 1–11. https://doi.org/10.1117/12.466539.

Witte, H.J., 1968. Airborne Observations of Cloud Particles and Infrared Flux Density in the Arctic (MS Thesis). University of Washington, Department of Atmospheric Sciences.

Yang, P., Liou, K.N., 1996. Geometric-optics–integral-equation method for light scattering by nonspherical ice crystals. Appl. Opt. 35, 6568. https://doi.org/10.1364/AO.35.006568.

Yang, P., Kattawar, G.W., Hong, Gang, Minnis, P., Hu, Yongxiang, 2008. Uncertainties associated with the surface texture of ice particles in satellite-based retrieval of cirrus clouds—part I: single-scattering properties of ice crystals with surface roughness. IEEE Trans. Geosci. Remote Sens. 46, 1940–1947. https://doi.org/10.1109/TGRS.2008.916471.

Zygmuntowska, M., Mauritsen, T., Quaas, J., Kaleschke, L., 2012. Arctic clouds and surface radiation – a critical comparison of satellite retrievals and the ERA-interim reanalysis. Atmos. Chem. Phys. 12, 6667–6677. https://doi.org/10.5194/acp-12-6667-2012.

PART 2

Modeling

CHAPTER 7

Simulations of Arctic Mixed-Phase Boundary Layer Clouds: Advances in Understanding and Outstanding Questions

Ann M. Fridlind, Andrew S. Ackerman
National Aeronautics and Space Administration, Goddard Institute for Space Studies, New York, NY, United States

Contents

1. INTRODUCTION

The objective of this chapter is to provide an overview of mixed-phase boundary layer cloud simulations. Our emphasis is on what detailed studies show—in particular what is

Mixed-Phase Clouds
https://doi.org/10.1016/B978-0-12-810549-8.00007-6

not relatively well understood or observed about the microphysical processes within such clouds—using analogous liquid-phase boundary layer clouds as a reference for the dynamical conditions. Since boundary layer clouds are characterized by turbulent mixing, the large-eddy simulation (LES) approach has been most widely used to represent the coupling between dynamical and mixed-phase microphysical processes, although there are limits to its ability to represent cloud-top entrainment and associated microphysical details at cloud top (e.g., Klingebiel et al., 2015; Mellado, 2016). Nevertheless, many LES studies of mixed-phase boundary layer clouds have been made over the past 20 years that the LES approach has been a relatively widely used technique. Among these are several model intercomparison studies that include results from differing LES models simulating the same case study.

Thus far nearly all detailed LES and intercomparison studies have been based on specific cloud systems observed during field campaigns. Whereas other chapters of this book broadly summarize observational findings, here we focus primarily on modeling results from three major field campaigns on which intercomparison studies have been based: the First International Satellite Cloud Climatology Project (ISCCP) Regional Experiment-Arctic Cloud Experiment (FIRE-ACE)/Surface Heat Budget in the Arctic (SHEBA) campaign (SHEBA; Curry et al., 2000), the Mixed-Phase Arctic Cloud Experiment (M-PACE; Verlinde et al., 2007), and the Indirect and Semi-Direct Aerosol Campaign (ISDAC; McFarquhar et al., 2011). Table 1 summarizes the general cloud-system properties for the respective intercomparison case studies based on observations from SHEBA (Morrison et al., 2011), M-PACE (Klein et al., 2009), and ISDAC (Ovchinnikov et al., 2014). Fig. 1 shows a satellite image representative of each case.

The case studies in Table 1 span a range of liquid water path (LWP), aerosol loading, and cloud temperatures. Considering the case studies as liquid-phase only for a moment and placing them in the context of LWP and droplet number concentration (N_d), they can be seen to span conditions from very thin, polluted clouds in the SHEBA case to very thick, clean clouds in the M-PACE case (Fig. 2). Drizzle can be expected to be an active process in liquid-phase clouds where $LWP/N_d \gg 0.1$ g m^{-2} cm^3 (Comstock et al., 2004), as in the M-PACE case. Drizzle drops are conspicuous in Cloud Particle Imager (CPI) data for that

Table 1 Mixed-phase boundary layer cloud model intercomparison case studies

Field campaign	Observation period (UTC)	Cloud top height (m)	Cloud temp. (°C)		Path (g m^{-2})		Conc. (cm^{-3})	
			Top	Base	Liquid	Ice	Drops	Ice
SHEBA	May 7, 1998	500	−20	−18	5–20	0.2–1	200	∼0.0001
M-PACE	Oct. 9–10, 2004	1000	−16	−9	110–210	8–30	40	∼0.01
ISDAC	Apr. 26, 2008	800	−15	−11	10–40	2–6	200	∼0.001

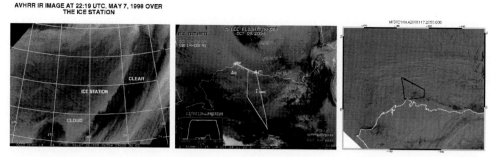

Fig. 1 Observation period satellite imagery from SHEBA (infrared; left; May 7 sea ice station location shown), M-PACE (infrared; middle; flight 9a track from Barrow, Alaska to Oliktok Point shown along coast), and ISDAC (mid-visible; right; flight 31 track from Barrow shown).

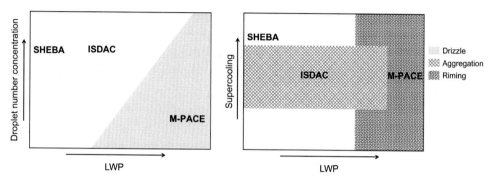

Fig. 2 Model intercomparison case studies ranked by liquid-phase (left) and ice-phase (right) cloud microphysical processes.

case, consistent with past evidence of drizzle formation under supercooled conditions (e.g., Cober et al., 1996). Drizzle in mixed-phase clouds is discussed further below.

Perhaps less easily deduced from the values listed in Table 1 are active ice-phase cloud microphysical processes. By definition all mixed-phase clouds contain ice crystals, which are formed evidently in part via heterogeneous nucleation (as discussed further below), and they grow most rapidly via vapor diffusion within the cloud layer, where humidity is saturated with respect to liquid and correspondingly supersaturated with respect to ice. As long as liquid cloud base is supercooled, as in all cases in Table 1, ice-supersaturated conditions extend below liquid cloud base. A deep layer of the cloud-topped boundary layer that extends from cloud top to below supercooled cloud base is thus a region where diffusional growth of ice is active. Ice throughout that zone is growing in both updrafts and downdrafts, both above and below cloud base.

Analogous to drizzle formation via a warm-phase collision-coalescence process, riming and ice aggregation may also be expected active ice growth processes via mixed-phase and ice-ice collection. Placing case studies in the context of LWP and supercooling, riming can be roughly understood to accompany high LWP (Fig. 2), although it should be noted that observations indicate that a mean droplet diameter of 10 μm is also required for the process to be an efficient growth mechanism (e.g., Lowenthal et al., 2011). Aggregation, on the other hand, may conceivably be roughly understood to accompany the presence of sufficiently numerous dendritic particles to undergo entanglement (e.g., Mitchell, 1988, and references contained therein). We represent this in Fig. 2 as a range of cloud supercooling around −15°C with a high LWP limit above which active riming may largely eliminate dendrites, as observed in the M-PACE case. At colder temperatures, the dendritic growth habit is not favored (e.g., Pruppacher and Klett, 1997), as consistent with observations from the SHEBA case study.

Taking the liquid- and ice-phase process occurrences in Fig. 2 together, SHEBA can be identified as the simplest case study insofar as no collisional processes are active. In the ISDAC case, aggregation is active. In the M-PACE case, drizzle and riming are active. For the purposes of illustrating model representation of these processes, we will discuss the case studies below in order of increasing complexity, drawing on related studies that accompany each intercomparison work. We then discuss open questions common to all case studies, and avenues for future progress.

2. SHEBA CASE STUDY

The conditions observed on May 7, 1998 over the SHEBA sea ice camp at roughly 76°N, 165°W occurred within a nine-day period dominated by shallow mixed-phase cloud over pack ice with a variable cloud-top height of 400–1200 m and LWP commonly exceeding $20 \, \mathrm{g \, m^{-2}}$ (Zuidema et al., 2005). The 12-h period selected for the Morrison et al. (2011) model intercomparison study, 12–24 UTC, exhibited weak winds of roughly $5 \, \mathrm{m \, s^{-1}}$ within a shallow cloud-topped boundary layer that was relatively well-mixed from surface to cloud top. Sustained radar reflectivity at all elevations below cloud top indicates persistent, continuous mixed-phase conditions (Fig. 3).

A notable feature of the SHEBA case study relative to others is that boundary layer air is supersaturated with respect to ice from cloud top (where liquid saturation defines substantial ice supersaturation) down to the surface, indicating that ice sublimation is not an active process (cf. Morrison et al., 2011, their Fig. 9a). This leaves three active liquid-phase processes (droplet activation, primarily within updrafts at cloud base; droplet growth, primarily within updrafts above cloud base; and droplet evaporation, primarily within downdrafts above cloud base) and two active ice-phase processes (ice nucleation, likely somewhere within the liquid-phase region; and ice diffusional growth, at all elevations throughout the cloud-topped turbulent boundary layer).

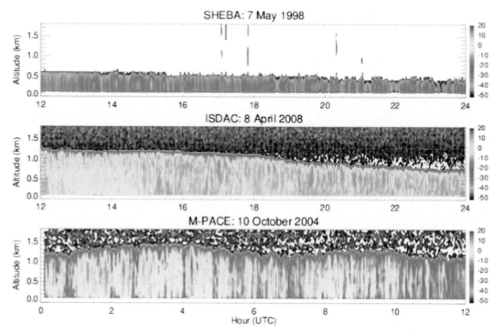

Fig. 3 Vertically pointing millimeter-wavelength cloud radar (MMCR) reflectivity (dBZ) observations during 12-h periods representative of the SHEBA, ISDAC, and M-PACE case studies.

2.1 Ice Crystal Number Concentration Budget

Fig. 4 illustrates the steady-state budget for ice crystal number concentration within a well-mixed boundary layer with steady cloud-top height and temperature in the case that ice crystals are formed exclusively via the activation of aerosol ice-freezing nuclei[1](IFN) that are entrained from the overlying free troposphere and rapidly nucleated, as commonly assumed (e.g., Pinto, 1998; Harrington and Olsson, 2001). Fig. 4 illustrates the steady-state budget for ice crystal number concentration within a well-mixed boundary layer with steady cloud-top height and temperature. Horizontal flux divergences are neglected. The only supply of new ice crystals to the boundary layer is via entrainment of heterogeneous ice-freezing nuclei at cloud top, which depends upon cloud-top entrainment rate (w_e). In the absence of aggregation and sublimation, the only sink of crystals is sedimentation to the surface, which can be cast in terms of number-weighted ice crystal fall speed (v_f). LES and observations both support approximating ice particle size distributions (PSDs) as vertically uniform in a well-mixed boundary layer (e.g., Fridlind et al., 2007; McFarquhar et al., 2007, 2011; Fridlind et al., 2012).

---

[1] Owing to lack of ambiguities that could occur in some literature (e.g., Vali et al., 2015), here we follow the traditional Pruppacher and Klett (1997) terminology for IFN, which parallels terminology for cloud condensation nuclei (CCN) in cloud microphysics literature.

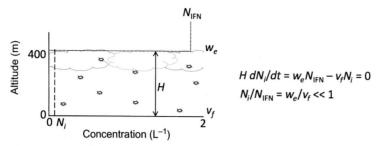

Fig. 4 Steady-state budget for ice crystal number concentration (N_i) in the SHEBA case, where H is boundary layer height, N_{IFN} is overlying ice nucleus number concentration, w_e of ~ 0.1 cm s^{-1} is cloud-top entrainment rate, and v_f of ~ 30 cm s^{-1} is number-weighted ice crystal fall speed at the surface, following Fridlind et al. (2012).

The salient result of the mixed-layer budget (cf. Fridlind et al., 2012) is that the ice crystal number concentration within the boundary layer (N_i) is found to be proportional to the product of the overlying IFN number concentration (N_{IFN}) and the ratio of cloud-top entrainment rate to the number-weighted ice crystal number concentration (w_e/v_f). Given LES estimates of v_f circa 30 cm s^{-1} and w_e c.0.1 cm s^{-1}, N_i is then two orders of magnitude smaller than N_{IFN}, which presents a stark contrast to $N_i \approx N_{IFN}$ near the leading edge of an orographic wave cloud (Eidhammer et al., 2010), for instance, where entrainment does not present a limitation to the supply of IFN.

Budgets based on LES of the SHEBA case study support this key result that $N_i \ll N_{IFN}$ under the assumptions just stated (Fridlind et al., 2012). The simulations also yield an ice crystal lifetime within the well-mixed boundary layer of roughly 1 h in this case, consistent with estimates obtained by other means for the ISDAC case study (Yang et al., 2013). Here we have assumed that all IFN are those that will activate essentially instantaneously in a measurable mode under cloud-top conditions, which are the coldest and most supersaturated within the boundary layer. LES results are insensitive to whether the IFN nucleation mode is assumed to be condensation, immersion, or deposition, but contact-mode nucleation is found to be in a separate class wherein the rate of collection of IFN by supercooled droplets does not yield rapid activation (Fridlind et al., 2012), consistent with assumptions that measurements made by a Counter-Flow Diffusion Chamber (CFDC; Rogers et al., 2001) instrument under cloud-top conditions do not include contact nucleation. Because the role of contact-mode nucleation remains uncertain (e.g., Ladino Moreno et al., 2013) we will return to this point later.

2.2 Intercomparison Specification

Model intercomparison studies commonly make a number of simplifying assumptions to limit the model components being tested and thereby the chances of being overwhelmed by diversity of model behaviors. For instance, surface turbulent heat fluxes may be specified and radiative transfer replaced with a simple parameterization (e.g., Ovchinnikov et al., 2014). In the case of the SHEBA model intercomparison, ice nucleation was also specified with a simple parameterization that maintains N_i approximately fixed at 1.7 L^{-1} (referred to as BASE), with sensitivity tests using 0.17 L^{-1} (LOWNI) and 5.1 L^{-1} (HIGHNI). The BASE value was selected to give N_i equal to the overlying N_{IFN} (at cloud-top temperature) based on observational evidence of approximate equivalence reported in field measurements (Prenni et al., 2009), in contrast to the budget argument provided above.

Large-scale subsidence and advective tendencies of heat and moisture must be applied to LES with periodic boundary conditions considered in an Eulerian rather than Lagrangian column framework, but such forcings are poorly constrained (e.g., Jiang et al., 2000; Morrison and Pinto, 2004). It is the rule rather than the exception that they are selected to produce observed conditions at least to some degree, as discussed by Vogelmann et al. (2015), who demonstrate the diversity of shallow cloud simulations that can result when large-scale forcings are adopted wholesale from a range of potential sources (global or regional reanalyses or mesoscale model simulations). In the case of the SHEBA case study specification, with N_i fixed at 1.7 L^{-1}, the large-scale forcings are selected to maintain LWP, cloud-top height, and thermodynamic profiles in quasi-equilibrium over the 12-h simulation time (cf. Morrison et al., 2011, their Fig. 4). If LOWNI or HIGHNI were adopted as the baseline, large-scale forcings would be selected to account for a lesser or greater desiccation rate. This relationship between specified large-scale forcings and specified N_i in sustaining a shallow mixed-phase cloud was well demonstrated by Jiang et al. (2000).

2.3 Intercomparison Results

It is seen in the intercomparison study that two-thirds of LES models maintain LWP in a quasi-steady state in the BASE case, as intended (cf. Morrison et al., 2011, their Fig. 4). The LOWNI case is reported to be similar to an ice-free state, and all models produce greater LWP by varying amounts. In the HIGHNI case, most models cannot maintain steady LWP. Using the estimates of w_e and N_{IFN} listed above, the HIGHNI case would correspond to a free troposphere concentration of rapidly nucleated IFN of ~ 1500 L^{-1}, which exceeds by roughly an order of magnitude those reported at any temperature in commonly used compilations of CFDC measurements (cf. DeMott et al., 2010).

Given extreme uncertainty in the actual concentration of ice based on in situ measurements within mixed-phase clouds (e.g., Fridlind et al., 2007, factor of 5) and the likelihood that unknown secondary ice multiplication mechanisms may exist with poorly defined limits (e.g., Yano and Phillips, 2011; Ackerman et al., 2015; Lawson et al.,

2015), it is vital to know quantitatively what ice loadings occur under commonly observed conditions. In their intercomparison case study Morrison et al. (2011) highlight the capability of high N_i to lead to complete cloud glaciation.

2.4 Additional Observational Constraints

Using the intercomparison case study as a foundation, Fridlind et al. (2012) sought additional constraints on ice loading by making two additional comparisons between the BASE simulation and observations. First, they compared radar reflectivity (Z) and mean Doppler velocity (V_D) measured by a Millimeter-wavelength Cloud Radar (MMCR) at the sea ice camp with that forward-simulated from the LES. Second, they compared in situ aircraft measurements of ice PSDs made below liquid cloud base, where conditions are relatively uniform with height, with those simulated. The last 2 h of the 12-h intercomparison period were selected to bound the aircraft sampling period.

Fridlind et al. (2012) found that their BASE intercomparison simulation overestimated median Z by roughly 12 dBZ, but quite accurately represented median V_D, indicating that ice PSDs were quite consistent with remote-sensing measurements but N_i was too great, all else being equal. Adjusting the case study specification, especially reducing both N_i and large-scale moisture convergence and increasing heat divergence, served to bring the LES results into line with both radar and in situ measurements simultaneously. The resulting simulation yielded a weak desiccation rate with N_i of circa 0.3 L^{-1}, allowing mixed-phase persistence over a 4-h simulation despite LWP of only 5 g m^{-2}. Weak desiccation can be considered as qualitatively consistent with the prevalence of long-lived mixed-phase conditions in observations (e.g., Shupe et al., 2006).

However, Fridlind et al. (2012) also found that a rapidly-nucleated overlying IFN concentration of roughly 50 L^{-1} would be required to explain in situ and radar observations of ice properties. At a cloud-top temperature of $-20°C$, this is still a very high concentration relative to global CFDC measurements (cf. DeMott et al., 2010). The occurrence of persistent ice precipitation from mixed-phase layer clouds that may not be greatly desiccating but still greatly exceeds the effect of observationally supported N_{IFN} values has been reported elsewhere (Westbrook and Illingworth, 2013), as discussed further below. Fridlind et al. (2012) also conclude that simulations are sensitive to the ice crystal physical properties that determine fall speed and growth rate, which are not quantitatively constrained by existing observational analyses. Thus while radar and in situ measurements provide constraints on obtaining a relatively realistic mixed-phase cloud state that is consistent with observations in many ways, two factors remained exceptionally poorly constrained: the mechanism(s) of new ice crystal formation and the ice crystal physical properties (see Fig. 5).

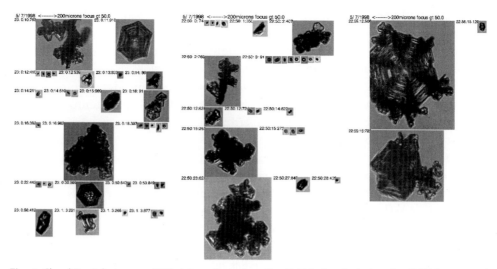

Fig. 5 Cloud Particle Imager (CPI) data collected on the C-130 aircraft during the SHEBA case study observation period show a variety of radiating plate shapes.

2.5 SHEBA Lessons

For the purposes of understanding the fundamentals of mixed-phase boundary layer clouds, the SHEBA case study has demonstrated that mixed-phase clouds can be very simple. In this case: a non-drizzling warm-phase stratocumulus type cloud plus the weak production of ice crystals that grow by vapor diffusion within the turbulent boundary layer until they sediment out. It is intuitive to consider the ice-free state and addition of a scarcely perceptible amount of ice, which is not dissimilar to observer experiences reported for this case study. If each crystal experiences the cloud conditions independently and LWP is unaffected, a characteristic ice size distribution can be considered to emerge from the results of crystals growing and sedimenting within a turbulent layer. If twice as many ice crystals are nucleated (anywhere in the cloud, it turns out), the horizontal-mean ice PSDs are shifted directly upward and the distribution of V_D is unaffected, as found in LES results (Fridlind et al., 2012). As more and more ice crystals are added, LWP will eventually be reduced and total desiccation could proceed as found for the Morrison et al. (2011) HIGHNI case.

It is notable that desiccation appears weak in the constrained case study of Fridlind et al. (2012). However, it is also notable that ice nucleation appears substantially stronger than would be expected from in situ measurements of overlying IFN during the case study (roughly 30 times greater) or globally. Fridlind et al. (2012) discuss conceivable causes for this, such as blowing snow despite weak winds. As shown further below, a lack of adequate IFN to explain observed ice appears to be the rule rather than exception in observed case studies for as yet undetermined reasons.

LES with broadly accepted (hereafter "known") physics are consistent with the observed coexistence of liquid and ice, and can reproduce close simulacra of observed conditions on many counts simultaneously. Degree of success depends on tuning ice production rate upwards in this case and probably also depends in part on tuning large-scale forcing terms, which could mask simulation errors, such as in cloud-top entrainment rate or ice crystal vapor growth rate. Requiring a model with relatively few internal tuning knobs to reproduce many observations at once within observationally determined uncertainties nonetheless provides some degree of a test of known physics. Within uncertainty in large-scale forcings, Fridlind et al. (2012) results indicate that known physics can go far. That said, it should be noted that different LES codes produce quite different results in the limit of fully liquid-phase conditions, consistent with warm cloud intercomparison studies (e.g., Ackerman et al., 2009, factor of 3 spread in LWP). Given fixed N_i in this case, it is not surprising that LES additionally predict differing ice production rates. Among models that predict LWP within a factor of two in the BASE intercomparison simulation, predicted IWP varies by less than a factor of two, a relatively close agreement owing at least in part to strongly constrained N_i.

Finally, ice formation mechanisms remain unclear, and quantitative ice crystal properties are a requirement unmet by observational analyses thus far, as discussed further below.

3. ISDAC CASE STUDY

As discussed by Avramov et al. (2011), flights 16 and 31 respectively on Apr. 8 and 26, 2008 during the ISDAC campaign both sampled widespread, single-layer, mixed-phase stratocumulus decks over pack ice with cloud bases colder than −8°C (Fig. 1). In both cases, soundings indicate a relatively well-mixed cloud layer of roughly 0.5 km depth with cloud-top temperatures around −15°C overlying a stable, moister surface layer of similar depth, and relatively uniform horizontal wind speeds of 7–9 m s^{-1} (Avramov et al., 2011; Ovchinnikov et al., 2014). The Apr. 8 case was selected by Avramov et al. (2011) owing to the availability of ground-based remote-sensing data in the same cloud deck over the US Department of Energy Atmospheric Radiation Measurement Program's North Slope of Alaska site, whereas the Apr. 26 case was selected by Ovchinnikov et al. (2014) for a model intercomparison study. Owing to the close similarity of the cases and the greater availability of observations for the Apr. 8 case, we will discuss both here, roughly interchangeably, and use examples from the earlier case to illustrate some points. For instance, the ground-based MMCR time series from the earlier case is shown in Fig. 3.

Microphysically, conditions during both flights were characterized by a wide range of dendritic crystals and their aggregates (Fig. 6 and Avramov et al., 2011, their Fig. 4). Dendrites in both cases exhibited a wide range of arm properties, ranging from needle-like to plate-like to highly branched. In both cases, single crystals transitioned to aggregates with increasing maximum dimension, over an estimated size range of 1–4 mm in the Apr. 8 case (Avramov et al., 2011). In other words, crystals smaller than 1 mm were predominantly unaggregated and those larger than 4 mm predominantly aggregated. The aircraft did not extensively sample elevations below roughly 0.5 km (e.g., Avramov et al., 2011, their Fig. 15), where sublimation was also an active process (e.g., Ovchinnikov et al., 2014, their Fig. 18).

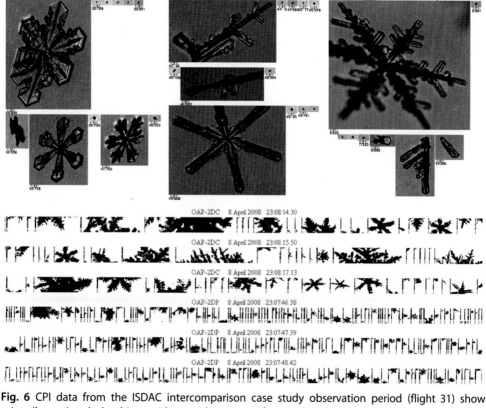

Fig. 6 CPI data from the ISDAC intercomparison case study observation period (flight 31) show primarily unrimed dendrites with a wide range of arm thickness and branch patterns. Two-Dimensional Cloud and Precipitation (2DC and 2DP) Optical Array Probe (OAP) data from similar conditions during flight 16 show similar dendrites and their aggregates. The vertical dimension of 2DC and 2DP images is ∼1 mm and 6 mm, respectively.

3.1 Intercomparison Specification

A bimodal PSD of ammonium bisulfate aerosol was specified for models that treat droplet activation, and a fixed droplet number concentration of 200 cm^{-3} was specified for the rest. As in the Morrison et al. (2011) intercomparison, Ovchinnikov et al. (2014) sought to constrain ice nucleation for the purposes of intercomparison by controlling N_i within cloud to be zero (liquid-phase only), 1 L^{-1}, based loosely on observations, and 4 L^{-1}. Going further than any previous specification to our knowledge, ice crystal properties were also specified in detail, including the relationships among mass, crystal maximum dimension, capacitance, and fall speed. Aggregates were neglected for simplicity. As recommended by the Morrison et al. (2011) study, the onset of ice formation was delayed until after a 2-h liquid-phase spin-up.

Unlike the other case studies, large-scale forcings were not specified based on reproducing evolution of boundary layer thermodynamics for a specified baseline. Rather, large-scale subsidence was selected in a manner that led to minimal cloud-top height evolution, whereas moisture and temperature, as well as zonal and meridional winds above the inversion, were nudged with 1 h and 2 h time scales, respectively. For thermodynamic quantities, 1-h nudging was not sufficient to avoid substantial evolution of the profile away from the observed decoupled state of the boundary layer toward a well-mixed state. LWP and dynamical properties of the cloud-topped and surface layers evolved in a manner similar to that also shown in the Avramov et al. (2011) ISDAC simulations, as discussed further below.

3.2 Intercomparison Results

Considering first liquid-phase only conditions, all participating LES models predicted a rapid increase of LWP from roughly 10 to 50 g m^{-2} over 8 h. The models that predicted a most rapid LWP increase were those that most rapidly deepened the cloud-topped layer downwards into the surface layer, reaching the surface and leading to a relatively well-mixed state within roughly 4 h. Those models with the slowest increase of LWP reached maximum LWP at roughly 7 h. Since these results are without ice, they indicate a relatively wide range of LES dynamical behavior with relatively simple microphysics (no drizzle, relatively high droplet number concentration).

In simulations with ice formation beginning at 2 h, $N_i = 1$ L^{-1} had a relatively weak desiccating effect for most models and 4 L^{-1} a greater effect. Despite the fact that they produce grossly differing rates of boundary layer coupling, the two models with independent size-resolved, bin microphysics schemes were shown to produce the greatest IWP and ice diffusional growth rates. In the case of $N_i = 4$ L^{-1} those two models also produced closely similar evolutions of IWP.

Through detailed comparison of bulk and bin microphysics and additional sensitivity tests, Ovchinnikov et al. (2014) demonstrate that the bin microphysics schemes prognose

PSD features that lead to systematically more ice than bulk schemes. By fitting gamma size distributions to the bin results, it is further demonstrated that ice size distributions are narrower than the exponential shape typically assumed in bulk schemes. When these factors were accounted for, for instance by specifying in the bulk schemes a gamma size distribution shape parameter calculated from the size distributions obtained using bin microphysics, then the bulk and bin schemes were brought into quite close agreement. Thus once ice individual-crystal properties are fully specified, it is found that ice size distribution shape is also an important determinant of mixed-phase cloud evolution. In this case over 6 h of ice formation, using an exponential size distribution led to roughly twice as much LWP as predicted with a bin scheme's fitted gamma shape parameter of roughly 3.

3.3 Related Studies

In a follow-on study focused on boundary layer dynamics for the Apr. 26 case, Savre et al. (2014) suggest a dominant role for near-surface large-scale advection of cold air in maintaining a decoupled cloud-topped boundary layer, consistent with ISDAC observations and in contrast to LES of both the Apr. 8 case (Avramov et al., 2011, their Fig. 20) and the Apr. 26 case (Ovchinnikov et al., 2014, their Fig. 2), which did not include such large-scale forcings. They explore factors that favor maintenance of cloud-topped layer decoupling commonly observed in the Arctic, including the role of a humidity inversion (i.e., increase with height) at cloud top (e.g., Curry, 1986), which is unknown in warm stratocumulus (e.g., Ackerman et al., 2004).

Avramov et al. (2011), Solomon et al. (2015), and Savre and Ekman (2015b) all examine the IFN budget in the context of LES and mesoscale modeling studies, each with a slightly differing emphasis and approach. In contrast to the SHEBA case conditions, ice sublimation is an active process in all of the ISDAC case studies, but the studies reach varying conclusions regarding the role of IFN recycling.

Avramov et al. (2011) seek to reconcile in situ IFN measurements via CFDC with in-cloud N_i. If cloud-top entrainment were the only source of rapidly activated IFN, they find N_{IFN}/N_i to be at least 50, reflecting the cloud-top entrainment limitation discussed above, inconsistent with the factor of 10 observed (roughly 10 L^{-1} IFN and 1 L^{-1} N_i). If N_{IFN} are similar in the surface layer below and that air is entrained substantially from the bottom of the cloud-topped mixed layer, they report that N_{IFN} measured could potentially explain N_i observed. However, as already noted, such rapid entrainment of below-cloud air in their simulations is not consistent with observations. They also report that sublimation of ice is a potential source of IFN, but only to a shallow surface layer that likely did not play a role during the observation period.

Solomon et al. (2015) consider a longer integration period of 40 h in mesoscale model simulations of the Apr. 8 conditions. They find that a subcloud layer that is well-stocked with IFN can serve as a persistent reservoir of IFN when the cloud is continually entraining from such a source beneath the mixed-layer base. In a simulation without recycling, the cloud layer deepening entirely depletes the surface layer as it mixes out. In a simulation with recycling, the surface layer is initially enriched in IFN, as in Avramov et al. (2011), and the layer serves as an efficient source if it is efficiently mixed out.

Savre and Ekman (2015b) reach an entirely different conclusion, namely that IFN recycling scarcely matters to predicted IFN in several ISDAC case studies. Rather than initializing N_{IFN} using CFDC measurements and assuming that such IFN are all rapidly nucleated, as did Avramov et al. (2011) and Solomon et al. (2015), Savre and Ekman (2015b) assign a distribution of contact angles to measured dust and soot aerosol concentrations. They conclude that cloud-top entrainment of IFN is responsible for only roughly 25%–40% of ice formed during 6 h simulations. In their simulations with a contact angle distribution, cloud top cooling that is accompanied by rising cloud top in all three of their case studies leads to nucleation of an increasing number of dust and soot IFN present initially within the cloud layer. In addition, they stress that the distribution of contact angles assigned to IFN is required to explain persistent ice formation because the more weakly active IFN help to sustain steady ice nucleation rates as discussed further below.

3.4 ISDAC Lessons

Relative to the other campaign cases, ISDAC studies were complicated by the representation of boundary layer decoupling and cooling of cloud top in simulations. Most simulations studied included cooling of cloud top but arguably none well demonstrated that they reproduced observed cloud-layer decoupling behavior compared with observations.

Given a cloud-topped boundary layer hosting ice production, Ovchinnikov et al. (2014) demonstrated that independent bin microphysics schemes agreed with one another when ice properties were completely specified, and demonstrated that bulk simulations could be brought into agreement with the bin simulations when the assumed gamma shape parameter was specified based on results from the bin simulations. Avramov et al. (2011) demonstrated that ice properties, such as those specified by Ovchinnikov et al. (2014), could not be readily derived from observations owing to a large diversity of dendrite shapes present. In short, no quantitative analyses were available to constrain the specified ice properties.

Regardless of dendrite habit selected by bracketing the habit range observed, Avramov et al. (2011) also demonstrated that a size-resolved microphysics scheme can accurately predict the transition from single crystals to aggregates with increasing size. Simulations agreed best with observed in situ measurements of ice PSD and remote-sensing measurements of X-band Z, and V_f and W-band Z, when ice crystals and their aggregates were assumed to be composed of low-density elements (with thin arms). Aggregates were important to proper prediction of Z, but scarcely reduced N_i and only modestly reduced IWP and slightly increased LWP.

Modeling studies arrived at diverse and conflicting conclusions regarding ice formation. Avramov et al. (2011) attempted a study of N_i–N_{IFN} closure using CFDC measurements, and came to the tentative conclusion that observed IFN could explain observed N_i if substantial entrainment of surface layer IFN were invoked, but observations were not available to establish what occurred and simulations likely overestimated entrainment of surface layer air. Savre and Ekman (2015b) also came to the tentative conclusion that observed aerosol properties could explain observed ice properties, but results were sensitive to assumed aerosol ice nucleating properties and were not constrained by CFDC measurements. Solomon et al. (2015) concluded that recycling was a chief factor sustaining N_i whereas Avramov et al. (2011) and Savre and Ekman (2015b) found otherwise for entirely differing reasons, as discussed further below.

Finally, it is notable that no studies considered a role for ice multiplication under ISDAC conditions, or, to our knowledge, for blowing snow or any source excepting primary ice nucleation via activation of IFN.

4. M-PACE CASE STUDY

The M-PACE case study observation period took place during an extended cold-air outbreak over the open Beaufort Sea (Fig. 1). During Oct. 8–12, cloud-top temperatures at Barrow fell from roughly −10°C to −17°C (cf. Fridlind et al., 2007, their Fig. 2). In contrast to the negligible surface turbulent heat fluxes over pack ice in the SHEBA and ISDAC cases, clouds rapidly approaching Barrow from the Beaufort Sea with horizontal wind speeds of roughly 13 m s^{-1} were fed by sensible and latent heat fluxes both in excess of 100 W m^{-2}, yielding LWP in excess of 100 g m^{-2} and correspondingly substantial IWP on the order of 10 g m^{-2} (Klein et al., 2009). Roll convection common to such cold-air outbreaks was evident in high-resolution imagery (cf. Klein et al., 2009, their Fig. 1), characterized by increasing roll aspect ratio with distance from the ice edge (e.g., Gryschka and Raasch, 2005, and references therein). Undulations in cloud top height seen by radar at Barrow may have been associated with meandering of roll structures (Fig. 3). The high LWP in combination with low droplet number concentration of 30–40 cm^{-3} led to active drizzle and riming processes evident in CPI data (Fig. 7).

4.1 Case Study Specification

As described by Klein et al. (2009), idealized initial thermodynamic and wind profiles were based on soundings at Barrow, with a fixed-temperature ocean surface. Idealized large-scale forcings were derived from reanalysis results 200 km upwind of Barrow. Models were to use their own interactive radiative fluxes with solar zenith angle varying realistically as a function of time, as in the SHEBA intercomparison and in contrast to the more constrained parameterized treatment used in the ISDAC case. Sensible and latent heat fluxes were fixed as in both SHEBA and ISDAC cases to increase constraint on models diverging for reasons other than cloud dynamics or microphysics, in this case

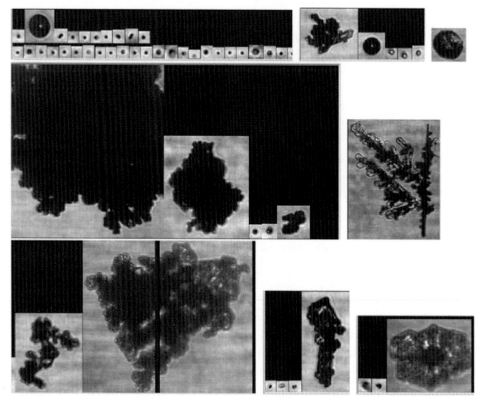

Fig. 7 CPI data from the Oct. 9–10 flights during M-PACE show a wide range of properties, from lightly to heavily rimed and from plate-like to spatially branched. Drizzle drops are relatively common near cloud base (top), but to our knowledge only one single image captured a pristine frozen drizzle drop (upper right).

to 110 and 140 W m^{-2}, respectively, based on reanalysis. A bimodal aerosol size distribution with accumulation and coarse modes was specified based on derivation from Barrow CCN data and a handheld particle counter mounted on an Aerosonde UAV (Morrison et al., 2008). An IFN concentration of 0.16 L^{-1} was reported (but its use not specified), based on CFDC measurements under varying above- and within-cloud conditions, which was noted to be close to the CFDC detection limit of roughly 0.1 L^{-1}.

4.2 Intercomparison Results

The intercomparison included a range of 2D eddy-resolving models and 3D LES and other cloud-resolving models (CRMs), as well as single-column models (SCMs). Results were not identified by model, and only classified by the microphysics scheme complexity.

From our knowledge of our submission based on code described in Fridlind et al. (2007), we can distinguish between the 2D and 3D bin microphysics simulations in the following discussion, but are otherwise limited in our ability to distinguish 2D from 3D bulk microphysics simulations.

Among the CRM results, in contrast to the SCM results, essentially all simulations maintained a mean mixed-phase boundary layer cloud depth of 1.5 km. However, that is where similarities ended. Median predicted LWPs ranged from roughly 0 to 175 g m^{-2} compared with an observational range of roughly 110–210 g m^{-2}. Only three of nine CRMs maintained LWP within roughly a factor of 2 of that observed, and most grossly underpredicted LWP. With an exception or two—one of which is a 2D model in the case of bin microphysics—simulated IWP was roughly anticorrelated with LWP in CRMs. Thus underprediction of LWP could be generally tied to overprediction of IWP, which can be understood as precipitation removal of LWP owing to overproduction of ice. Klein et al. (2009) emphasize that complexity of microphysics did not necessarily lead to improved performance, at least partly related to the fact that CRMs reported an astonishing five order of magnitude range of predicted N_i. The extreme diversity of results when specifying N_{IFN} in this study led directly to the strong constraints applied essentially directly to N_i in the SHEBA and ISDAC intercomparison studies.

We note that our LES submission to the intercomparison was based on simulations that included production of ice associated with evaporation of droplets, which we identified out of many proposals in the literature as one possible means by which N_i might be maintained within the range of observations (Fridlind et al., 2007). A possible surface source of IFN or an unknown ice multiplication mechanism were identified as other possibilities. Without that ad hoc ice production mechanism, our simulations would have reported essentially negligible IWP, as discussed further below, and LWP similar to the no-ice sensitivity test discussed by Klein et al. (2009). It is notable that CRM simulations without the ice phase already differed by more than a factor of three; 2D versus 3D could play some role in that. This diversity can be considered quite surprising since the case study is relatively simple from the standpoint that turbulence is robust and the boundary layer is relatively well-mixed throughout.

4.3 Related Studies

Here we will focus on two related studies that included detailed microphysics with prognostic IFN in 3D simulations. The significance of a prognostic instead of diagnostic treatment of IFN is that the former allows consumption of IFN upon nucleation of an ice crystal. In contrast, the latter does not deplete the abundance of IFN available for further nucleation and thus will tend to result in greater ice crystal formation rates by virtue of ignoring the basic fact that once an IFN is within an ice crystal it is no longer available for further primary nucleation. Aggregation of ice crystals that form on IFN will reduce the number of IFN returned to the atmosphere upon complete sublimation of the aggregate, and sedimentation of ice formed on IFN will also serve as a sink of IFN.

Fridlind et al. (2007) focused primarily on the gross inability of observed IFN to explain observed N_i. In that study, when specifying thermodynamic soundings and winds based on a Barrow sounding and sea surface temperature offshore with an ocean surface and predicted latent and sensible heat fluxes, it was first found that the liquid phase cloud properties could be roughly reproduced without difficulty. Drizzle was not well constrained by observations but was predicted and seen in observations as discussed above. However, IFN that were based on CFDC measurements and were, by extension, considered to be rapidly nucleated, were quickly consumed. As noted by Fridlind et al. (2007), this IFN consumption process was well described by Harrington and Olsson (2001) based on their earlier simulations of similar conditions. Fridlind et al. (2007) furthermore reported that assuming CFDC-observed IFN to be fully restored and available for reactivation upon sublimation made little difference. Given that N_i was estimated at 10 L^{-1} and was clearly visible precipitating to the surface in radar measurements, it is not surprising that IFN could not build up far beyond the 0.2 L^{-1} value observed and could not possibly account for 10 L^{-1} ice, even when considering an estimated factor of 5 uncertainty in observed N_i.

In response to this, Fridlind et al. (2007) sought possible mechanisms to explain the M-PACE ice observations in decades of literature, which also documented evidence of such discrepancies in both stratiform and cumuliform clouds. Chief among these were several ice multiplication mechanisms, which Fridlind et al. (2007) found insufficiently effective in simulations, conceivably in part owing to a lack of properly specified ice properties, as discussed further below. Other possibilities identified were an ocean surface source of IFN or potential physicochemical changes in droplet residuals, for instance associated with collision-coalescence of droplets containing biogels (Leck and Bigg, 2005) with those containing sulfate, which could lead to exposure of ice-nucleating solids. Now as then, all of these possible processes remain unproven, although the ocean is increasingly viewed as a relatively weak source of IFN (e.g., Demott et al., 2016). An ice multiplication mechanism involving the coexistence of fragile and dense ice has been further investigated (Yano and Phillips, 2011), and the potential of large freezing droplets to produce more ice splinters than previously established may also emerge from new laboratory measurements (e.g., Lawson et al., 2015).

Using an independent LES code with an independent size-resolved microphysics scheme, Fan et al. (2009) largely confirmed the basic findings of Fridlind et al. (2007) insofar as the gross inability of observed IFN to account for observed ice. However, Fan et al. (2009) make the additional point that IFN recycling can become important when IFN are as abundant as required to sustain observed N_i, for instance via a within-droplet physicochemical process. Fan et al. (2009) also illustrate that even when N_i in simulations is substantially increased by some additional mechanism, forward-simulated Z remains low compared with MMCR. This is important to consider since observations of ice crystal number size distribution remain extremely uncertain owing to poorly established artifacts from crystals shattering on probes as well as difficulties establishing fundamental probe calibration (e.g., Baumgardner et al., 2011; Korolev et al., 2011). In other words, comparison of forward simulations of Z from measured PSDs and Z and V_D from model results are helpful to support conclusions regarding consistency between simulated and observed ice loading when in situ measurement uncertainties are great (e.g., Fan et al., 2009; Avramov et al., 2011; Fridlind et al., 2012).

4.4 M-PACE Lessons

M-PACE demonstrated the potential for severe lack of model skill in simulating, and perhaps also observing, mixed-phase boundary layer clouds. On the observation side, at least M-PACE conditions closely conformed with those encountered on earlier airborne surveys of moderately supercooled stratiform clouds with large droplets and copious ice (Rangno and Hobbs, 2001), as discussed further below. On the modeling side, on the other hand, an intercomparison study produced five orders of magnitude difference in N_i and the most detailed studies with prognostic IFN reported essentially no ability to explain N_i far exceeding N_{IFN} with known microphysical mechanisms. Perhaps underemphasized in this earliest of three case studies was the role of ice properties. Avramov and Harrington (2010) demonstrated that their simulations were strongly sensitive to assumed ice habit, but did not attempt to constrain their habit assumptions with in situ observations. As evidenced by the variety of shapes shown in Fig. 7, doing so would not have been an easy task.

5. DISCUSSION

We take the non-controversial view that parameterization efforts are hampered by lack of understanding of fundamental microphysical processes. In the following discussion, we therefore discuss the greatest microphysics knowledge gaps across these several case studies and identify outstanding questions.

5.1 Persistence and Strength of Ice Production

A central feature of all three observed case studies summarized here is persistent ice formation, as evidenced clearly by cloud radar (Fig. 3). Essentially none of the simulations with detailed microphysics and prognostic IFN fail to reproduce persistent ice formation, with the possible exception of some reported by Klein et al. (2009). However, the simulations tend to not produce sufficient ice, most extremely so in the M-PACE case.

Although observed ice loadings often appear to be substantially greater than can be explained by collocated IFN measurements and known physics, the evidence that it is sufficient to rapidly glaciate available LWP is not strong. Ice consistent with that observed appears to be playing a relevant role in the mixed-layer water budget, but it is not the dominant player that is seen when N_i is made extraordinarily greater than observed in SHEBA intercomparison sensitivity test simulations discussed above, for instance.

Prior to all the case study work described above, this central feature of continuous ice formation was already well identified in the Morrison et al. (2005) study of SHEBA conditions where a role for contact freezing was proposed to explain it. The role of IFN consumption in limiting persistent ice production was also previously identified by Harrington and Olsson (2001).

In short, what the modeling and analysis of these case studies suggest is that CFDC measurements of IFN and known physics going into models seem inadequate to explain observed ice. Results are most inconsistent in the M–PACE case, when drizzle and riming are also active and a poorly known or even unknown multiplication process appears likely to play a powerful role. That said, ice consistent with observations then plays a greater role in the water budget, but still remains far from glaciating such clouds because the conditions consistent with multiplication in the observed M–PACE case also happen to be those where there is no shortage of water vapor with such a strong surface vapor flux. The two conditions are related since a latent heat flux in the M–PACE case leads to a substantial LWP, contributing to active drizzle and riming, which appear implicated in multiplication (e.g., Rangno and Hobbs, 2001). It also seems possible that such a multiplication process could be self-limiting in the sense that if explosive ice formation were to substantially reduce LWP, then the ice formation itself would also be slowed.

5.2 Primary Ice Formation

It is natural to begin discussion of ice formation with the idea that there are cases where ice multiplication is not active. The limiting case of very low LWP and relatively high N_d as in the SHEBA case could be representative of such conditions. Under such conditions, de Boer et al. (2011) have convincingly argued that ice formation appears to accompany the presence of liquid water, as in condensation, immersion, or contact freezing. They base that on analyses of collocated lidar and radar measurements that show reflectivity associated with ice formation and growth increasing only after supercooled cloud water is seen by lidar. Thus available IFN under moderate supercooling appear to require a liquid phase or at least water saturation to be activated, which is consistent with the laboratory finding that IFN are orders of magnitude more active above water saturation than below in CFDC measurements (e.g., Sullivan et al., 2010, their Fig. 1).

Using prognostic IFN with a singular treatment (in which activation occurs instantaneously upon attaining specified conditions (cf. Phillips et al., 2008)), Fridlind et al.

(2012) found in simulations that contact IFN behaved fundamentally differently from IFN acting in other modes. Namely, a boundary layer full of contact IFN that could be activated under cloud-top conditions would be only slowly depleted from the boundary layer because contact between IFN and droplets served as a rate-limiting step that proceeds so slowly an initial boundary layer reservoir of IFN present before cloud formation could produce ice continuously and steadily for tens of hours. This contrasted with IFN assumed to act in other modes where conditions within the boundary layer would guarantee activation and depletion of the whole boundary layer IFN reservoir within roughly 1 h and sedimentation of all ice crystals on a similar time scale.

Fridlind et al. (2012) therefore refer to non-contact IFN as rapidly nucleated, consistent with measurements made with a CFDC instrument, which has an effective residence time of roughly 4 s (Paul DeMott, personal communication). Fig. 8 illustrates the differing behaviors of a reservoir of rapidly activated versus slowly activated IFN, using contact

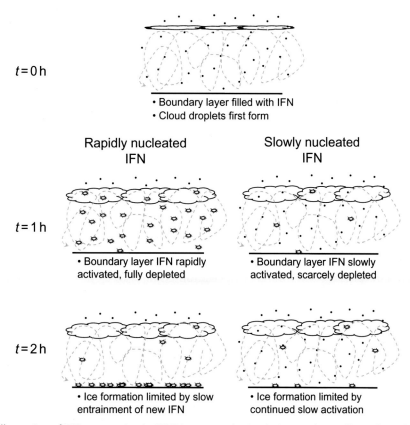

Fig. 8 Illustration of IFN progression in SHEBA case study simulations with rapidly nucleated IFN (left) and slowly nucleated IFN (right), as in the "Prognostic IN" and "Contact IN only" simulations reported by Fridlind et al. (2012).

IFN as example of slowly nucleated particles. It should be noted that if an initial boundary layer reservoir of IFN are entirely and quickly depleted, water vapor will be substantially depleted if the initial concentration is high enough. For instance, in order to avoid rapid LWP depletion in Fridlind et al. (2012) with 50 L^{-1} IFN in the SHEBA case, simulations had to be initialized with steady-state depleted levels within the boundary layer, allowing the entrainment source of IFN to slowly balance its surface loss rate.

Savre and Ekman (2015b) report simulations of ISDAC case studies, discussed above, where slow IFN nucleation proceeds via the immersion mode rather than the contact mode. Using a time-dependent (nonsingular) treatment of IFN based on classical nucleation theory (Savre and Ekman, 2015a), they assign a distribution of contact angles to the dust and soot particles estimated from in situ single-particle observations, further subject to population fractions of two-thirds and one-third able to act as IFN, respectively. They arrive at a substantial reservoir of potential IFN in the boundary layer, at least two orders of magnitude greater than observed by CFDC with a 4 s residence time. By following the contact angle distribution prognostically, they find that that substantial reservoir sustains N_i similar to that observed in all three cases, aided by cooling of cloud top, whereas a singular treatment of the IFN fails to do so.

The idea that many IFN may be relatively weakly active or may be active only in the contact mode—in either case slowing their depletion and providing a steady source of ice—was also suggested by Westbrook and Illingworth (2013). In a well observed decoupled mixed-phase cloud layer with stable cloud-top temperature, they argue that cloud top and cloud base entrainment are both negligible and that some type of slow activation process must be required to sustain the ice observed. Based on budgets of N_i flux estimated from observations, they also report that sustaining ice for tens of hours would require more IFN than ever reported by CFDC measurements in the immersion mode. In other words, not only was slow activation required to explain their data, but an IFN source beyond that commonly understood to be present in the atmosphere was also required.

One reason it may be difficult to distinguish how nucleation is proceeding is that very different spatial distributions of nucleation could lead to similar occurrence of a relatively uniform distribution of N_i when averaged horizontally. For instance, in SHEBA case study simulations where nucleation is concentrated at cloud top or is distributed throughout the liquid cloud layer (Fig. 9), the horizontal mean N_i fields are similarly uniform vertically owing to turbulent mixing, as demonstrated in Fridlind et al. (2012, their Fig. 10). Yang et al. (2013) proposed that examining the underlying relationship of N_i and ice mass mixing ratio (q_i) could give insight into where the nucleation is occurring, as well as the underlying rate. Fig. 10 shows such the differing patterns of N_i versus q_i below cloud base for a SHEBA case study simulation, which are similar to those within-cloud but likelier easier to observe. They differ somewhat from those reported by Yang et al. (2014) for an ISDAC case study with cloud-top seeding.

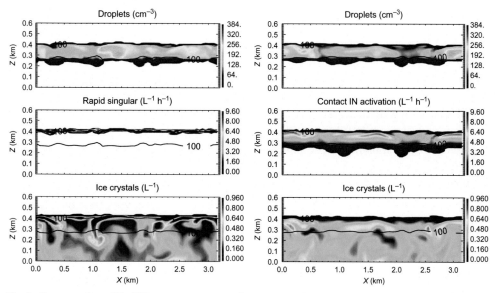

Fig. 9 Cross-section of N_d, IFN activation rate, and N_i in simulations with rapidly nucleated IFN (left) and slowly nucleated IFN in the contact mode (right), as in the "Prognostic IN" and "Contact IN only" simulations reported by Fridlind et al. (2012).

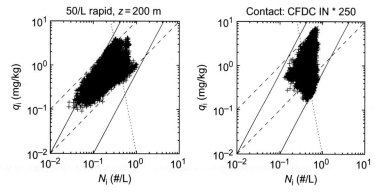

Fig. 10 Ice mass mixing ratio (q_i) versus number concentration (N_i) at an elevation of 0.2 km in simulations with rapidly nucleated IFN (left) and slowly nucleated IFN in the contact mode (right) shown in Fig. 9, as in the "Prognostic IN" and "Contact IN only" simulations reported by Fridlind et al. (2012). Lines following slopes of 1, 2.5, and −5 following Yang et al. (2013, 2014) drawn for comparison.

Recently it has been reported that CFDC measurements may underestimate IFN by a factor 2–10 owing to their exclusion of aerosol with diameter exceeding roughly 1.5 μm (Mason et al., 2016), and the bias increases at warmer temperatures. This under-counting could actually exacerbate rather than diminish a curious feature of Arctic IFN from CFDC measurements, namely a pronounced weak dependence on temperature (e.g., Prenni et al., 2007). Unlike Savre and Ekman (2015b), in part owing to a focus on CFDC measurements as an observational constraint, Fridlind et al. (2012) and Avramov et al. (2011) neglected or underestimated the contribution of increasing IFN available owing to progressive cooling of cloud top. In the ISDAC cases, aircraft observations clearly indicate that cooling of cloud top is relatively rapid on Lagrangian flight legs. We are unaware of analysis of the downwind evolution of cloud-top temperatures for the M-PACE or SHEBA cases. It is also not clear the degree to which assumptions made by Savre and Ekman (2015b) in ISDAC cases are consistent with ISDAC CFDC measurements when applying the CFDC residence time and aerosol diameter sampling limitations.

While it is beyond the scope of this chapter to review the somewhat recent explosion of studies of ice nucleation since the first case study discussed here, we can at least identify some aspects of recent debate. For instance, recent conclusions regarding the importance of time dependence in measuring, reporting, and modeling primary ice nucleation have been notably diverse. For instance, Wright et al. (2013) concluded that neglecting time dependence in measurements leads to minimal error in reported IFN properties and that models will be correspondingly insensitive to inclusion of time dependence. On other hand, Herbert et al. (2014) concluded that time dependence is more important for some nucleating materials than others and recommend an approach to (i) reconcile measurements with differing cooling rate or residence time procedures and (ii) include it in models.

Some have also emphasized that IFN diversity within a given well-defined class (e.g., Arizona test dust) may be of negligible importance compared with adequately constraining sample aerosol surface area (Alpert and Knopf, 2016) whereas others argue that diversity is required to explain the freezing behavior of some materials but not others (Herbert et al., 2014). These factors become entwined when considering that isothermal experiments with diverse particles lead to extended IFN lifetimes owing to the slower freezing behavior of less efficient particles (e.g., Herbert et al., 2014, their Fig. 7). Savre and Ekman (2015b) emphasize this latter factor in explaining persistent ice formation in their ISDAC simulations that assume diverse nucleating efficiencies of dust and soot particles based on laboratory measurements. Since tracking such diversity introduces a non-negligible cost to simulations, it would present a major benefit if such schemes could be suitably reconciled with CFDC measurements under field conditions and collocated field measurements with instruments using substantially longer residence times than the CFDC.

In a climate-model study reporting implementation of a relatively complex time-dependent IFN nucleation scheme, it was concluded that sensitivity to time dependence is weak compared with other IFN properties whose values remain profoundly uncertain (Wang and Liu, 2014). For instance, recent intercomparisons of IFN measurement techniques for a material reported very weakly time dependent remains three orders of magnitude apart in activity or 8°K, and differences are most pronounced at the warmer temperatures similar to those in the case studies discussed here (Hiranuma et al., 2015).

Finally, it is relevant that a consensus seems to be building across the community that immersion freezing is the dominant ice formation process in supercooled clouds (Vali and Snider, 2015), but it could be premature to rule out a role for contact freezing specifically in very long-lived supercooled clouds owing to uncertainties in contact rates (e.g., Ladino Moreno et al., 2013).

5.3 Secondary Ice Formation

By definition secondary ice formation must be initiated by some primary ice nucleation, and is thought to increase N_i by an order of magnitude or more (e.g., Pruppacher and Klett, 1997, and references therein).

The only ice multiplication mechanism widely included in current microphysics schemes is Hallett-Mossop rime-splintering. Based on laboratory experiments, that process requires riming by droplets with diameter greater than 24 µm on an ice hydrometeor with a surface temperature of $-3°C$ to $-8°C$ (Pruppacher and Klett, 1997). Across these case studies, it is notable that ice formation is greatest by far in the M-PACE case, with large droplets and an active riming process, despite cloud base temperature colder than the Hallett-Mossop temperature range. This common scenario has led many to conclude that an unknown or poorly constrained ice multiplication mechanism is likely active under such conditions (e.g., Morrison et al., 2008), as originally surmised by Rangno and Hobbs (2001) in explaining their survey of slightly to moderately supercooled clouds of varying droplet number concentration and varying active processes.

In particular, Rangno and Hobbs (2001) identify riming and splintering as processes that accompany the production of a droplet effective radius exceeding 12 or 20 µm within a stratiform cloud that is supercooled by 10 or 20°C, respectively, and find contrasting conditions of weak ice formation otherwise. It could be the case that ice multiplication under such conditions is either dependent upon the production of splinters by the freezing process of large droplets, as suggested by Lawson et al. (2015), or is dependent on the production of splinters by ice-ice collisions, as suggested by Yano and Phillips (2011). In the latter case, the process could be substantially dependent upon ice properties. For instance, a detailed observational study found that at least 20% of dendrites had undergone natural fragmentation as evidenced by partial regrowth under conditions similar to the ISDAC case study (Schwarzenboeck et al., 2009). It is notable that relatively warm-temperature ice multiplication mechanisms are now being invoked to explain ice distributions emanating from updrafts within growing tropical cumulus (Lawson et al., 2015) and mature mesoscale convective systems (Ackerman et al., 2015; Fridlind et al., 2017), where conditions could be rather similar from the perspective of coexisting large droplets and ice with an active riming process.

It may be useful to separate the M-PACE conditions with likely ice multiplication from the SHEBA conditions of relatively negligible ice formation putting ISDAC perhaps closer to SHEBA than M-PACE. In the phase space of Fig. 2, fragmentation can be considered to increase with LWP, perhaps always accompanying aggregation or riming to some degree. Field studies of ice formation and multiplication require faithful measurements of PSDs, which brought to the fore concerns of severe contamination over several decades of N_i measurements by ice shattering (Korolev et al., 2011). Perhaps suggesting that such concerns have been adequately addressed through instrument design and post-measurement analysis, arguments that observed IFN cannot explain N_i at the level of repeated measurements and climatologies are again being put forth (e.g., Petters and Wright, 2015). Such conclusions are being driven by renewed emphasis on ice formation measurements (e.g., DeMott et al., 2011), which is leading to the advancement of measurement techniques for both N_i and IFN, if not substantially greater confidence in either as of yet (e.g., Lawson, 2011; Baumgardner et al., 2012; Hiranuma et al., 2015).

5.4 Ice Crystal Properties

Variations among well-defined ice crystal habits can lead to substantial differences in water vapor budgets owing to differences in shape factor and sedimentation rate (e.g., Avramov and Harrington, 2010). In addition, it has been noted that ice crystal shapes observed in stratiform mixed-phase clouds are often highly irregular (Korolev et al., 2000), consistent with increasing polycrystalline forms and particle shape complexity, which has been observed to be associated with increasing ice supersaturation in the laboratory (e.g., Bailey and Hallett, 2004). In the context of supercooled mixed-phase clouds, increasing supersaturation is a natural consequence of decreasing temperature, where the relative humidity with respect to ice is increasingly supersaturated while relative humidity with respect to water remains essentially at the saturation value.

One way to understand the challenge for models is at the case study level. There, model setup can exploit CPI data, with a characteristic habit used for a given case study, and use crystal properties from the literature for that habit (e.g., Fridlind et al., 2012; Savre and Ekman, 2015b). However, as encountered by Avramov et al. (2011), habits such as dendrites may occur within the same cloud system in a wide range of forms that have quite dramatically different properties according to the literature. In short, at the case study level, there is currently an absence of quantitative information that can be used to assign relevant ice properties in a model that is designed to have those flexibly assigned or, much more challenging, predicted (e.g., Harrington et al., 2013). Furthermore, individual crystal mass was not systematically measured by any means during any of the field studies used here nor, to our knowledge, has it become a routine measurement in such studies since.

5.5 Climatology and Climate Sensitivity

We have discussed three case study conditions of shallow, single-layer mixed-phase cloud decks, all observed in the Arctic—one a cold-air outbreak over the Beaufort Sea in autumn and two over pack ice in spring. LES with detailed microphysics are able to reproduce basic aspects of all the case studies, including continuous ice formation within the context of a well-mixed liquid-cloud-topped layer. Microphysically, a chief open question is how the observed amount of ice is forming within such clouds. Assumed ice properties may also bear a significant influence on the water vapor budget and reflectivity properties. Active microphysical properties are related to ice properties insofar as riming and aggregation affect ice morphology.

Looking towards climate model parameterization, a leading question is whether these cases are representative of the shallow clouds that contain most of the radiatively important liquid at high latitudes and elsewhere that such clouds may be climatically relevant. Recent preliminary analyses of radar and lidar measurements at the North Slope of Alaska support the tentative conclusion that these case studies are representative of the most commonly occurring shallow mixed-phase clouds at that site. Namely, single-layer liquid-topped clouds represent about 72% of all shallow mixed-phase clouds observed during 2006–11, with the remaining occurrences being primarily characterized by more than one liquid-topped layer (23%) or multi-layer without liquid identifiable at the top of the uppermost layer (Katia Lamer, personal communication). Future work is aimed at establishing the frequency of active drizzle, riming, and aggregation processes. How often are these occurring and how important are they to the occurrence statistics for the radiatively important supercooled water?

If the North Slope of Alaska is a representative site for climatological significance, then it appears important to resolve immediately outstanding questions about ice formation. Is time-dependence important to include in models or not? Is it important to interpretation of observations? Does neglecting time dependence in a model serve to place the cloud state correctly or incorrectly into one of the two quasi-steady state conditions shown in Fig. 8, namely into a cloud-top entrainment limited state where recycling of IFN is more likely to be important, or into a state where a large initial reservoir of IFN is relatively slowly depleted? The role of cooling of cloud top in a deepening mixed-layer in a Lagrangian framework, as well as the understanding of whether and why immersion IFN may be so weakly temperature dependent under Arctic conditions, also deserves additional climatological evaluation.

REFERENCES

Ackerman, A.S., Kirkpatrick, M.P., Stevens, D.E., Toon, O.B., 2004. The impact of humidity above stratiform clouds on indirect aerosol climate forcing. Nature 432 (7020), 1014–1017.

Ackerman, A.S., Stevens, B., Savic-Jovcic, V., Bretherton, C., Chlond, A., Hamburg, G., et al., 2009. Large-eddy simulations of a drizzling, stratocumulus-topped marine boundary layer. Mon. Weather Rev. 137, 1083–1110.

Ackerman, A.S., Fridlind, A.M., Grandin, A., Dezitter, F., Weber, M., Strapp, J.W., et al., 2015. High ice water content at low radar reflectivity near deep convection–Part 2: evaluation of microphysical pathways in updraft parcel simulations. Atmos. Chem. Phys. 15, 11729–11751.

Alpert, P.A., Knopf, D.A., 2016. Analysis of isothermal and cooling-rate-dependent immersion freezing by a unifying stochastic ice nucleation model. Atmos. Chem. Phys. 16 (4), 2083–2107. https://doi.org/10.5194/acp-16-2083-2016-supplement.

Avramov, A., Harrington, J.Y., 2010. Influence of parameterized ice habit on simulated mixed phase Arctic clouds. J. Geophys. Res. 115 (D3), D03205. https://doi.org/10.1029/2009JD012108.

Avramov, A., Ackerman, A.S., Fridlind, A.M., van Diedenhoven, B., Botta, G., Aydin, K., et al., 2011. Toward ice formation closure in Arctic mixed-phase boundary layer clouds during ISDAC. J. Geophys. Res. 116. https://doi.org/10.1029/2011JD015910.

Bailey, M., Hallett, J., 2004. Growth rates and habits of ice crystals between –20° and –70°C. J. Atmos. Sci. 61 (5), 514–544.

Baumgardner, D., Brenguier, J.-L., Bucholtz, A., Coe, H., DeMott, P., Garrett, T.J., et al., 2011. Airborne instruments to measure atmospheric aerosol particles, clouds and radiation: a cook's tour of mature and emerging technology. Atmos. Res. 102 (1-2), 10–29. https://doi.org/10.1016/j.atmosres.2011.06.021.

Baumgardner, D., Avallone, L., Bansemer, A., Borrmann, S., Brown, P., Bundke, U., et al., 2012. In situ, airborne instrumentation: addressing and solving measurement problems in ice clouds. Bull. Am. Meteorol. Soc. 93 (2), ES29–ES34. https://doi.org/10.1175/BAMS-D-11-00123.1.

Cober, S.G., Strapp, J.W., Isaac, G., 1996. An example of supercooled drizzle drops formed through a collision–coalescence process. J. Appl. Meteorol. 35, 2250–2260.

Comstock, K., Wood, R., Yuter, S., Bretherton, C., 2004. Reflectivity and rain rate in and below drizzling stratocumulus. Q. J. R. Meteorol. Soc. 130 (603), 2891–2918. https://doi.org/10.1256/qj.03.187.

Curry, J.A., 1986. Interactions among turbulence, radiation and microphysics in arctic stratus clouds. J. Atmos. Sci. 43 (1), 90–106. https://doi.org/10.1175/1520-0469(1986)043<0090:IATRAM>2.0.CO;2.

Curry, J.A., Hobbs, P.V., King, M.D., Randall, D., Minnis, P., Isaac, G.A., et al., 2000. FIRE Artic clouds experiment. Bull. Am. Meteorol. Soc. 81 (1), 5–29.

de Boer, G., Morrison, H., Shupe, M.D., Hildner, R., 2011. Evidence of liquid dependent ice nucleation in highlatitude stratiform clouds from surface remote sensors. Geophys. Res. Lett. 38 (1), L01803. https://doi.org/10.1029/2010GL046016.

DeMott, P.J., Prenni, A.J., Liu, X., Kreidenweis, S.M., Petters, M.D., Twohy, C.H., et al., 2010. Predicting global atmospheric ice nuclei distributions and their impacts on climate. Proc. Natl. Acad. Sci. U. S. A. 107 (25), 11217–11222. https://doi.org/10.1073/pnas.0910818107.

DeMott, P.J., Möhler, O., Stetzer, O., Vali, G., Levin, Z., Petters, M.D., et al., 2011. Resurgence in ice nuclei measurement research. Bull. Am. Meteorol. Soc. 92 (12), 1623–1635. https://doi.org/10.1175/2011BAMS3119.1.

Demott, P.J., Hill, T.C.J., McCluskey, C.S., Prather, K.A., Collins, D.B., Sullivan, R.C., et al., 2016. Sea spray aerosol as a unique source of ice nucleating particles. Proc. Natl. Acad. Sci. U. S. A. 113 (21), 5797–5803. https://doi.org/10.1073/pnas.1514034112.

Eidhammer, T., DeMott, P., Prenni, A., Petters, M., Twohy, C., Rogers, D., et al., 2010. Ice initiation by aerosol particles: measured and predicted ice nuclei concentrations versus measured ice crystal concentrations in an orographic wave cloud. J. Atmos. Sci. 67, 2417–2436.

Fan, J., Ovtchinnikov, M., Comstock, J.M., Mcfarlane, S.A., Khain, A., 2009. Ice formation in Arctic mixed-phase clouds: insights from a 3-D cloud-resolving model with size-resolved aerosol and cloud microphysics. J. Geophys. Res. 114, D04205. https://doi.org/10.1029/2008JD010782.

Fridlind, A.M., Ackerman, A.S., McFarquhar, G.M., Zhang, G., Poellot, M.R., DeMott, P.J., et al., 2007. Ice properties of single-layer stratocumulus during the mixed-phase arctic cloud experiment: 2. Model results. J. Geophys. Res. 112 (D24), D24202. https://doi.org/10.1029/2007JD008646.

Fridlind, A.M., van Diedenhoven, B., Ackerman, A.S., Avramov, A., Mrowiec, A., Morrison, H., et al., 2012. A FIRE-ACE/SHEBA case study of mixed-phase Arctic boundary layer clouds: entrainment rate limitations on rapid primary ice nucleation processes. J. Atmos. Sci. 69 (1), 365–389. https://doi.org/10.1175/JAS-D-11-052.1.

Fridlind, A.M., Li, X., Wu, D., van Lier-Walqui, M., Ackerman, A.S., Tao, W.-K., et al., 2017. Derivation of aerosol profiles for MC3E convection studies and use in simulations of the 20 May squall line case. Atmos. Chem. Phys. 17, 5947–5972. https://doi.org/10.5194/acp-17-5947-2017.

Gryschka, M., Raasch, S., 2005. Roll convection during a cold air outbreak: a large eddy simulation with stationary model domain. Geophys. Res. Lett. 32 (14), 1–5. https://doi.org/10.1029/2005GL022872.

Harrington, J., Olsson, P.Q., 2001. On the potential influence of ice nuclei on surfaceforced marine stratocumulus cloud dynamics. J. Geophys. Res. 106, 27473–27484.

Harrington, J.Y., Sulia, K., Morrison, H., 2013. A method for adaptive habit prediction in bulk microphysical models. Part I: theoretical development. J. Atmos. Sci. 70 (2), 349–364. https://doi.org/10.1175/JAS-D-12-040.1.

Herbert, R.J., Murray, B.J., Whale, T.F., Dobbie, S.J., Atkinson, J.D., 2014. Representing time-dependent freezing behaviour in immersion mode ice nucleation. Atmos. Chem. Phys. 14 (16), 8501–8520. https://doi.org/10.5194/acp-14-8501-2014-supplement.

Hiranuma, N., Augustin-Bauditz, S., Bingemer, H., Budke, C., Curtius, J., Danielczok, A., et al., 2015. A comprehensive laboratory study on the immersion freezing behavior of illite NX particles: a comparison of 17 ice nucleation measurement techniques. Atmos. Chem. Phys. 15 (5), 2489–2518. https://doi.org/10.5194/acp-15-2489-2015-supplement.

Jiang, H., Cotton, W.R., Pinto, J., Curry, J., Weissbluth, M.J., 2000. Cloud resolving simulations of mixed-phase Arctic stratus observed during BASE: Sensitivity to concentration of ice crystals and large-scale heat and moisture advection. J. Atmos. Sci. 57 (13), 2105–2117.

Klein, S.A., Mccoy, R.B., Morrison, H., Ackerman, A.S., Avramov, A., Boer, G.D., et al., 2009. Intercomparison of model simulations of mixed-phase clouds observed during the ARM Mixed-Phase Arctic Cloud Experiment. I: single-layer cloud. Q.J.R. Meteorol. Soc. 135, 979–1002. https://doi.org/10.1002/qj.416.

Klingebiel, M., de Lozar, A., Molleker, S., Weigel, R., Roth, A., Schmidt, L., et al., 2015. Arctic low-level boundary layer clouds: in situ measurements and simulations of mono- and bimodal supercooled droplet size distributions at the top layer of liquid phase clouds. Atmos. Chem. Phys. 15 (2), 617–631. https://doi.org/10.5194/acp-15-617-2015.

Korolev, A., Isaac, G., Hallett, J., 2000. Ice particle habits in stratiform clouds. Q. J. R. Meteorol. Soc. 126 (569), 2873–2902.

Korolev, A.V., Emery, E.F., Strapp, J.W., 2011. Small ice particles in tropospheric clouds: fact or artifact? Airborne icing instrumentation evaluation experiment. Bull. Am. Meteorol. Soc. 92 (8), 967–973.

Ladino Moreno, L.A., Stetzer, O., Lohmann, U., 2013. Contact freezing: a review of experimental studies. Atmos. Chem. Phys. 13 (19), 9745–9769. https://doi.org/10.5194/acp-13-9745-2013.

Lawson, R.P., 2011. Effects of ice particles shattering on optical cloud particle probes. Atmos. Meas. Tech. 4 (1), 1361–1381. https://doi.org/10.5194/amtd-4-939-2011.

Lawson, R.P., Woods, S., Morrison, H., 2015. The microphysics of ice and precipitation development in tropical cumulus clouds. J. Atmos. Sci. 72 (6), 2429–2445. https://doi.org/10.1175/JAS-D-14-0274.1.

Leck, C., Bigg, E.K., 2005. Source and evolution of the marine aerosol—a new perspective. Geophys. Res. Lett. 32 (19), L19803. https://doi.org/10.1029/2005GL023651.

Lowenthal, D.H., Borys, R.D., Cotton, W., Saleeby, S., Cohn, S.A., Brown, W.O.J., 2011. Atmospheric environment. Atmos. Environ. 45 (2), 519–522. https://doi.org/10.1016/j.atmosenv.2010.09.061.

Mason, R.H., Si, M., Chou, C., Irish, V.E., Dickie, R., Elizondo, P., et al., 2016. Size-resolved measurements of ice-nucleating particles at six locations in North America and one in Europe. Atmos. Chem. Phys. 16, 1637–1651. https://doi.org/10.5194/acp-16-1637-2016-supplement.

McFarquhar, G.M., Zhang, G., Poellot, M.R., Kok, G.L., McCoy, R., Tooman, T., et al., 2007. Ice properties of single-layer stratocumulus during the mixed-phase arctic cloud experiment (MPACE): part I observations. J. Geophys. Res. 112, D24201. https://doi.org/10.1029/2007JD008633.

McFarquhar, G.M., Ghan, S., Verlinde, J., Korolev, A., Strapp, J.W., Schmid, B., et al., 2011. Indirect and semi-direct aerosol campaign: the impact of arctic aerosols on clouds. Bull. Am. Meteorol. Soc. 92, 183–201. https://doi.org/10.1175/2010BAMS2935.1.

Mellado, J.P., 2016. Cloud-top entrainment in stratocumulus clouds. Annu. Rev. Fluid Mech. 49 (1), 145–169. https://doi.org/10.1146/annurev-fluid-010816-060231.

Mitchell, D., 1988. Evolution of snow-size spectra in cyclonic storms: Part I: snow growth by vapor deposition and aggregation. J. Atmos. Sci. 45 (22), 3431–3452.

Morrison, H., Pinto, J.O., 2004. A new approach for obtaining advection profiles: application to the SHEBA column. Mon. Weather Rev. 132 (3), 687–702.

Morrison, H., Shupe, M., Pinto, J., Curry, J., 2005. Possible roles of ice nucleation mode and ice nuclei depletion in the extended lifetime of Arctic mixed-phase clouds. Geophys. Res. Lett. 32, L18801.

Morrison, H., Pinto, J., Curry, J., McFarquhar, G.M., 2008. Sensitivity of modeled Arctic mixed-phase stratocumulus to cloud condensation and ice nuclei over regionally varying surface conditions. J. Geophys. Res. 113, D05203.

Morrison, H., Zuidema, P., Ackerman, A.S., Avramov, A., De Boer, G., Fan, J., et al., 2011. Intercomparison of cloud model simulations of Arctic mixed-phase boundary layer clouds observed during SHEBA/FIRE-ACE. J. Adv. Model. Earth Syst. 3, 1–23. https://doi.org/10.1029/2011MS000066.

Ovchinnikov, M., Ackerman, A.S., Avramov, A., Cheng, A., Fan, J., Fridlind, A.M., et al., 2014. Intercomparison of large-eddy simulations of Arctic mixed-phase clouds: importance of ice size distribution assumptions. J. Adv. Model. Earth Syst. 6 (1), 223–248. https://doi.org/10.1002/2013MS000282.

Petters, M.D., Wright, T.P., 2015. Revisiting ice nucleation from precipitation samples. Geophys. Res. Lett. 42, 8758–8766. https://doi.org/10.1002/(ISSN)1944-8007.

Phillips, V.T.J., DeMott, P., Andronache, C., 2008. An empirical parameterization of heterogeneous ice nucleation for multiple chemical species of aerosol. J. Atmos. Sci. 65, 2757–2783.

Pinto, J., 1998. Autumnal mixed-phase cloudy boundary layers in the Arctic. J. Atmos. Sci. 55 (11), 2016–2038.

Prenni, A.J., Demott, P.J., Kreidenweis, S.M., Harrington, J.Y., Avramov, A., Verlinde, J., et al., 2007. Can ice-nucleating aerosols affect Arctic seasonal climate? Bull. Am. Meteorol. Soc. 88 (4), 541–550. https://doi.org/10.1175/BAMS-88-4-541.

Prenni, A.J., Demott, P.J., Rogers, D.C., Kreidenweis, S.M., McFarquhar, G.M., Zhang, G., et al., 2009. Ice nuclei characteristics from M-PACE and their relation to ice formation in clouds. Tellus Ser. B Chem. Phys. Meteorol. 61 (2), 436–448. https://doi.org/10.1111/j.1600-0889.2009.00415.x.

Pruppacher, H.R., Klett, J.D., 1997. Microphysics of Clouds and Precipitation, second ed. Kluwer Academic Publishers, Boston, MA.

Rangno, A., Hobbs, P.V., 2001. Ice particles in stratiform clouds in the Arctic and possible mechanisms for the production of high ice concentrations. J. Geophys. Res. 106 (D14), 15065–15075.

Rogers, D., DeMott, P., Kreidenweis, S., Chen, Y., 2001. A continuous-flow diffusion chamber for airborne measurements of ice nuclei. J. Atmos. Ocean. Technol. 18, 725–741.

Savre, J., Ekman, A., 2015a. A theory based parameterization for heterogeneous ice nucleation and implications for the simulation of ice processes in atmospheric models. J. Geophys. Res. 120, 4937–4961. https://doi.org/10.1002/2014JD023000.

Savre, J., Ekman, A.M.L., 2015b. Large-eddy simulation of three mixed-phase cloud events during ISDAC: conditions for persistent heterogeneous ice formation. J. Geophys. Res. 120 (15), 7699–7725. https://doi.org/10.1002/2014JD023006.

Savre, J., Ekman, A.M.L., Svensson, G., Tjernström, M., 2014. Large-eddy simulations of an Arctic mixed-phase stratiform cloud observed during ISDAC: sensitivity to moisture aloft, surface fluxes and large-scale forcing. Q. J. R. Meteorol. Soc. 141 (689), 1177–1190. https://doi.org/10.1002/qj.2425.

Schwarzenboeck, A., Shcherbakov, V., Lefevre, R., Gayet, J.F., Pointin, Y., Duroure, C., 2009. Indications for stellar-crystal fragmentation in Arctic clouds. Atmos. Res. 92, 220–228.

Shupe, M., Matrosov, S., Uttal, T., 2006. Arctic mixed-phase cloud properties derived from surface-based sensors at SHEBA. J. Atmos. Sci. 63 (2), 697–711.

Solomon, A., Feingold, G., Shupe, M.D., 2015. The role of ice nuclei recycling in the maintenance of cloud ice in Arctic mixed-phase stratocumulus. Atmos. Chem. Phys. 15 (18), 10631–10643. https://doi.org/10.5194/acp-15-10631-2015.

Sullivan, R.C., Miñambres, L., Demott, P.J., Prenni, A.J., Carrico, C.M., Levin, E.J.T., et al., 2010. Chemical processing does not always impair heterogeneous ice nucleation of mineral dust particles. Geophys. Res. Lett. 37(24). https://doi.org/10.1029/2010GL045540.

Vali, G., Snider, J.R., 2015. Time-dependent freezing rate parcel model. Atmos. Chem. Phys. 15 (4), 2071–2079. https://doi.org/10.5194/acp-15-2071-2015.

Vali, G., DeMott, P.J., Möhler, O., Whale, T.F., 2015. Technical note: a proposal for ice nucleation terminology. Atmos. Chem. Phys. 15 (18), 10263–10270. https://doi.org/10.5194/acp-15-10263-2015-corrigendum.

Verlinde, J., Harrington, J., Yannuzzi, V.T., Avramov, A., Greenberg, S., Richardson, S.J., et al., 2007. The mixed-phase arctic cloud experiment. Bull. Am. Meteorol. Soc. 88 (2), 205–221. https://doi.org/10.1175/BAMS-88-2-205.

Vogelmann, A.M., Fridlind, A.M., Toto, T., Endo, S., Lin, W., Wang, J., et al., 2015. RACORO continental boundary layer cloud investigations: 1. case study development and ensemble large-scale forcings. J. Geophys. Res. 120, 5962–5992.

Wang, Y., Liu, X., 2014. Immersion freezing by natural dust based on a soccer ball model with the community atmospheric model version 5: climate effects. Environ. Res. Lett. 9(12). https://doi.org/10.1088/1748-9326/9/12/124020.

Westbrook, C.D., Illingworth, A.J., 2013. The formation of ice in a long-lived supercooled layer cloud. Q. J. R. Meteorol. Soc. 139 (677), 2209–2221. https://doi.org/10.1002/qj.2096.

Wright, T.P., Petters, M.D., Hader, J.D., Morton, T., Holder, A.L., 2013. Minimal cooling rate dependence of ice nuclei activity in the immersion mode. J. Geophys. Res. 118 (18), 10535–10543. https://doi.org/10.1002/jgrd.50810.

Yang, F., Ovchinnikov, M., Shaw, R.A., 2013. Minimalist model of ice microphysics in mixed-phase stratiform clouds. Geophys. Res. Lett. 40 (14), 3756–3760. https://doi.org/10.1002/grl.50700.

Yang, F., Ovchinnikov, M., Shaw, R.A., 2014. Microphysical consequences of the spatial distribution of ice nucleation in mixedphase stratiform clouds. Geophys. Res. Lett. 41, 5280–5287. https://doi.org/10.1002/2014GL060657.

Yano, J.-I., Phillips, V.T.J., 2011. Ice–ice collisions: an ice multiplication process in atmospheric clouds. J. Atmos. Sci. 68 (2), 322–333. https://doi.org/10.1175/2010JAS3607.1.

Zuidema, P., Baker, B., Han, Y., Intrieri, J., Key, J., Lawson, P., et al., 2005. An Arctic springtime mixed-phase cloudy boundary layer observed during SHEBA. J. Atmos. Sci. 62 (1), 160–176.

CHAPTER 8

Subgrid Representation of Mixed-Phase Clouds in a General Circulation Model

Kalli Furtado
Met Office, Exeter, United Kingdom

Contents

1. INTRODUCTION

The representation of mixed-phase clouds in weather and climate models is important because these clouds are related to model biases on a range of spatial and time scales. In weather forecasting models, mixed-phase processes are important for accurate forecasting of cloud cover and precipitation (Illingworth et al., 2007; Williams and Brooks, 2008; Forbes and Ahlgrimm, 2014). This is particularly the case in the midlatitudes, where mixed-phase processes play a role in the breakup of stratocumulus clouds in the cold-sectors of cyclones (Field et al., 2014a; Huang et al., 2014), but mixed-phase processes are also important in the Arctic (Klein et al., 2009; Morrison et al., 2012), where the climate is determined by a delicate interplay of cloud and radiation processes (Curry et al., 1996; Verlinde et al., 2007).

Deficiencies in the representation of mixed-phase clouds can therefore lead to model errors, such as too rapid dissolution of mixed-phase layer cloud, which cause biases in the radiative balance of climate models (Trenberth and Fasullo, 2010; Williams et al., 2013; Bodas-Salcedo et al., 2012, 2014, 2016a,b). For example, a lack of liquid water in cold-sector boundary layers can lead to too little short-wave radiation being reflected to space at high latitudes (Bodas-Salcedo et al., 2014). For many climate models, this bias is particularly evident over the Southern Ocean where it is accompanied by excessive heat transfer to the surface (Bodas-Salcedo et al., 2012). In coupled models this is a possible cause of large biases in sea surface temperature, and sea-ice extent, which, amongst other

Mixed-Phase Clouds
https://doi.org/10.1016/B978-0-12-810549-8.00008-8

185

consequences, cause uncertainties in climate projection (Sen Gupta et al., 2009; Meijers, 2014) severe enough to raise questions about the reliability of models for some applications.

There is now a good deal of evidence from climatological, synoptic-scale, and small-scaler (process-based) analyses, that many models have difficulty producing and maintaining sufficient quantities of supercooled liquid water (Illingworth et al., 2007; Klein et al., 2009). This raises the question as to why models struggle to represent mixed-phase clouds correctly. The answer to this is twofold. First, the errors could result from the representation of the *source* of liquid water by condensation of water vapor. Secondly, the sinks of liquid water due to, for example, interaction with the ice-phase, may be poorly represented. In both cases, it should be hoped that if parametrizations containing enough of the essential physics of the production and depletion processes can be developed, then realistic mixed-phase environments can be simulated. In this chapter, we will focus on reviewing recent approaches to modeling the *source* of supercooled liquid water.

The accurate modeling of liquid water condensation is, of course, the long-standing goal of cloud macrophysics schemes (so-called to distinguish the prediction of liquid water content and cloud fraction from the subsequent evolution of these variables by small-scale (*micro*physical) processes inside clouds). Cloud schemes range vastly in complexity from the simple diagnostic schemes (see, e.g., Smith, 1990), to schemes that make use of increasing numbers of prognostic variables to keep track of information about model cloudiness (Forbes and Ahlgrimm, 2014; Tiedtke, 1993; Tompkins et al., 2007; Golaz et al., 2002; Wilson et al., 2008).

However, all these schemes have some features in common. First, they all exploit the fact that liquid water condensation occurs very rapidly compared to the model time step. Second, all cloud schemes attempt to model a *probability density function* (PDF) for subcloud-scale temperature and water-substance variations, and then use information about this PDF to specify processes in the model.

The first assumption (rapid condensation of liquid water) allows the PDF to be related to familiar cloud variables such as cloud fraction, ϕ_l, and mean liquid water content, $\overline{q_l}$. Throughout this chapter, water contents will be defined as mass mixing-ratios, relative to dry air. For example, $q_l = \rho_l / \rho_a$, where ρ_l is the mass of liquid water per unit volume, and ρ_a is the mass per unit volume of dry air. Because condensation is a fast process, an excess of the *water content*

$$q_t = q_v + q_l, \tag{1}$$

over the water saturated mass mixing-ratio, $q_{s,w}(p, T)$, of water vapor in air, is equivalent to a mass of liquid water

$$q_l = q_t - q_{s,w}(p, T). \tag{2}$$

This means, for example, that if the PDF, $P(q_t, T)$, for total water, q_t, and temperature, T, is known then:

$$\phi_l = \int dT \int_{q_{s,w}}^{\infty} dq_t P, \qquad (3)$$

$$\overline{q_l} = \int dT \int_{q_{s,w}}^{\infty} dq_t (q_t - q_{s,w}) P. \qquad (4)$$

In other words, the first two moments of the PDF specify the most important cloud properties. Simple diagnostic cloud-schemes assume a functional form for the PDF and then use this to determine the liquid water content and cloud fraction. More complex prognostic schemes are based on evolution equations for the low-order moments of the PDF. These evolution equations contain terms for any model processes that affect the prognostic PDF moments. Hence the challenge is usually the development of closure-relations for all these terms.

In this chapter we will describe the development of a novel approach to constructing a diagnostic cloud scheme for mixed-phase clouds. Rather than assuming a functional form for the PDF of water-content variations, the scheme is based on analytical solutions to an exact but simplified model for the cloud-scale dynamics of relative-humidity variations. Moreover, the dynamical model that underlies the PDF is stochastic in nature and therefore samples the states of a model of a turbulent atmosphere. The model also includes, in a physically based manner, the effects of the presence of the ice crystals on water variations. We will see that, whereas turbulence is the main process driving the *production* of humidity variations, the ice phase acts as a sink of variability that makes condensation of liquid water less likely to occur. We may paraphrase the philosophy behind this approach as an attempt to address the following question:

> given enough information about (a) the turbulent state of the atmosphere, (b) the state of any pre-existing ice cloud, is it possible to diagnose the liquid-cloud properties, using an approach that is based on exact solutions to the dynamical equations for atmospheric water-content variations?

By turbulent state we mean the physical parameters that describe the bulk properties of a turbulent fluid. These might include the turbulent kinetic energy (TKE), dissipation rate, and information about the geometry of the flow (for example, the depth of a stratus cloud layer). Similarly, a description of the bulk properties of the ice cloud might include quantities such as the ice water content, ice crystal number concentration, and the mean size of the ice crystals. Because the model is physically based it is broadly applicable to many types of mixed-phase clouds, for example: supercooled stratoscumulus, but also mid-level clouds (altostratus and altocumulus), and pockets of mixed-phase conditions embedded in frontal clouds. It could also be applied to shallow- and deep-convective clouds, provided the TKE used to describe the turbulent state is representative of such environments.

The approach to be presented evolved from a sequence of papers investigating the competing roles of turbulence and ice microphysics in mixed-phase clouds. This begins with the work of Heymsfield (1977) and Mazin (1986), in the 1970s and 1980s, and concludes more recently with the development of a parametrization for general circulation models that is based on their ideas (Korolev and Field, 2008; Field et al., 2014b; Furtado et al., 2016). We will begin in Section 2 with an overview of the background literature. This also introduces some of the basic physics of liquid condensation in mixed-phase clouds. In Section 3 we describe the development of the background theory into an exactly-solvable model that can form the basis of a cloud-scheme parametrization. In Section 4, we discuss the evaluation of the model against realistic simulations of turbulent mixed-phase clouds. In the same section, we describe the implementation of the model as a parametrization in a general circulation model, and demonstrate the effects that this has on longstanding, radiative-flux biases over the Southern Hemisphere.

2. HISTORICAL OVERVIEW

It has been appreciated for a long time that microphysical processes must play an important role in controlling the longevity of mixed-phase clouds. Because the saturated vapor pressure over an ice surface, $e_{s,i}(T)$, is different from that over a liquid-water surface, $e_{s,w}(T)$, at the same temperature, T, an isothermal three-phase mixture of ice, liquid water, and moist air is not thermodynamically stable. For temperatures below 0°C, this instability manifests itself via a transfer of water molecules from the liquid phase onto the ice phase. This process is referred to as the Wegener-Bergeron-Findeisen (WBF) process, after the work of these scientists on understanding the importance of its consequences for the coexistence of liquid droplets and ice crystals in mixed-phase clouds (Wegener, 1911; Bergeron, 1935; Findeisen, 1938). In fact, Bergeron proposed that the thermodynamically driven glaciation of mixed-phase clouds is so efficient a mechanism for producing precipitation-sized particles that the majority of surface rainfall was the result of the melting of the ice particles thereby formed (Field and Heymsfield, 2015).

Much more recently, Korolev and Isaac (2003) used simplified parcel modeling to derive rigorous bounds on the glaciation time of mixed-phase cloudy volumes due to the WBF process alone. They obtained an informative relationship, $t_g \sim (q_{l0}/N_i)^{2/3}/S_{i,w}$, between the time-to-glaciation, t_g, the initial liquid water content, q_{l0}, the ice number concentration, N_i, and the supersaturation difference $S_{i,w} = e_{s,w}/e_{s,i} - 1$. From this they concluded, for example, that at the temperature of $-10°C$ a stationary air parcel, with an initial liquid water content of 0.1 g kg^{-1}, could remain mixed-phase for between 10 and 10^4 s, depending on the number of ice crystals present. In other words, based on purely thermodynamic considerations, mixed-phase cloud lifetimes of between a few seconds and a few hours are all that we should expect to observe in nature. In fact, this estimate could be expected to a be an *upper* bound on t_g, because other microphysical processes also occur which can further limit the persistence of liquid water. For example, the presence of ice nucleating aerosol particles can lead to the freezing of liquid droplets, and ice crystals can accrete liquid onto their surface via interparticle collisions (a process known as riming). Korolev and Isaacs also showed that a mixed-phase air parcel moving with a constant vertical velocity cannot maintain liquid water indefinitely, and will always glaciate fully in a relatively short amount time. For conditions typical of stratiform clouds, glaciation was due to the WBF mechanism. Only for parcels undergoing rapid accents, could glaciation be delayed until spontaneous freezing occurs at $T = -40°C$.

Contrary to these estimates, observations show that mixed-phase cloud layers can persist for several days (Verlinde et al., 2007; Rauber and Tokay, 1991; Pinto, 1998). It is therefore clear that microphysics alone is not sufficient to explain the properties of mixed-phase environments. The microphysical processes that deplete liquid water need to be considered together with the dynamical processes producing liquid condensate. Moreover, the finding, by Korolev and Isaac, that ascent with a constant velocity cannot produce long-lived liquid water clouds shows that the dynamical circulations maintaining liquid water must be nonuniform in nature. For example, Korolev and Isaacs investigated air parcels driven by a periodically oscillating vertical velocity and showed that such motions can produce cycles of periodic condensation and subsequent reevaporation of liquid water. They showed that a parameter regime exists where the mean liquid water content, averaged over a cycle, becomes constant.

Previously, Heymsfield (1977) had also recognized the importance of dynamical processes for explaining aircraft measurements of mixed-phase stratiform clouds. He noted that mixed-phase conditions were observed in regions where adiabatic lifting motions were sufficiently rapid to maintain water-saturated conditions in the presence of the sink of water molecules to the ice particles. Heymsfield considered the dynamics of supersaturation over ice, which is defined by $S_i = e_v/e_{s,i} - 1$, where e_v is the partial pressure of water vapor in air. He showed that, for an ascending motion to maintain liquid water, the updraft speed, w, must exceed a critical value, w_c, defined by $w_c = \alpha(dq_i/dt)_w$, where $(dq_i/dt)_w$ is the rate of change of the ice water content *evaluated at water-saturation* and the proportionality constant, α, is a function of pressure, p, and temperature, T (Heymsfield, 1977).

Heymsfield's criteria is a first step toward formulating dynamical constraints on the existence of mixed-phase clouds. It raises the possibility of addressing the following questions: Given a large population of air parcels constituting a cloud, are there conditions which can be applied to determine the fraction of parcels which are mixed-phase at any given instant? Similarly, could the mean liquid water content of these parcels also be predicted? Answering these questions provides a conceptual route toward a subgrid parametrization that could be a used in a GCM. If we envisage the subgrid state as being composed of a large ensemble of individual air parcels, we would like to predict, given the values of the resolved fields, how many of these parcels contain liquid water and how much liquid is present on average in the grid box. How much information about the dynamical and microphysical state of the air parcels is required to carry out this program? For example, Heymsfield's criteria allow mixed-phase parcels to be identified if the instantaneous vertical velocities, supersaturations, and ice-particle growth rates are known.

Korolev and Mazin (2003) further investigated the significance of Heymsfield's criterion. They noted that for the liquid water and vapor components of a three-phase cloud to be in steady state, whilst the ice phase is growing, the following conditions must be met. Firstly, the air must be saturated with respect to water (to ensure that the liquid droplets are neither evaporating nor growing). Secondly, the production of supersaturation by vertical lifting must equal the sink the supersaturation due to growth of the ice crystals. Expressed mathematically, these conditions are $S_w = 0$ and $dS_w/dt = 0$, where $S_w = e_v/e_{s,w} - 1$ is the supersaturation with respect to liquid water.[1] Hence, given a population of mixed-phase air parcels, which all have $S_w = 0$, we can identify the subset which also have $dS_w/dt = 0$ as having steady-state liquid water contents in the presence of growing ice crystals. Moreover, if the time derivative of S_w is nonzero for an air parcel then steady-state mixed-phase conditions are not possible. If $dS_w/dt > 0$ then the adiabatic lifting dominates, leading to production of supersaturation and further condensation of liquid water. If $dS_w/dt < 0$ then the sink of water molecules to the ice phase causes the air to become subsaturated with respect to water and the liquid droplets quickly evaporate, glaciating the cloud parcel. By applying these criteria, Korolev and Mazin showed that Heymsfield's threshold velocity could be expressed as

$$w > w_c^* = \alpha^* N_i \bar{r}_i \tag{5}$$

where N_i is the number concentration of ice crystals, $\bar{r}_i$ is an appropriately defined mean size for the ice crystals and $\alpha^*(p, T)$ is another known function of p and T (see Korolev and Mazin for an exact definition of this parameter).

The condition $w > w_c^*$ identifies those mixed-phase parcels that will maintain their liquid water contents with their current vertical velocities. Similarly, mixed-phase parcels, which have $w < w_c^*$ at a given instant, will subsequently become ice-only. However, the criteria do not identify those parcels in a cloud that are currently ice-only but will

---

[1] Korolev and Mazin applied these criteria by considering the so-called "quasisteady" (or quasistatic, c.f., Squires, 1952) supersaturation, S_{wqs}. The quasisteady supersaturation is attained *exactly* only when $dS_w/dt = 0$, which occurs instantaneously when an ascending air parcel encounters its maximum supersaturation. However, Korolev and Mazin showed that S_{wqs} is approximately attained at "late times," in the sense that $\lim t \to \infty (S_w(t)/S_{wqs}(t)) = 1$. The reason for this is that the rate of change of S_w becomes slow, and tends gradually to zero, as time progresses. Korolev and Mazin applied $S_{wqs} = 0$ to derive their expression for w_c, which is equivalent to requiring $dS_w/dt = 0$ for a water saturated air parcel.

become mixed-phase at a later time. Korolev and Field (2008) recognized the importance of the single-to-mixed transition for understanding the longevity of mixed-phase clouds in nature, and for building a subgrid parametrization. Returning to the conceptual model of an ensemble of subgrid air parcels, we can imagine a highly simplified picture of "turbulent" dynamics as follows. Suppose that the time evolution of the system is composed of a sequence of "snapshots," separated in time by some interval, Δt. At each instant, each air parcel "jumps" a distance $w\Delta t$, and then resets its velocity to a new value $w(t + \Delta t)$. At any given instant some of the parcels are mixed-phase and will remain so, provided $w(t) > w_c^*$, and some are mixed-phase but will become ice-only $\left(w(t) < w_c^*\right)$. Other parcels are ice-only but have the potential to activate liquid water during the time-step. The population is therefore apparently composed of distinct subgroups of parcels, according to phase-composition, with individual parcels transitioning between these states as time progresses. For a steady-state mixed-phase cloud to exist, the net rates of transition between the different subgroups of parcels must be zero. So, for example, the rate at which mixed-phase parcels glaciate must be balanced by the supply of new mixed-phase parcels via activation of liquid water.[2]

Formulating an exact condition for an initially ice-only parcel to activate liquid water is difficult because in principle it involves solving for the time evolution of the parcel variables. At present, it is only possible to do this analytically for a few special cases. For example, if there is no condensate in the air parcel, then changes in the supersaturation over ice, S_i, are due entirely to adiabatic lifting and $S_i(t)$ can be determined from the vertical displacement, Δz, of the air parcel using $S_i(t) \approx S_i(0) + a_i\Delta z(t)$, where a_i is a function of p and T. If such an air parcel is to activate liquid water then its displacement must be at least large enough to bring it to water saturation: $\Delta z > \Delta Z_c \approx (S_{i,w} - S_i(0))/a_i$. Korolev and Field formulated ΔZ_c more precisely as $\Delta Z_c = \ln(e_{s,w}/e_0)/a_i$, where e_0 is the initial vapor pressure in the air parcel. They also noted that this special case provides a necessary condition for the activation of liquid water. If an air parcel contains ice, then the change in supersaturation is *less* than that experienced by an ice-free parcel for the same vertical displacement. Hence the required parcel displacement must be at least as large as ΔZ_c.

Korolev and Field combined their displacement criteria with Korolev and Mazin's velocity threshold to state two conditions which must be satisfied for conversion of a parcel to mixed-phase: (1) the parcel must be lifted until it becomes saturated with respect to water; (2) having attained water-saturation, it must be ascending fast enough to overcome the sink of water-molecules to the ice phase. They examined how these criteria operated in two dynamically interesting scenarios: periodically oscillating vertical motion of the

---

[2] This is actually a condition of constant mixed-phase fraction. The condition of steady-state liquid water content is more subtle because increasing LWC in ascending mixed-phase parcels would need to be accounted for.

kind shown by Korolev and Isaacs to produce time-averaged steady states (Korolev and Isaac, 2003), and a simplified model of a turbulent cloud for which the dynamics are similar to the "jumpy turbulence" described above, with $w(t)$ at each time-step selected from a specified probability distribution. In both cases, they found that steady-state liquid water contents could be maintained indefinitely.

The conditions formulated by Korolev and Field were based on idealized parcel modeling. A drawback of these models is that many effects are neglected. For example, particle size distributions are treated as mono disperse, growth of ice by vapor deposition and riming are omitted and ice particles cannot sediment. The dynamical processes are also simplified because of the purely Lagrangian description of the prescribed vertical velocities in the model. To evaluate the criteria stated by Korolev and Field in a more realistic setting, Hill et al. (2014) used large-eddy simulations (LESs) of mixed-phase clouds. Fig. 1 shows two-dimensional cross sections of TKE and liquid water content from two of their simulations. A layer of mixed-phase cloud was generated by applying a prescribed shear rate to the early stages of the simulations. The turbulence generated then decayed over the course of the simulations, which allowed a range of domain-mean energies and dissipation rates to be sampled. A suite of such simulations were performed, spanning a range of temperatures, initial shear-rates, ice water contents and ice number concentrations. The number

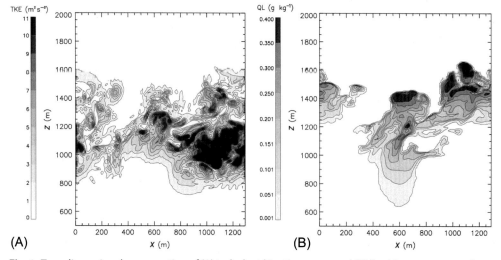

Fig. 1 Two-dimensional cross-section of (A) turbulent kinetic energy and (B) liquid water content, from one of the large-eddy simulations, performed by Hill et al. (2014) (their "BASE(-6C)" simulation) with $N_i = 10\,L^{-1}$ and an initial shear rate of $0.2\,s^{-1}$. *(Hill, A.A., Field, P.R., Furtado, K., Korolev, A., Shipway, B.J., 2014. Mixed-phase clouds in a turbulent environment. Part 1: Large-eddy simulation experiments. Q. J. R. Meteorol. Soc. 140, 855–869. © 2013 Crown copyright, the Met Office, and Her Majesty the Queen in Right of Canada. Quarterly Journal of the Royal Meteorological Society © 2013 Royal Meteorological Society. Modified from the original: only two panels from the original figure are reproduced here).*

concentrations where selected to span a wide range of values, from relatively pristine conditions (1 L^{-1}) representative of Arctic boundary layers, to values typical of more ice-nuclei–rich environments at lower-latitudes, or over continental sites. However, the simulated clouds were necessarily idealized and served mainly to generate a range of values, in space of underlying physical parameters, against which theoretical ideas of Korolev and Field could be evaluated. In addition, idealized simulations with ice growth and sedimentation turned off were compared to simulations with full ice microphysics. These simulations provided a test of Korolev and Field's criteria against realistic turbulent flow statistics and ice cloud microphysics. To test the criteria, Hill et al. assumed that approximate steady states existed where the number of air parcels transitioning to mixed-phase balanced those returning to single-phase, at any instant. Under these assumptions the liquid cloud properties could be estimated from the LES fields by counting the number of model grid boxes satisfying Korolev and Field's criteria. Fig. 2 shows the results of this analysis for two of the simulations. In both cases the agreement between the LES time-series of cloud fraction and the theoretical estimates is reasonably good.

Evaluating the criteria for the existence of mixed-phase conditions against LES, gives some confidence that they could form the basis of subgrid parametrization of liquid water condensation in the presence of ice. The next steps in this direction were taken by Field et al. (2014b), who developed an analytically solvable model for steady-state mixed-phase clouds maintained by turbulence. A description of their model and its implementation in a GCM is described in the next section.

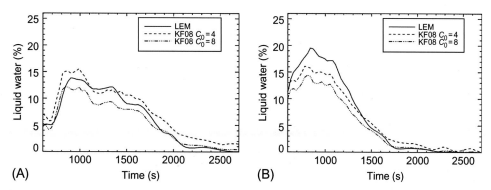

Fig. 2 Time series of liquid cloud fraction from a simulation by Hill et al. with $N_i = 100 \text{ L}^{-1}$. (A) Low initial shear rate, 0.2 s^{-1}, (B) high initial shear rate, 0.3 s^{-1}. *Dashed lines* are theoretical estimates using Korolev and Field's conditions for two different assumed values of the Lagrangian structure constant, C_0 (see Section 3): $C_0 = 4$ *(dashed line)* and $C_0 = 8$ *(dot-dashed line)*. (Hill, A.A., Field, P.R., Furtado, K., Korolev, A., Shipway, B.J., 2014. Mixed-phase clouds in a turbulent environment. Part 1: Large-eddy simulation experiments. Q. J. R. Meteorol. Soc. 140, 855–869. © 2013 Crown copyright, the Met Office, and Her Majesty the Queen in Right of Canada. Quarterly Journal of the Royal Meteorological Society © 2013 Royal Meteorological Society. Modified from the original: only two panels from the original figure are reproduced here).*

3. THEORETICAL MODELS OF MIXED-PHASE CLOUDS

In this section we review a model of turbulent mixed-phase clouds developed by Field et al. (2014b). The challenge in constructing a cloud-scheme is to model the PDF of water content, q_t, and temperature, T, well enough to obtain good estimates of the cloud fraction, ϕ_l, and mean liquid water content, $\overline{q_l}$. To obtain information about the PDF the most fundamental approach is to start from the *statistical dynamics* of the underlying variables, q_t and T. At this point, it is helpful to borrow some terminology from statistical physics in calling the underlying dynamics the microscale and the overlying description of the evolution of the PDF of the dynamical quantities the macroscale (Español, 2004). The process of deriving macroscale properties from microscale details is often called coarse graining, because the finer details of the individual dynamical trajectories are averaged out to obtain a model for the statistical properties of a large-ensemble of real-izations of the dynamics (Español, 2004). The approach of coarse graining microscale atmospheric dynamics to derive information about an ensemble of cloud systems has been employed by a number of authors, particular Larson et al. (2001a), Larson et al. (2001b), Larson et al. (2002). Prior to these studies, Russian scientists developed a range of coarse-graining techniques for studying processes in clouds, focusing on the role of supersatu-ration fluctuations in determining the droplet size spectrum in clouds (Sedunov, 1974; Stepanov, 1975; Voloshchuk and Sedunov, 1977) (see Khvorostyanov and Curry, 1999 for a review). Below we will describe a simple approach to coarse graining the microscale dynamics of supersaturation to derive expressions for $\overline{q_l}$ and ϕ_l. However, before doing so we will review a simple problem from statistical physics, which, it turns out, is highly informative for understanding the statistics of supersaturation. The problem is a classical one from the theory of Brownian motion (Uhlenbeck and Ornstein, 1930).

A mass attached to a spring and subjected to an external force, $f(t)$, has the following equation of motion

$$\frac{dl_s(t)}{dt} = -\kappa(l_s(t) - l_0) + f(t), \tag{6}$$

where $l_s(t)$ is the length of the spring at time t, and l_0 is the natural length of the spring when no force is acting on it. The terms on the right-hand side of Eq. (6) represent the effects of a linear restoring-force, which tries to pull the spring back to its natural length, and an external driving force, f, which injects energy into the system. These dynamical components are somewhat analogous to the dynamics of supersaturation in clouds, where the microphysical growth of the condensed phase attempts to bring the air to a water-saturated state and adiabatic vertical motions provide a source of supersaturation counteracting the microphysical damping. The case where f fluctuates randomly with time is therefore analogous to supersaturation fluctuations in the presence of a random (turbulent) velocity field. We will make this analogy precise, but for now we note only that if f is a randomly fluctuating force then Eq. (6) is called a stochastic differential equation (SDE) (Gardiner, 2004; Arnold, 1974).

We will use the notation $\overline{(\cdot)}$, to denote the average of a quantity over realizations of the random force. Suppose that the force satisfies the following relationship:

$$\overline{f(t)f(t')} = 2\Gamma\delta(t-t'), \tag{7}$$

where δ is the Dirac distribution (Lighthill, 1980), Γ is a constant, and t and t' are any two times. In this case, f is called Gaussian white noise and $l_s(t)$ is called the Ornstein-Uhlenbeck process (Gardiner, 2004). Because $l_s(t)$ is a random quantity, we can investigate its probability density, $P(l, t)$, where $P(l, t)dl$ is the probability that the length of the spring at time t is within dl of l. We can move from the microscale description of the dynamics of $l_s(t)$, given by Eq. (6), to a macroscale description for the PDF, $P(l, t)$, by using a standard method, taken from the statistical mechanics of many-particle systems, for reformulating a SDE in terms of an equivalent partial differential equation (Fokker-Planck equation) for the PDF on the phase-space of the system. (Details of the procedure can be found in, e.g., Gardiner, 2004.) Conceptually, one can imagine a large ensemble of individual trajectories each governed by the SDE, Eq. (6). At time t, each trajectory is defined by its position, $l = l_s(t)$, in the one-dimensional phase-space defined by the length-coordinate, l. As time progresses the individual trajectories disperse in terms of the values of l that are occupied by ensemble members. Roughly speaking, the PDF, $P(l, t)$, at time t, can be constructed by subdividing l into a large number of small intervals and counting the number of trajectories in each of intervals at time t. Similarly, the evolution equation for the PDF can be constructed by considering the rate at which the trajectories cross into, and out of, each of these intervals. For the example of the Ornstein-Uhlenbeck process, this procedure yields the following equation of time-evolution of the PDF:

$$\frac{\partial P}{\partial t} = \kappa\frac{\partial}{\partial l}(l - l_0)P + \Gamma\frac{\partial^2 P}{\partial l^2}. \tag{8}$$

The terms on the right-hand side of Eq. (8) represent the macroscale effects of the microscale processes in the underlying stochastic differential (Eq. 6). The linear damping of variations in length and the stochastic forcing both contribute to the net flux of probability along the coordinate, l, which corresponds to the length of the spring. The effect of the linear damping term on the PDF is to *advect* the probability distribution with the "velocity" $\kappa(l_0 - l)$. On the other hand, the stochastic forcing gives rise to diffusion of the probability density. This process corresponds to the dispersion of the individual microscale trajectories due to the action of the stochastic force, f, which is why the diffusion constant, Γ, is related to the intensity of the fluctuations in the external force.

Given an initial probability distribution, $P_0(l)$, the PDF will evolve, according to Eq. (8), and eventually tend toward a steady-state for which $\partial P/\partial t = 0$. It will be useful below that the eventual steady state solution of Eq. (8) is given by

$$P(l) = \left(\frac{\kappa}{2\pi\Gamma}\right)^{1/2} \exp\left(-\kappa(l - l_0)^2/2\Gamma\right), \tag{9}$$

which is a Gaussian distribution with mean l_0 and variance $\sigma_l^2 = \Gamma/\kappa$.

For clouds with sufficiently low liquid contents, the full statistical-dynamics of coupled $q_t - T$ fluctuations contains more information than is required because, in this regime, temperature fluctuations are only important only insofar as they affect the *supersaturation over ice*, defined by

$$S = \frac{q_v - q_{s,i}(p, T)}{q_{s,i}(p, T)}, \tag{10}$$

(we will omit the subscript "i" from now on). The cloud properties can be estimated from the PDF, $P(S)$, of supersaturation:

$$\phi_l = \int_{S_{i,w}} dS\, P(S), \tag{11}$$

$$\overline{q_l} = q_{s,i}(p, T) \int_{S_{i,w}} dS\,(S - S_{i,w}) P(S), \tag{12}$$

where $q_{s,i}$ is the saturated value of q_v with respect to ice. These equations are only approximations to the more precise formulation given by Eqs. (3) and (4), but it can be shown that the approximation is reasonably good provided the liquid water contents diagnosed (or, equivalently, the typical values of $S - S_{i,w}$ achieved) are small. The idea of using the PDF of ice supersaturation, in place of the full PDF of q_t and T, originates in from the investigations by Heymsfield (1977), Korolev and Field (2008) and, subsequently, Field et al. (2014a), who showed that liquid water contents derived from $P(S)$ agree well with the results of cloud-resolving simulations. Based on these considerations, what is sought is therefore a physically based model for the PDF of S, from which the liquid cloud properties can be derived.

To establish such a model, we can turn to the equations for the dynamics of supersaturation. If we assume that the variations in supersaturation are not large, compared to unity, then the dynamics is given by a linearized version of Squires' equation (Squires, 1952), for the supersaturation, $S(t) = S(\mathbf{x}(t), t)$, of an air parcel with the position, $\mathbf{x}(t)$:

$$\frac{dS}{dt} = -\frac{S}{\tau_p} + a_i w, \tag{13}$$

$$\frac{d\mathbf{x}}{dt} = \mathbf{w}, \tag{14}$$

where $\mathbf{w} = (u, v, w)$ is the velocity of the parcel and τ_p is a time-scale characterizing the growth of ice crystals by water vapor deposition,

$$\frac{1}{\tau_p} = \beta(p, T) \int dD D f(D),$$ (15)

and $f(D)$ is the particle size distribution (PSD) of the ice crystal population and β is a function of p and T (Field et al., 2014a). The first term on the right-hand side of Eq. (13) represents the sink (or source) of supersaturation due to ice crystal growth, the second term is production (or depletion) of supersaturation due to adiabatic lifting (or sinking) air motions. In clear-sky conditions, the second term alone determines the dynamics of $S(t)$. In this case, the time-dependence of $S(t)$ is given in terms of the air parcel displacement: $S = S_0 + a_i \Delta z(t)$, where S_0 is the initial supersaturation of the air parcel. In the presence of ice cloud, it is the competition between ice microphysics and dynamics that determines the observed supersaturations.

If supersaturation variations that are comparable to unity need to be considered then a nonlinear factor, $1/(S+1)$, should multiply the time derivative on the left-hand side of Eq. (13) (Squires, 1952; Korolev and Isaac, 2003; Field et al., 2014a). Although including this factor extends the applicability of the model, it also significantly complicates the course graining of the microscale dynamics and therefore makes it difficult to calculate the PDF of S. The parametrization developed will therefore only apply in regimes where the missing nonlinearity does not strongly influence the dynamics. Similarly, the factor a_i in the adiabatic-lifting term is function of pressure and temperature (as is the phase time-scale, τ_p). This means that, in principle, the dynamical description should include equations for p and T variations. This extends the microscale dynamics in the direction of the full-blown, parcel modeling considered by Korolev and Isaac (2003), Korolev and Mazin (2003), and Korolev and Field (2008). However, the basic properties of the supersaturation distribution can be understood from Eq. (13) if constant values of p and T are assumed in the coefficients.

An apparently more serious omission is an evolution equation for the ice particle size distribution, $f(D)$. The fact that the first moment of the PSD influences supersaturation fluctuations means that, in principle, the ice PSD should be included in the dynamical description. However, this extra complexity can be circumvented if the phase time-scale for ice growth, τ_p, is long compared to the other time scales in Eq. (13). Since the only other time scale that appears is the time scale intrinsic to the adiabatic lifting-motions, this restriction implies that depositional growth of the ice phase must be slow compared to characteristic time for vertical motions. If this separation of time scales is achieved then the mass of ice (and, hence, $\int dD D f$) can be assumed to remain constant.

Because the adiabatic lifting term Eq. (13) involves the vertical velocity, it is not possible to model the statistics of S without introducing a model for w. Unlike the ice mass, w cannot be assumed to be constant in a realistic model. In fact, using a typical turbulent dissipation rate, $\varepsilon \sim 0.002 \text{ m}^2 \text{ s}^{-3}$ and TKE $K \sim 1.5 \text{ m}^2 \text{ s}^{-2}$, gives a turbulent characteristic eddy duration of $t \sim 100$ s, for isotropic turbulence. Here, we have taken, as the characteristic duration, the Lagrangian decorrelation time scale, τ_d, of the turbulent fluctuations. This is defined as the time scale over which the two-point Lagrangian correlation function, $C(t - t') = \overline{w(t)w(t')}$, decays (the Lagrangian integral time scale). The estimate given is obtained from the following formula, which can be found in the monograph by Rodean (1997):

$$\tau_d = \frac{2\sigma_w^2}{\varepsilon C_0}, \tag{16}$$

where σ_w^2 is the variance of the w-fluctuations and C_0 is a dimensionless constant whose exact value is uncertain for atmospheric flows, but is of the order 1-to-10 (Rodean, 1997; Rizza et al., 2006).

A sophisticated model for S-fluctuations would therefore involve coupling Eq. (13) to the fluid-dynamical equations for nonhydrostatic, buoyancy-driven momentum transport. Equations of this kind can be solved numerically at a cloud-resolving scale by LESs, and can be used to determine the statistical properties of supersaturation for a given cloud regime (as was done by Hill et al., for mixed-phase clouds; Hill et al., 2014). Another possible model for w, often used in atmospheric dispersion modeling, is to treat w as a random process (Rodean, 1997; Wilson and Sawford, 1996). A detailed dynamical description of the turbulence would be to model the fluctuating vertical velocity as an Ornstein-Ulhenbeck process:

$$\frac{dw}{dt} \quad -\frac{w}{\tau_d} + f(t). \tag{17}$$

where, as in Eq. (6), the random driving force $f(t)$ is Gaussian white noise with a correlation function $\overline{f(t)f(s)} = C_0 \varepsilon \delta(t - s)$. The choice, $C_0 \varepsilon$, for the intensity of the noise is required for consistency with the similarity theory of turbulence (see Rodean, 1997, or Wilson and Sawford, 1996, for a review). The model given by Eq. (17) is often used in atmospheric dispersion modeling, to model transport of pollutants by boundary-layer turbulence. We note that, by applying Eq. (9), along with Eq. (16) for τ_d, the steady-state distribution of w is a Gaussian with zero mean and standard deviation, σ_w.

Eq. (17) offers a relatively detailed way of coupling random velocity fluctuations to Squires' equation for the supersaturation dynamics. However, Field et al. proposed a simpler dynamical model, which also treats w as a random variable (Field et al., 2014a). Rather than introduce an equation of motion for w, they modeled w itself (rather than the fluid acceleration) as δ-correlated noise:

$$\overline{w(t)w(t')} = \frac{2\Gamma}{a_i^2} \delta(t - t'), \tag{18}$$

where Γ is a constant that determines the intensity of turbulent velocity fluctuations. To inform the choice of Γ we can consider the parcel displacements, given by integrating

Eq. (14). If **w** is Gaussian white noise then the displacement is a continuous random walk (also called a Weiner process Gardiner, 2004; Arnold, 1974). The vertical displacement is given by $\Delta z = \int_0^t ds w$, which can be thought of as the continuous limit of a series of random "jumps" of length $\Delta t w$, where Δt is small discrete time-interval. If the underlying vertical velocity distribution had a variance of σ_w^2, then a typical value of Δz would be $\tau_d \sigma_w$, where τ_d is the eddy-duration given by Eq. (16). For consistency with these typical step-lengths, Field et al. made the following choice of Γ:

$$\Gamma = a_i^2 \frac{\tau_d \sigma_w^2}{2}. \tag{19}$$

In this case, Eq. (13) again defines S to be an example of the Ornstein-Uhlenbeck process. Modeling w as white noise, avoids the need for introducing an equation of motion for vertical velocity into the microscale dynamics. Moreover, Field et al. showed that for typical mixed-phase environments, with a range of levels of turbulence and concentrations of ice crystals, this stochastic model for w-dynamics is sufficient to reproduce the statistics of supersaturation generated by LES.

Subject to these assumptions about the statistical properties of the w-fluctuations, Field et al. were therefore able to proceed by using the steady-state statistics of the Ornstein-Uhlenbeck process to model the statistics of $S(t)$. Before doing so they introduced an additional term into Eq. (13) to model the effect of nonadiabatic exchange between the fluctuating air-parcel and its surroundings. In introducing this mixing term, we will adopt a different approach, which brings out the relationship of their assumptions to the processes governing the marcoscale PDF.

We can move from the statistical-dynamically description given by Eqs. (13) and (14), to an equation for the evolution for the PDF, $P(\mathbf{x}, S, t)$, by using the same procedure employed in the analysis of the stochastically forced spring described by Eq. (6). In this case, the probability density is distributed over a four-dimensional space with coordinates given by spatial position, $\mathbf{x}$, and supersaturation, S. As with the example of the spring, we can imagine a large ensemble of air parcels and investigate the fraction of those parcels which, at a given time, t, inhabit a state close to a point with coordinates $(\mathbf{x}, S)$. Using this method, the PDF of S satisfies

$$\frac{\partial P(\mathbf{x}, S, t)}{\partial t} = \frac{\partial}{\partial S} \frac{S}{\tau_p} P + \frac{\Gamma}{a^2} \nabla_\mathbf{x}^2 P + \Gamma \frac{\partial^2}{\partial S^2} P, \tag{20}$$

The terms in Eq. (20) arise from the microscale processes that influence the individual air parcels in the microscale ensemble. The terms on the right-hand side arise from the supersaturation dynamics and turbulence. The first term is analogous to the advection of the length, l, of the spring by the drift-velocity, $\kappa(l_0 - l)$, in Eq. (8). It represents the effects of the damping of supersaturation fluctuations by depositional growth of the ice particles. The supersaturation that is equivalent to the natural length, l_0, of the spring is ice saturation, $S=0$. The damping influences the probability distribution by imposing a constant drift toward $S=0$, with a drift-velocity of $-S/\tau_p$. The second term is due to real diffusion of air parcels in the physical-space coordinate, $\mathbf{x}$. It arises because the position of an air parcel undergoes a random walk and therefore the ensemble disperses in terms of the spread of positions of the air parcels. The diffusion constant is the intensity, Γ/a^2, of the vertical velocity fluctuations. The third term is diffusion of probability in the supersaturation coordinate, and is due to the stochastic forcing by adiabatic motions.

It would be expensive computationally to solve this equation for the PDF. In fact, from a parametrization point of view, this equation is too general: it specifies the full, space-time evolution of the PDF, including memory of previous states and communication between neighboring grid boxes. To localize the PDF, we can assume a steady state in each grid box and make some additional assumptions about how neighboring grid boxes interact. To apply these assumptions we can proceed by integrating Eq. (20) over a control volume, V, which can represent, for example, a model grid box. We can then define the PDF, $P_V(S)$, of S within the volume v to be

$$P_V(S) = \int_v P d\mathbf{x} \tag{21}$$

and it follows that

$$\frac{\partial P_V(S,t)}{\partial t} - \frac{\Gamma}{a^2}\int_S d\hat{\mathbf{n}}_S \cdot \nabla_{\mathbf{x}}P = \frac{\partial}{\partial S}\frac{S}{\tau_p}P_V + \Gamma\frac{\partial^2}{\partial S^2}P_V \tag{22}$$

where $\hat{\mathbf{n}}_S$ is the outward-pointing unit normal vector to the surface, S, of the control volume. The second term on the right-hand side represents the flux of supersaturation variability out of the control volume due to turbulent velocity fluctuations, i.e., mixing of the air in the control volume with environmental air from the surroundings. The model for nonadiabatic mixing, introduced by Field et al., corresponds to adopting the following model for the effect of turbulent mixing with the environment:

$$\frac{\Gamma}{a^2}\int_S d\hat{\mathbf{n}}_S \cdot \nabla_{\mathbf{x}}P = \frac{\partial}{\partial S}\frac{S-S_E}{\tau_E}P_V, \tag{23}$$

where S_E and τ_E are parameters which represent the mean supersaturation of the environment and a time scale for turbulent exchange. The model given by Eq. (23) treats the effects of mixing on the PDF as a damping term. These assumptions result in the following equation for the volume-integrated PDF:

$$\frac{\partial P_V(S,t)}{\partial t} = \frac{\partial}{\partial S}\left\{\frac{S}{\tau_p} + \frac{S-S_E}{\tau_E}\right\}P_V + \Gamma\frac{\partial^2}{\partial S^2}P_V \tag{24}$$

This equation corresponds to Eq. (8) for the PDF of the Ornstein-Ulhenbeck process. In particular, the solution given by Eq. (9) can be applied to obtain the steady-state PDF of S, within the volume V:

$$P_V(S) = \left(\frac{1}{2\pi\sigma_S^2}\right)^{1/2} \exp\left(-(S-\overline{S})^2/2\sigma_S^2\right). \tag{25}$$

where

$$\overline{S} = S_E \frac{1/\tau_E}{1/\tau_p + 1/\tau_E}, \tag{26}$$

$$\sigma_S^2 = \Gamma \frac{1}{1/\tau_p + 1/\tau_E}, \tag{27}$$

are the mean and variance of the supersaturation distribution.

The model of nonadiabatic mixing, given by Eq. (23), is equivalent to adding a term of the form

$$-\frac{S - S_E}{\tau_E} \tag{28}$$

to the right-hand side of Eq. (13), yielding the following modification of Squires' equation:

$$\frac{dS}{dt} = -\frac{S}{\tau_p} - \frac{S - S_E}{\tau_E} + a_i w(t) \tag{29}$$

In fact, this is how it was introduced by Field et al. (2014a). Again, we can note that Eq. (29) defines S to be an example of the Ornstein-Ulhenbeck process. For sufficiently large times, the solutions tend to a statistically steady state in which $S(t)$ looses all memory of the initial supersaturation. In this limit, the mean and variance of $S(t)$, over an ensemble of realizations of the noise, w, become independent of time. In this section, we have seen that the assumption that w is white noise significantly simplifies the properties of this steady state. Moreover, the Gaussian PDFs specified by Eqs. (25)–(27) are simple enough to form a useful basis for a diagnostic subgrid cloud scheme. However, to do this, closure relations are needed for the parameters that specify the mean and variance of the PDF. This is taken up in the next section.

4. CLOSURE RELATIONS

The aim is to use the steady-state solution given by Eq. (25) as the basis for a subgrid scale parametrization. To do this, closure relations are required that relate the parameters τ_p, τ_E, S_E, and Γ to the resolved-scale variables in the GCM. The ice-microphysical time scale, τ_p, can be calculated from the ice particle size distribution using Eq. (15). Field et al. proposed the following closure relations for the parameters describing the turbulence. Following the monograph on atmospheric turbulence by Rodean (1997), they related the intensity, Γ, of the velocity fluctuations to two fundamental properties of the turbulence: the TKE, K, and the Lagrangian decorrelation time scale, τ_d, of the turbulent eddies. Firstly, as discussed above, they chose the intensity, Γ, of the vertical velocity fluctuations to given by Eq. (19), so that typical air-parcel displacements are of the order $\tau_d\sigma_w$. They also assumed that the turbulence was isotropic, in which case the TKE, K, is given by $K = 3\sigma_w^2/2$ and therefore

$$\Gamma = a_i^2 \frac{\tau_d K}{3}. \tag{30}$$

For the Lagrangian decorrelation time scale, τ_d, of the turbulence, they used the formula (16). This implies that, if w were modeled as the Ornstein-Ulhenbeck process, specified by Eq. (17), then the steady-state distribution of w would be Gaussian with variance σ_w^2. This gives the following expression for Γ, in terms of the TKE and dissipation rate:

$$\Gamma = a_i^2 \frac{4K^2}{9\varepsilon C_0}. \tag{31}$$

A closure relation is also needed for the turbulent mixing time scale, τ_E. For this, Field et al. used the time scale, given by Pope (2000), for the time needed to homogenize a region of extent L_E, by turbulent mixing:

$$\tau_E = \left(\frac{L_E}{\varepsilon}\right)^{1/3}. \tag{32}$$

4.1 Comparison to LES

For given values of τ_p, K, ε and the vertical extent, L_E, for the cloud layer, the theoretically predicted PDF can be compared to the results of LESs. This comparison was undertaken by Field et al. (2014a), who tested the statistical model against the suite of LESs of turbulent mixed phase clouds produced by Hill et al. (2014). A turbulent mixed-phase cloud layer was generated by initially forcing the simulation with a range of a prescribed horizontal shear-rates. The energy injected then decayed, over the course of the simulations, allowing the simulations to sample a range of energies within the cloud layer. The cloud layer depth also evolved with time, so the simulations also sampled a range of values of the mixing length scale, L_E. The simulations were also configured with a range ice water contents and ice crystal number concentrations. In addition, simulations including full ice-phase microphysics (e.g., ice nucleation, riming and depositional growth of ice), were compared to simulations with idealized ice microphysics (e.g., no riming, and no

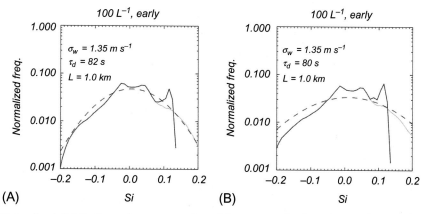

Fig. 3 Normalized distributions of supersaturation with respect to ice from the simulations of Hill et al. (2014) with low initial shear rates (0.2 s^{-1}), at early times (900 s), for two different prescribed ice number concentrations: (A) 100 L^{-1} and (B) 1 L^{-1}. The *black solid lines* show the distributions calculated from the LES, the *dashed lines* are the theoretical predictions and the *gray solid lines* show the distributions of total supersaturation (including condensed water), $S+q_l/q_{s,i}$, from the LES simulations. (*Reproduced from Field, P.R., Hill, A., Furtado, K., Korolev, A., 2014. Mixed-phase clouds in a turbulent environment. Part 2: analytic treatment. Q. J. R. Meteorol. Soc. 21, 2651–2663. © 2013 Crown copyright, the Met Office, and Her Majesty the Queen in Right of Canada. Quarterly Journal of the Royal Meteorological Society © 2013 Royal Meteorological Society. Modified from the original: only two panels from the original figure are reproduced here*).

growth of ice). In all the cases examined, their results showed that the individual PDFs of S produced by the theoretical model closely resembled those generated by the LESs.

Fig. 3 compares the normalized frequency distributions of supersaturation from two of the simulations to the theoretically calculated distributions. The two cases have the same levels of turbulence (as specified by σ_w^2 and τ_d) and the same cloud-layer depth, L_E, but differ in terms of the number concentration of ice crystals. The distributions of S calculated from the simulations are shown by the solid black lines and the dashed lines show the theoretical predictions. The two distributions agree well, up to the water-saturated value of S, at which the LES distribution of S is curtailed due to condensation of liquid water in the simulations. The dashed line continues above water-saturation because there is no condensation of water droplets in the theoretical model. Instead, the moments of the part of the PDF above water saturation can be converted into liquid cloud properties. If the liquid water contents of the LES are converted to equivalent supersaturations over ice, then the distributions obtained (shown by the gray lines) are in good agreement with the predictions of the theoretical model.

Further evidence of the close correspondence between the simulated and modeled statistics of supersaturation can be found in Field et al. Of particular interest is their Fig. 3, reproduced here as Fig. 4. It shows the ratio of the predicted and simulated values

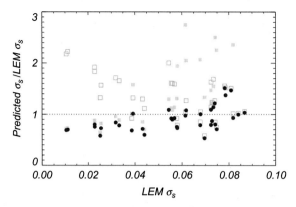

Fig. 4 The ratio of the predicted standard deviation of the supersaturation distribution to the standard deviation calculated from the LES (using the total supersaturation, including liquid water, $S+q_l/q_{s,w}$). The *solid circles* show the theoretical estimates made using Eq. (27), with the closures described in the in the text. The *solid squares* show the prediction obtained if mixing with the environment is neglected, by assuming that $1/\tau_E=0$ in Eq. (27). The *open squares* show the estimate obtained if the microphysical time-scale is neglected ($1/\tau_p=0$). *(Reproduced from Field, P.R., Hill, A., Furtado, K., Korolev, A., 2014. Mixed-phase clouds in a turbulent environment. Part 2: analytic treatment. Q. J. R. Meteorol. Soc. 21, 2651–2663. © 2013 Crown copyright, the Met Office, and Her Majesty the Queen in Right of Canada. Quarterly Journal of the Royal Meteorological Society © 2013 Royal Meteorological Society).*

of the standard deviation, σ_S^2, of the supersaturation fluctuations for the full range of parameters. Also shown is the effect, on the theoretical predictions, of assuming that either ice depositional growth or nonadiabatic mixing with environment can be neglected. It is interesting to note that *both* processes need to be modeled to capture the full range of variability produced by the simulations. If the theoretical model of the supersaturation PDF omits either of these processes then the agreement is poor for limiting values of parameters. For example, if the effects of ice-microphysical growth on the PDF are neglected (by setting $1/\tau_p=0$) then the predicted PDF widths are too broad for cases where the LES PDF is narrow. This is because the narrow LES PDF are due to the damping of supersaturation fluctuations by high ice crystal number concentrations. Similarly, if the effects of nonadiabatic mixing with the environment are neglected the predicted PDF are too dispersed when the simulated PDF are broad. This is because the broad PDF are generated when the turbulence is intense enough to overcome the ice microphysical sink. In these cases, it is mixing with the nonturbulent environment acting to damp-out large excursions in supersaturation. Evidently, any parametrization of condensation in mixed-phase clouds should include both processes, in order to cover the full range of physical behavior.

As a consequence of the good agreement between the simulated and predicted PDF of supersaturation, the bulk (averaged) properties of the simulated mixed-phase cloud layers

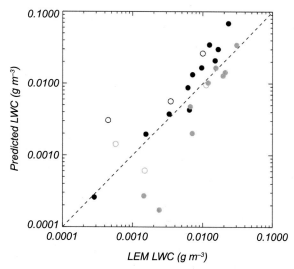

Fig. 5 Scatter plot comparing the predicted and LES-calculated liquid water contents for early, intermediate, and late times (and hence different domain mean turbulent properties), from all the LES simulations performed. *Black symbols*: high initial shear rate. *Gray symbols*: low initial shear. *Solid symbols*: idealized simulations without full ice microphysics. *Open symbols*: simulations with full ice microphysics. *(Reproduced from Field, P.R., Hill, A., Furtado, K., Korolev, A., 2014. Mixed-phase clouds in a turbulent environment. Part 2: analytic treatment. Q. J. R. Meteorol. Soc. 21, 2651–2663. © 2013 Crown copyright, the Met Office, and Her Majesty the Queen in Right of Canada. Quarterly Journal of the Royal Meteorological Society © 2013 Royal Meteorological Society).*

could be represented by the moments of the predicted PDF to a reasonable degree of accuracy. Figs. 5 and 6 show comparisons of the predicted and simulated liquid water contents and liquid cloud fractions. Overall, the theoretical predictions perform well, over almost three orders of magnitude of liquid cloud water.

5. IMPLEMENTATION AS A SUBGRID SCHEME

For the comparison to LES, Field et al. used the simulated turbulent kinetic energies and dissipation rates as inputs to the PDF model. In a GCM, a method is needed for obtaining estimates of these variables. Furtado et al. (2016) adapted the scheme of Field et al. for use in the Met Office Unified Model. They proposed the following closure relations. In each model grid box, the ice microphysical time-scale, τ_p, can be obtained from the ice-mass prognostic, q_i, using the parametrization for the PSD from the cloud microphysics scheme. In the Unified Model the first moment, $\int dD D f(D)$, of the ice PSD is given as a function of q_i and T using the moment–estimation parametrization of the ice PSD developed by Field et al. (2007). In a different GCM, a different PSD parametrization

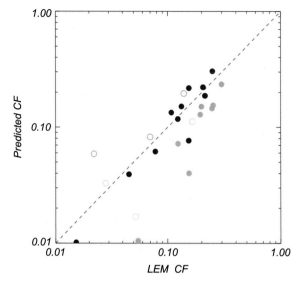

Fig. 6 As for Fig. 5 but for liquid cloud fraction. *(Reproduced from Field, P.R., Hill, A., Furtado, K., Korolev, A., 2014. Mixed-phase clouds in a turbulent environment. Part 2: analytic treatment. Q. J. R. Meteorol. Soc. 21, 2651–2663. © 2013 Crown copyright, the Met Office, and Her Majesty the Queen in Right of Canada. Quarterly Journal of the Royal Meteorological Society © 2013 Royal Meteorological Society).*

could be used for consistency with the microphysics scheme in that model. The TKE, K, can be obtained from a diagnostic in the boundary-layer scheme, which, in turn, is based on scaling arguments. Although the boundary layer scheme in the Unified Model also provides an estimate of eddy dissipation rate, ε, based on steady-state arguments, this was not used. Instead, an additional consistency requirement was introduced between the characteristic turbulent length-, time-, and velocity-scales:

$$L_E = \sigma_w \tau_d = \frac{2\sigma_w^3}{C_0 \varepsilon}; \tag{33}$$

which can be used to determine the dissipation rate, ε, if L_E and σ_w are given. The length scale for entrainment was taken to be the grid-box depth, although an equally plausible choice would be the estimated turbulent mixing length provided by the boundary-layer scheme. The supersaturation of the environment, S_E, can be taken to be the grid-box mean supersaturation. The resulting expressions for the parameters can be substituted into Eqs. (26) and (27) to obtain a closed expression for the PDF, $P(S)$. The first two moments of $P(S)$ can then be evaluated in each model grid box to give the values of the liquid cloud fraction and grid-box mean liquid water content, using Eqs. (11) and (12).

In a GCM with a purely diagnostic cloud scheme, the liquid cloud properties obtained by the above method define the grid box mean values of q_l and φ_l. In a model, such as the Unified Model, which uses a cloud scheme with prognostic variables for q_l and φ_l some care must be taken to ensure consistency with other model processes which can create or remove liquid cloud water. The philosophy behind the Unified Model's prognostic cloud scheme (PC2) is that each process in the model that affects liquid cloud should generate a prescribed increment to the model liquid cloud prognostics (Wilson et al., 2008). For example, there are increments to the liquid water contents and subgrid scale cloud fractions arising from convection, boundary-layer mixing, and microphysical processes. To some extent this means that the PDF of subgrid moisture and temperature fluctuations, which we shall call P_{PC2}, is not explicitly defined, and instead is determined only up to its first two moments.

However, the prognostic cloud scheme does not allow the functional form of the PDF to be completely free and imposes some constraints on the PDF in the limits of completely clear skies and completely overcast conditions. The PDF is constrained, at the end of each model time step, by using a diagnostic cloud scheme to ensure that the evolving subgrid liquid cloud properties are consistent with an assumed width of the subgrid PDF (Wilson et al., 2008). Moreover, this assumed PDF width is not necessarily consistent with the PDF calculated from the parametrization of Field et al. To circumvent this inconsistency, the most satisfactory approach would be a wholesale redefinition of P_{PC2} to be consistent with Field et al. scheme. Instead, a hybrid approach was used that is partially consistent with both the Field et al. parametrization and the PC2 initialization scheme. To do this, Furtado et al. used the observation by Morcrette (2012) that liquid water content increments, Δq_l, and cloud-fraction increments, $\Delta\phi_l$, should be related by

$$\Delta\phi_l = \frac{Q_c P_{PC2}(-Q_c)}{\phi_l Q_c - q_l}\Delta q_l, \tag{34}$$

where Q_c is a function of the grid-box mean quantities. The approach taken to implementing the Field et al. diagnostic PDF scheme, alongside a prognostic cloud scheme, was therefore to use the Field et al. PDFs to determine the increment Δq_l and then use this increment, together with Eq. (34), to determine a cloud-fraction increment that is consistent the PC2 scheme.

Fig. 7 is from a global simulation with the Met Office Unified Model using the Field et al. cloud-scheme. Fig. 7A shows the vertical velocity variance, σ_w^2, diagnosed by the boundary-layer scheme, at a height of 1 km, a couple of hours after the simulation was initialized. Fig. 7B and C show the resulting increments to the liquid water content and cloud fraction. Because condensation is driven by the subgrid-scale vertical velocity fluctuations, the liquid cloud increments produced by the scheme are confined to the regions of high TKE.

Fig. 8 shows the net effect of these cloud increments averaged over a 20-year integration of the global climate. It shows the biases in outgoing short-wave (SW) radiation at

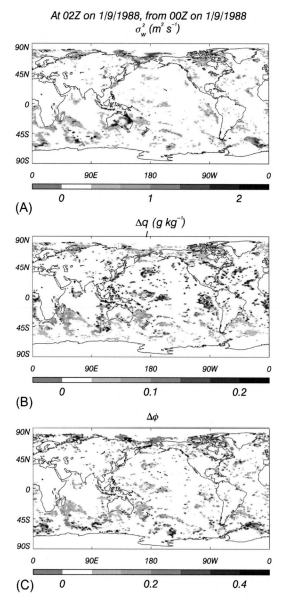

Fig. 7 Instantaneous model fields at 1 km above the surface, from a low-resolution global model simulations using the Field et al. parametrization. (A) the vertical velocity variance, σ_w^2, diagnosed by the GCM boundary-layer scheme, (B) the resulting increment to the liquid water content, due to the mixed-phase scheme, (C) the increment in cloud fraction. *(Furtado, K., Field, P.R., Boutle, I.A., Morcrette, C.R., Wilkinson, J., 2016. A physically-based, subgrid parametrization for the production and maintenance of mixed-phase clouds in a general circulation model. J. Atmos. Sci. 73, 279–291. © American Meteorological Society. Used with permission).*

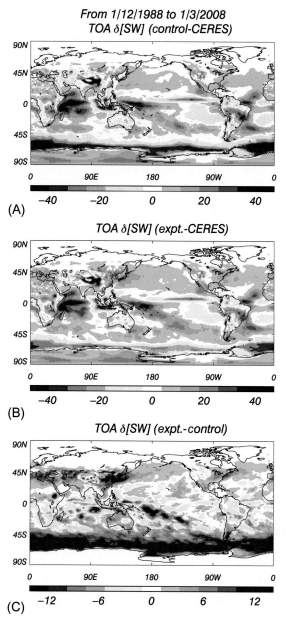

Fig. 8 Biases in the 20-year-mean, DJF outgoing SW flux, at the top of the atmosphere, compared to CERES-EBAF for: (A) control simulation; (B) an experiment, including the mixed-phase parametrization. (C) The difference in flux between the simulations. *(Furtado, K., Field, P.R., Boutle, I.A., Morcrette, C.R., Wilkinson, J., 2016. A physically-based, subgrid parametrization for the production and maintenance of mixed-phase clouds in a general circulation model. J. Atmos. Sci. 73, 279–291. © American Meteorological Society. Used with permission).*

the top of the atmosphere compared to the remotely sensed measurement from CERES-EBAF (Loeb et al., 2009). Fig. 8A shows the 20-year mean bias, from December to January in a control simulation performed with the Global Atmosphere 6 configuration of the Unified Model (Walters et al., 2016) running in AMIP-mode at a resolution of N96 (192 equally-spaced grid points along each latitude circle). The area of red colors over the Southern Ocean correspond to a deficit of outgoing SW flux in that region. Fig. 8B shows the result from a corresponding simulation that includes the Field et al. parametrization, and Fig. 8C shows the difference in outgoing SW between the two simulations. The enhanced liquid water paths in the Southern Ocean region, generated by the scheme, cause more SW flux to be reflected back to space and therefore reduce the SW bias. Changes in the top-of-atmosphere radiative fluxes are accompanied by changes in the radiative balance at the surface. The increased liquid cloud reduces transmission of SW radiation and increases the surface downwelling long-wave radiation (Furtado et al., 2016). The competition between these effects was shown to lead to an overall reduction in the net heat flux into the surface. Subsequent experiments with a fully coupled, atmosphere-ocean model have shown that this improves the sea-surface temperature biases in the Southern Ocean (Hyder et al., 2016).

Furtado et al. also evaluated the global distribution of liquid water path, in weather-forecast simulations, against remotely sensed estimates from the Advanced Microwave Scanning Radiometer (AMSR-E) 3-day global composite (Wentz et al., 2014). They found that liquid water path was improved at high-latitudes, when the Field et al. parametrization was used. In addition, a case study of Arctic stratus showed improvements in liquid water path and downwelling long-wave flux at the surface compared to aircraft and ground-based observations from the MPACE flying campaign (Klein et al., 2009; Verlinde et al., 2007)

6. CONCLUSIONS

A possible mechanism for the observed longevity of mixed-phase clouds is that turbulent air motions maintain a statistically steady state in which water-saturated conditions are maintained in a constant fraction of the cloudy volume (Field et al., 2014a; Korolev and Field, 2008). This process competes with the ice phase, which acts to remove liquid water by microphysical processes and damps out variations in relative humidity by depositional growth of ice crystals. The conditions for an air motion to produce liquid water in the presence of ice crystals can be formulated mathematically as the requirement that an air parcel cool sufficiently to attain water saturation during adiabatic ascent and that the ascent rate be rapid enough to overcome depletion of supersaturation due to ice (Field et al., 2014a; Heymsfield, 1977; Korolev and Field, 2008; Korolev and Isaac, 2003; Korolev and Mazin, 2003). This suggests a conceptual approach to subgrid-scale parametrization of mixed-phase clouds in which the subgrid state is viewed as consisting of a statistical ensemble to air parcels undergoing random, turbulence-induced motions (Korolev and Field, 2008). If the rates at which the ensemble members transition between mixed-phase and ice-only states are balanced then a steady state exists in which a constant fraction of the grid box contains liquid water. This approach can be developed further by introducing a stochastic vertical velocity into Squires' equation for the dynamics supersaturation in an air parcel (Field et al., 2014a). If the stochastic term is modeled as Gaussian white noise, the problem of determining the probability density of supersaturation fluctuations reduces to a classical problem in the theory of Brownian motion, for which analytical solutions are known (Uhlenbeck and Ornstein, 1930).

The solution PDF is a Gaussian with a mean and variance that are determined by the bulk properties of the turbulence (TKE and dissipation rate) and any preexisting ice cloud (number concentration and particle size) (Field et al., 2014a). When these parameters are matched to those from LESs of mixed-phase clouds, the predicted PDF (and derived liquid-cloud properties) were shown to be in good agreement with the simulation results for a wide range of liquid water contents (Field et al., 2014a). The parametrization can be developed into a subgrid cloud scheme and implemented in a GCM by coupling the parameters in the PDF to the boundary-layer and cloud-microphysics schemes. The new parametrization has been shown to improve the realism of Arctic stratus, compared to *in situ* observations and improve the distribution of liquid water at high latitudes (Furtado et al., 2016). This leads to a reduction in long-standing radiative biases over the Southern Ocean and improves sea-surface temperatures in fully coupled climate simulations (Furtado et al., 2016; Hyder et al., 2016).

The scheme is, however, based on a number of simplifications, the effects of which have been only partially analyzed (Field et al., 2014a). Among these, the assumptions that the ice phase does not grow and that the subgrid state is an approximate steady state stand-out as offering potential to extend the scheme. Moreover, the scheme is based on the dynamics of supersaturation over ice, a limitation that can be shown to limit the applicability of scheme to regimes where liquid condensation is relatively small (Furtado et al., 2016). In future work it would be interesting to understand the implications of these assumptions and the extent to which they can be lifted, while retaining a model that is simple enough to act as a diagnostic cloud scheme. It would also be interesting to evaluate the effects of the parametrization on mixed-phase cloud feedbacks of the kinds discussed in Chapters 9 and 10 of this volume.

REFERENCES

Arnold, L., 1974. Stochastic Differential Equations: Theory and Applications. John Wiley and Sons, New York, NY.

Bergeron, T., 1935. On the physics of clouds and precipitation. In: Proc. Ve Assemblée Générale de l'Union Gé odésique et Geophysique Internationale, Lisbon, Portugal, International Union of Geodesy and Geophysicspp. 156–180.

Bodas-Salcedo, A., Williams, K.D., Field, P.R., Lock, A.P., 2012. The surface downwelling solar radiation surplus over the Southern Ocean in the met office model: the role of midlatitude cyclone clouds. J. Clim. 25, 7467–7486. https://doi.org/10.1175/JCLI-D-11-00702.1.

Bodas-Salcedo, A., et al., 2014. Origins of the solar radiation biases over the Southern Ocean in CFMIP2 models. J. Clim. 27, 41–56. https://doi.org/10.1175/JCLI-D-13-00169.1.

Bodas-Salcedo, A., et al., 2016a. Large contribution of supercooled liquid clouds to the solar radiation budget of the Southern Ocean. J. Clim. 29, 4113–4228. https://doi.org/10.1175/JCLI-D-15-0564.1.

Bodas-Salcedo, A., Andrews, T., Karmalkar, A.V., Ringer, M.A., 2016b. Cloud liquid water path and radiative feedbacks over the Southern Ocean. Geophys. Res. Lett. 43, 10,938–10,946. https://doi.org/10.1002/2016GL070770.

Curry, J.A., Randall, D., Rossow, W.B., Schramm, J.L., 1996. Overview of arctic cloud and radiation characteristics. J. Clim. 9, 1731–1764.

Español, P., 2004. Statistical mechanics of coarse-graining. Novel Methods in Soft Matter Simulations. Lecture Notes in Physics, vol. 640. Springer, Berlin, Heidelberg, pp. 69–115.

Field, P.R., Heymsfield, A.J., 2015. Importance of snow to global precipitation. Geophys. Res. Lett. 42, 9512–9520. https://doi.org/10.1002/2015GL065497.

Field, P., Heymsfield, A., Bansemer, A., 2007. Snow size distribution parameterization for midlatitude and tropical ice clouds. J. Atmos. Sci. 64, 4346–4365. https://doi.org/10.1175/2007JAS2344.1.

Field, P.R., Cotton, R.J., McBeath, K., Lock, A.P., Webster, S., Allan, R.P., 2014a. Improving a convection-permitting model simulation of a cold air outbreak. Q. J. R. Meteorol. Soc. 140, 124–138. https://doi.org/10.1002/qj.2116.

Field, P.R., Hill, A., Furtado, K., Korolev, A., 2014b. Mixed-phase clouds in a turbulent environment. Part 2: analytic treatment. Q. J. R. Meteorol. Soc. 21, 2651–2663. https://doi.org/10.1002/qj.2175.

Findeisen, W., 1938. Kolloid-meteorologische vorgänge bei neiderschlagsbildung. Meteorol. Z. 55, 121–133.

Forbes, R.M., Ahlgrimm, M., 2014. On the representation of high-latitude boundary layer mixed-phase cloud in the ECMWF global model. Mon. Weather Rev. 144, 3445–4324.

Furtado, K., Field, P.R., Boutle, I.A., Morcrette, C.R., Wilkinson, J., 2016. A physically-based, subgrid parametrization for the production and maintenance of mixed-phase clouds in a general circulation model. J. Atmos. Sci. 73, 279–291. https://doi.org/10.1175/JAS-D-15-0021.

Gardiner, C., 2004. Handbook of Stochastic Methods: For Physics, Chemistry and the Natural Sciences. third ed. Springer Series in SynergeticsSpringer Verlag, Berlin Heidelberg.

Golaz, J.-C., Larson, V.E., Cotton, W.R., 2002. A PDF-based model for boundary layer clouds. Part I: method and model description. J. Atmos. Sci. 59, 3540–3551.

Heymsfield, A.J., 1977. Precipitation development in stratiform ice clouds – microphysical and dynamical study. J. Atmos. Sci. 34, 367–381.

Hill, A.A., Field, P.R., Furtado, K., Korolev, A., Shipway, B.J., 2014. Mixed-phase clouds in a turbulent environment. Part 1: large-eddy simulation experiments. Q. J. R. Meteorol. Soc. 140, 855–869. https://doi.org/10.1002/qj.2177.

Huang, Y., Siems, S.T., Manton, M.J., Thompson, G., 2014. An evaluation of WRF simulations of clouds over the Southern Ocean with A-train observations. Mon. Weather Rev. 142, 647–667.

Hyder, P., et al., 2016. Atmosphere heat flux errors control Southern Ocean near-surface biases in coupled climate models. Nat. Geosci. in preparation.

Illingworth, A.J., et al., 2007. Cloudnet—continuous evaluation of cloud profiles in seven operational models using ground-based observations. Bull. Am. Meteorol. Soc. 88, 883–898.

Khvorostyanov, V.I., Curry, J.A., 1999. Toward the theory of stochastic condensation in clouds. Part I: a general kinetic equation. J. Atmos. Sci. 56, 3985–3996.

Klein, S.A., et al., 2009. Intercomparison of model simulations of mixed-phase clouds observed during the ARM Mixed-Phase Arctic Cloud. Q. J. R. Meteorol. Soc. 135, 979–1002.

Korolev, A., Field, P.R., 2008. The effect of dynamics on mixed-phase clouds: theoretical considerations. J. Atmos. Sci. 65, 66–86.

Korolev, A.V., Isaac, G.A., 2003. Phase transformation in mixed phase clouds. Q. J. R. Meteorol. Soc. 129, 19–38.

Korolev, A.V., Mazin, I.P., 2003. Supersaturation of water vapor in clouds. J. Atmos. Sci. 60, 2957–2974.

Larson, V.E., Wood, R., Field, P.R., Golaz, J.-C., Vonder Haar, T.H., Cotton, W.R., 2001a. Small-scale and mesoscale variability of scalars in cloudy boundary layers: one-dimensional probability density functions. J. Atmos. Sci. 58, 1978–1994.

Larson, V.E., Wood, R., Field, P.R., Golaz, J.-C., Vonder Haar, T.H., Cotton, W.R., 2001b. Systematic biases in the microphysics and thermodynamics of numerical models that ignore subgrid-scale variability. J. Atmos. Sci. 58, 1117–1128.

Larson, V.E., Golaz, J.-C., Cotton, W.R., 2002. Small-scale and mesoscale variability in cloudy boundary layers: joint probability density functions. J. Atmos. Sci. 59, 3519–3539.

Lighthill, M.J., 1980. An Introduction to Fourier Analysis and Generalised Functions. Cambridge University Press, Cambridge. 79 pp.

Loeb, N.G., Wielicki, A., Doelling, D.R., LouisSmith, G., Keyes, D.F., Kato, S., Manalo-Smith, N., Wong, T., 2009. Toward optimal closure of the earth's top-of-atmosphere radiation budget. J. Clim. 22, 748–766. https://doi.org/10.1175/2008JCLI2637.1.

Mazin, I.P., 1986. Relation of clouds phase structure to vertical motion. Sov. Meteorol. Hydrol. N11, 27–35.

Meijers, A.J., 2014. The Southern Ocean in the coupled model intercomparison project phase 5. Philos. Trans. R. Soc. A. 372 (2019), 20130296. https://doi.org/10.1098/rsta.2013.0296.

Morcrette, C.J., 2012. Improvements to a prognostic cloud scheme through changes to its cloud erosion parametrization. Atmos. Sci. Lett. 13, 95–102. https://doi.org/10.1002/asl.374.

Morrison, H., de Boer, G., Feingold, G., Harrington, J., Shupe, M.D., Sulia, K., 2012. Resilience of persistent Arctic mixed-phase clouds. Nat. Geosci. 5, 11–17. https://doi.org/10.1038/ngeo1332.

Pinto, J.O., 1998. Autumnal mixed-phase cloudy boundary layers in the Arctic. J. Atmos. Sci. 55, 2016–2038.

Pope, S.B., 2000. Turbulent Flows. Cambridge University Press, Cambridge.

Rauber, R.M., Tokay, A., 1991. An explanation for the existence of supercooled water at the top of cold clouds. J. Atmos. Sci. 48, 1005–1023.

Rizza, U., Mangia, C., Carvalho, J.C., Anfossi, D., 2006. Estimation of the Lagrangian velocity structure function constant C_0 by large-eddy simulation. Bound.-Layer Meteorol. 120, 25–37.

Rodean, H.C., 1997. Stochastic Lagrangian models of turbulent diffusion. In: Meteorological Monographs. vol. 48. American Meteorological Society, Boston, MA.

Sedunov, Y.S., 1974. Physics of Drop Formation in the Atmosphere. Wiley.

Sen Gupta, A., Santoso, A., Taschetto, A., Ummenhofer, C., Trevena, J., England, M., 2009. Projected changes to the Southern Hemisphere ocean and sea ice in the IPCC AR4 climate models. J. Clim. 22, 3047–3078. https://doi.org/10.1175/2008JCLI2827.1.

Smith, R.N.B., 1990. A scheme for predicting layer clouds and their water contents in a general circulation model. Q. J. R. Meteorol. Soc. 116, 435–460.

Squires, P., 1952. The growth of cloud drops by condensation. Aust. J. Sci. Res. 5, 66–86.

Stepanov, A.S., 1975. Condensational growth of cloud droplets in a turbulized atmosphere. Izv. Atmos. Oceanic Phys. 11, 27–42.

Tiedtke, M., 1993. Representation of clouds in large-scale models. Mon. Weather Rev. 121, 3040–3061.

Tompkins, A.M., Gierens, K., Rädel, G., 2007. Ice supersaturation in the ECMWF integrated forecast system. Q. J. R. Meteorol. Soc. 133, 53–63.

Trenberth, K., Fasullo, J., 2010. Simulation of present-day and twenty-first-century energy budgets of the Southern Oceans. J. Clim. 23, 440–454. https://doi.org/10.1175/2009JCLI3152.1.

Uhlenbeck, G.E., Ornstein, L.S., 1930. On the theory of Brownian motion. Phys. Rev. 36, 823–841.

Verlinde, J., et al., 2007. The mixed-phase Arctic Cloud Experiment. Bull. Am. Meteorol. Soc. 88, 205–221.

Voloshchuk, V.M., Sedunov, Y.S., 1977. A kinetic equation for the evolution of the droplet spectrum in a turbulent medium at the condensation stage of cloud development. Sov. Meteorol. Hydrol. 3, 3–14.

Walters, D., et al., 2016. The Met Office Unified Model Global Atmosphere 6.0/6.1 and JULES Global Land 6.0/6.1 configurations. Geosci. Model Dev. 10 (4), 1487–1520. https://doi.org/10.5194/gmd-2016-194.

Wegener, A., 1911. Thermodynamik der Atmosph. Verlag Von Johann Ambrosius Barth, Leipzig 331 pp. (in German).

Wentz, F.J., Meissner, T., Gentemann, C., Brewer, M., 2014. Remote Sensing Systems AQUA AMSR-E 3-Day Environmental Suite on 0.25 deg Grid, Version 7.0. Available online at:www.remss.com/missions/amsre (Accessed 4 November 2014).

Williams, K.D., Brooks, M.E., 2008. Initial tendencies of cloud regimes in the met Office Unified Model. J. Clim. 21, 833–840.

Williams, K.D., et al., 2013. The transpose-AMIP II experiment and its application to the understanding of Southern Ocean cloud biases in climate models. J. Clim. 26, 3258–3274. https://doi.org/10.1175/JCLI-D-12-00429.1.

Wilson, J.D., Sawford, B.L., 1996. Review of stochastic Lagrangian models for trajectories in the turbulent atmosphere. Bound.-Layer Meteorol. 78, 191–210.

Wilson, D.R., Bushell, A.C., Kerr-Munslow, A.M., Price, J.D., Morcrette, C.J., 2008. PC2: a prognostic cloud fraction and condensation scheme I: scheme description. Q. J. R. Meteorol. Soc. 134, 2093–2107. https://doi.org/10.1002/qj.333.

CHAPTER 9

Mixed-Phase Cloud Feedbacks

Daniel T. McCoy*, Dennis L. Hartmann†, Mark D. Zelinka‡
*University of Leeds, Leeds, United Kingdom
†University of Washington, Seattle, WA, United States
‡Lawrence Livermore National Laboratory, Livermore, CA, United States

Contents

1. INTRODUCTION

Oceanic planetary boundary layer (PBL) cloud cover strongly affects reflected shortwave (SW) radiation, but has relatively little effect on the outgoing longwave (LW). This leads to a negative cloud radiative effect that strongly affects the Earth's radiative balance (Hartmann and Short, 1980). Because of this it is important to represent the response of PBL cloud to warming accurately to calculate 21st century climate change. Unfortunately, PBL clouds must be parameterized in global climate models (GCMs). This is because turbulent motions with length scales smaller than a GCM grid cell create boundary-layer cloud. The ability of the PBL cloud parameterizations to reproduce cloud behavior in the current climate can be evaluated using observations, however it is difficult to use the existing observational record to evaluate the accuracy of the response of PBL cloud to warming. This results in a cloud feedback that is highly uncertain, even in the most recent generation of GCMs (Bony et al., 2006; Caldwell et al., 2013; Vial et al., 2013; Webb et al., 2013) and accounts for most of the uncertainty in the estimation of equilibrium climate sensitivity (ECS) (Vial et al., 2013; Webb et al., 2006).

Even though the global cloud feedback varies widely across GCMs, the spatial structure of GCM cloud feedbacks is relatively similar (Zelinka et al., 2012, 2013, 2016). One particularly striking feature is the similarity in the latitudinal pattern of the response of cloud SW reflection to warming. We will refer to this change in the reflection of SW due to changes in cloud optical depth and amount with warming as the SW cloud feedback. The SW cloud feedback over oceans in the fifth climate model intercomparison project (CMIP5) is shown in Fig. 1. Across GCMs the SW cloud feedback transitions

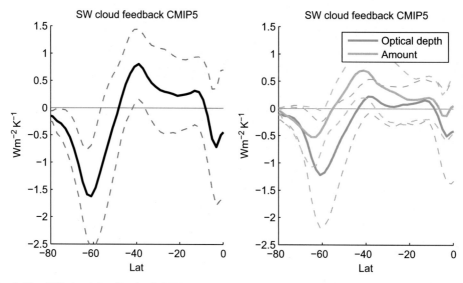

Fig. 1 The SW cloud feedback of GCMs participating in CMIP5. The figure on the left shows the multimodel mean SW cloud feedback with one standard deviation across the GCMs shown as a *dashed line*. The same figure is shown on the right, but with SW cloud feedback decomposed into contributions from optical depth and amount feedbacks. *(See Zelinka, M.D., Klein, S.A., Hartmann, D.L., 2012. Computing and partitioning cloud feedbacks using cloud property histograms. Part II: attribution to changes in cloud amount, altitude, and optical depth. J. Climate 25(11), 3736–3754).*

from positive to negative in the subtropics poleward of around 50 degrees. This is particularly pronounced in the Southern Hemisphere, but also occurs in the Northern Hemisphere. The SW cloud feedback may be decomposed into contributions from cloud optical depth and amount (Zelinka et al., 2012). This is shown in Fig. 1. It is clear that the majority of the positive subtropical feedback originates from cloud area decreasing and revealing the relatively dark ocean beneath, while the negative midlatitude feedback is due to increasing cloud optical depth.

Decreasing cloud cover with warming has been studied extensively and is a robust feature of both large-eddy simulation and observational analysis (Blossey et al., 2013; Bretherton, 2015; Bretherton and Blossey, 2014; Bretherton et al., 2013; Clement et al., 2009; Klein et al., 1995; Myers and Norris, 2013, 2015, 2016; Norris and Leovy, 1994; Norris et al., 2016; Qu et al., 2014, 2015a,b; Rieck et al., 2012; Seethala et al., 2015). It is well known that increasing boundary layer stability increases cloud cover and that boundary layer stability increases as the planet warms (Klein and Hartmann, 1993; Myers and Norris, 2015; Qu et al., 2015a; Webb et al., 2013; Wood and Bretherton, 2006). However, the increase in cloud cover due to increasing stability seems to be overwhelmed by decreases driven by thermodynamic mechanisms linked to sea surface temperature increases (Bretherton and Blossey, 2014). This positive subtropical cloud amount feedback increases ECS, and the negative feedback at high latitudes has a counterbalancing effect on ECS. The robustness of the positive cloud amount feedback in the subtropics makes it particularly important to understand whether the negative feedback in the midlatitudes is physical, and if so, if its strength is accurately represented.

The potential for a pronounced change in cloud optical depth due to mixed-phase clouds (MPCs) transitioning to a relatively more liquid-dominated state was first noted by Mitchell et al. (1989) and Li and Le Treut (1992). Over the last decade this so-called MPC feedback has been of increasing interest in the climate modeling community (Ceppi et al., 2016a; Choi et al., 2014; Kay et al., 2016; McCoy et al., 2016; Naud et al., 2006; Tan and Storelvmo, 2016; Tan et al., 2016; Tsushima et al., 2006) and has been recently featured in review articles (Gettelman and Sherwood, 2016; Storelvmo et al., 2015). It appears that representing MPC behavior in a way that is both physically robust and tractable from a modeling standpoint is becoming a widely acknowledged challenge in accurately predicting 21st century climate change. As discussed in Mitchell et al. (1989) and Li and Le Treut (1992), the increase in cloud optical depth in the midlatitudes appears to be due to transitions of MPC cover to a relatively less ice-dominated and more liquid-dominated state. At zeroth order this is simply because ice crystals tend to be larger than liquid droplets and thus less reflective for a constant amount of condensate (McCoy et al., 2014; Tsushima et al., 2006; Zelinka et al., 2012). In addition to this effect it is probable that the cloud water mass will increase with warming because ice precipitates much more efficiently than liquid (Ceppi et al., 2016a; Field and Heymsfield, 2015; Heymsfield et al., 2009; McCoy et al., 2015a; Mitchell et al., 1989; Morrison et al., 2011). This *mixed-phase cloud feedback* is the subject of this chapter.

The MPC feedback is particularly difficult to constrain in GCMs for several reasons. These may be generally grouped into bottom-up and top-down uncertainties. From the bottom up, the MPC feedback is uncertain because it is governed by ice nucleation; and other MPC physics, which are a complex interplay of different mechanisms, many of which still lack a strong constraint (Atkinson et al., 2013; Hoose and Möhler, 2012; Morrison et al., 2011; Murray et al., 2012; Tan and Storelvmo, 2016). From the top down the feedback is uncertain because we cannot accurately measure the amount of cloud ice mass, making it difficult for models to be rigorously evaluated (Carro-Calvo et al., 2016; Hu et al., 2010; Jiang et al., 2012). Together, these top-down and bottom-up uncertainties yield a wide variety of mixed-phase behaviors in climate models and have led to MPC feedbacks being one of the major contributors to uncertainty in the cloud feedback, and thus climate sensitivity (McCoy et al., 2016; Zelinka et al., 2016). In this chapter we will discuss the origins, mechanisms, and possible constraints on this feedback.

2. THE MPC FEEDBACK IN GCMs

As we discussed in Section 1, understanding the robustness and strength of the MPC feedback in GCMs is important to better constraining ECS and offering better predictions of 21st century climate change (Tan et al., 2016). Because MPC physics operate at a length scale smaller than GCM resolution their behavior in GCMs must be parameterized. Readers interested in a more in-depth discussion of how MPCs are parameterized in GCMs should read the chapter on MPC parameterization by Furtado (2017) in this volume.

As noted above, parameterization of MPC physics is not the focus of this chapter, but it is useful to discuss it briefly. When confronted with the need to represent MPCs, GCMs may either attempt to represent the nucleation of ice by aerosol and the growth of ice particles in MPCs, or may simply diagnose the partitioning of ice and liquid based on a function of temperature (Cesana et al., 2015; Tsushima et al., 2006). Both approaches are problematic. Diagnosing liquid fraction as a function of atmospheric temperature is a very stable method of describing MPCs, and can be implemented based on aircraft sampling of clouds (see Bower et al.,1996), however it cannot represent the impacts of regional variability in ice nuclei (IN) on supercooled liquid clouds (Atkinson et al., 2013; Kanitz et al., 2011; Murray et al., 2012). Indeed, clouds in regions that have access to IN (particularly dust) have noticeably less supercooled liquid (Hu et al., 2010; Kanitz et al., 2011; Tan et al., 2014). Sources of IN, particularly feldspar, are much more common in the Northern Hemisphere than in the Southern Hemisphere, leading to Northern Hemisphere clouds being more glaciated (Atkinson et al., 2013; Morrison et al., 2010; Murray et al., 2012).

While MPC processes are complex, when a MPC is warmed it should transition from a more ice-dominated to a more liquid-dominated state as sinks of cloud water through ice phase precipitation are suppressed (Ceppi et al., 2016a; McCoy et al., 2015a; Mitchell et al., 1989; Morrison et al., 2011). Because of the difference in the radiative properties of ice and liquid this results in an increase in upwelling SW and a negative optical depth feedback, providing that the size of ice crystals and liquid droplets are reasonably represented in a given GCM.

Do GCMs all agree on the MPC temperature range? By examining the behavior of MPCs in GCMs as a function of atmospheric temperature it becomes clear that climate models disagree strongly as to the temperature range inhabited by MPCs. This is shown in Fig. 2 for GCMs participating in CMIP5. To create the curves shown in Fig. 2 for each GCM the fraction of liquid condensate is calculated within each model-level and latitude-longitude grid box. The fraction of liquid condensate is then averaged as a function of atmospheric temperature. This yields a gross statistical representation of the partitioning of ice and liquid as a function of temperature within each model. Examination of these curves reveals substantial disagreement between GCMs in terms of their mixed-phase condensate partitioning behavior. Some GCMs maintain liquid water to temperatures as low as 220 K, well below the homogeneous freezing temperature, while some models are entirely composed of ice at temperatures as high as 260 K. Overall, there is a

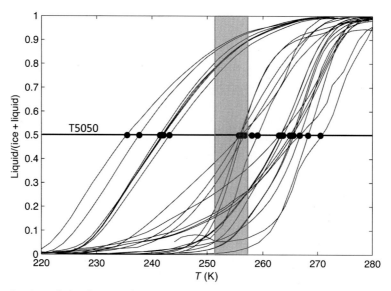

Fig. 2 The fraction of cloud water that is liquid as a function of atmospheric temperature from a selection of GCMs participating in CMIP5 (for a full list of GCMs and details of the calculation see McCoy et al., 2015a). The midpoint of the curves, where ice and liquid are equally mixed (T5050), is shown highlighted by *dots*. The range in of T5050 that would be inferred based on the CALIPSO cloud top phase (Hu et al., 2010) combined with comparison between paritioning and simulated lidar data from Cesana et al. (2015) and Hu et al. (2010) is shown as a solid shaded area.

nearly 35 K range across models where ice and liquid are equally prevalent. While it is useful to discuss the temperature range for which MPCs exist in a particular GCM, we will utilize the temperature at which ice and liquid are equally mixed for the remainder of this chapter. This is useful for brevity and characterizing the MPC temperature range of each model by a single number and still has the capability to explain a significant amount of intermodel variability. We will refer to this quantity, the atmospheric temperature at which ice and liquid each make up 50% of existing condensate, as T5050 (McCoy et al., 2016; Naud et al., 2006).

We have shown that in the models participating in CMIP5 there is an approximately 35 K range in the temperature where ice and liquid are equally prevalent (Fig. 2). Can the range of GCM ice to liquid partitioning shown in Fig. 2 be constrained using observations? As noted above, the curves in Fig. 2 show the temperature dependent partitioning of ice and liquid for vertical averages over GCM model levels (see McCoy et al., 2015a for calculation details). Because of this it is hard to evaluate this model behavior with observations. Evidently this is not directly comparable to in situ measurements made from an airplane, because airplane measurements are made in specific cloud regimes and at high temporal and spatial resolutions (Bower et al., 1996; Cober et al., 2001;

Isaac and Schemenauer, 1979; Korolev and Isaac, 2003; Moss and Johnson, 1994; Mossop et al., 1970; Storelvmo et al., 2015).

A more direct comparison may be made between GCM phase partitioning and ground- and space-based remote sensing. Naud et al. (2006) utilized Moderate Resolution Imaging Spectroradiometer (MODIS) (King et al., 2003) measurements of cloud top phase in northern hemisphere cyclones to show that cloud tops were equally partitioned between ice and liquid at roughly 255 K in the Northern Hemisphere. Surface-based lidar estimates made by Kanitz et al. (2011) showed a T5050 that varied between 242 K for pristine maritime regions and 260 K for a continental site in Leipzig, Germany. This contrast between pristine maritime regions away from dust sources and continental sites is echoed by studies conducted using space-based lidar (Hu et al., 2010; Tan et al., 2014). Komurcu et al. (2014) evaluated a selection of state-of-the-art GCMs that do not treat ice and liquid partitioning as a function of temperature alone. The simulated cloud lidar output from these models showed that all six GCMs produced clouds that were much more glaciated than observed by the CALIPSO lidar (Winker et al., 2009). This result is reinforced by the analysis performed by Cesana et al. (2015) and McCoy et al. (2015a) who diagnosed the effective ice to liquid partitioning curve used by several of the models participating in CMIP5 (Fig. 2). However, it was shown by Cesana et al. (2015) using simulated lidar output from GCMs that lidar-diagnosed ice to liquid partitioning is not directly comparable to the curves shown in Fig. 2. This makes using space-borne observations to constrain ice in MPCs in models problematic. McCoy et al. (2016) offered a rough estimate of the range where ice and liquid are equally mixed based on results from Cesana et al. (2015) and Hu et al. (2010). This range was estimated at 254–258 K, in the global mean. This is a much smaller range than the range of temperatures from CMIP5 models (shown as a shaded area in Fig. 2), and supports the idea that the current generation of GCMs tends to freeze liquid at temperatures that are too high (Cesana et al., 2015; Komurcu et al., 2014; McCoy et al., 2016).

The most apparent effect of this diversity in model parameterization manifests itself in a wide variety of climatological cloud properties in GCMs. GCMs that maintain liquid down to colder temperatures tend to both have more liquid and less ice, as one would naïvely expect. This is shown in Fig. 3 by examining how the T5050 temperature relates to the intermodel spread in historical LWP and IWP in CMIP5 GCMs. In addition, GCMs with a higher T5050 appear to have less overall cloud water (ice and liquid combined), which is generally consistent with the idea of enhanced precipitation efficiency in more glaciated clouds (see Morrison et al., 2011; Ceppi et al., 2016a; McCoy et al., 2015a).

Mixed-phase parameterizations have the capability to substantially influence the climate mean-state ice and liquid content in the mixed-phase regions. This variety in GCM climate mean-state ice and liquid water content may potentially be due to the weak observational constraint on ice phase condensate in the current climate (Jiang et al., 2012).

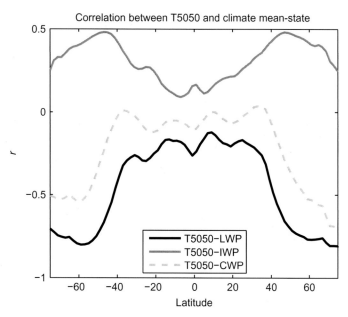

Fig. 3 The across model correlation between T5050 (see Fig. 2) and zonal-mean climate-mean state cloud properties over oceans: liquid water path (LWP), ice water path (IWP), and ice and liquid, or condensed water path (CWP). In the midlatitudes, GCMs that freeze liquid at warmer temperatures (high T5050) have less liquid and more ice. They also have less overall ice and liquid water path. *(Adapted from McCoy, D.T., Tan, I., Hartmann, D.L., Zelinka, M.D., Storelvmo, T., 2016. On the relationships among cloud cover, mixed-phase partitioning, and planetary albedo in GCMs. J. Adv. Model. Earth Syst. 8, 650–668, https://doi.org/10.1002/2015MS000589).*

Only the MODIS and Cloudsat instruments offer estimates of the cloud ice water content through the vertical extent of the atmosphere. MODIS only retrieves IWP while the sun is up, which excludes nighttime and high-latitude winter. It is difficult to estimate an error in this retrieval beyond errors engendered by the assumed particle size distribution used in the retrieval and intercomparison of GCMs and observations by Jiang et al. (2012) assigned a factor of 2 uncertainty in the IWP retrieval from MODIS. The Cloudsat radar is highly sensitive to the partitioning of cloud ice and precipitation (Eliasson et al., 2011) as well as to the assumed particle size distribution (Jiang et al., 2012). The uncertainty range in Cloudsat IWP assigned by Jiang et al. (2012) is between 50% and a factor of 2 depending whether or not columns that the cloud radar has identified as precipitating are excluded from the dataset. Ultimately, this wide variability in the IWP that can be consistent with observations means that GCMs are left with relatively little observational constraint in the creation of cloud parameterizations.

Evidently GCM mixed-phase parameterizations play an important role in determining the column-integrated ice and liquid in mixed-phase regions. Does this matter to the SW cloud feedback? In general, it appears that models whose MPCs contain a greater

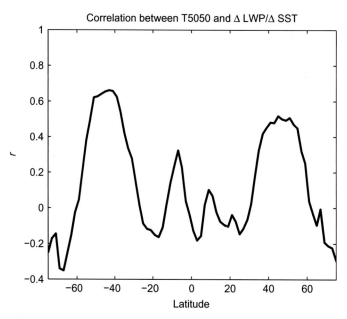

Fig. 4 As in Fig. 3, but showing the correlation between T5050 and change in zonal-mean LWP over oceans normalized by change in SST between historical and RCP8.5 scenarios in CMIP5. GCMs that have a mixed-phase scheme that has generated a large amount of susceptible ice in the climate mean state increase their liquid water path more with warming.

amount of ice that is susceptible to transitioning to water will have a larger increase in liquid water with warming. This is shown in Fig. 4 by examining the change in LWP for a CO_2-induced warming, where the change in LWP has been normalized by surface temperature change. Fig. 4 shows that the intermodel spread in T5050 is strongly correlated with warming-induced increases in LWP. That is to say, models that glaciate their cloud cover more at warmer temperatures also increase their LWP more strongly in a warming climate and have a more pronounced negative optical depth feedback.

The first intercomparison of GCM MPC feedbacks was performed by Tsushima et al. (2006), who analyzed five of the GCMs participating in the third climate model intercomparison project (CMIP3) and showed that there was a strong relationship between the phase partitioning in MPCs and warming-induced increases in LWP. This dependence of the optical depth feedback on ice and liquid partitioning was also demonstrated by Choi et al. (2014), who created several versions of the CAM3 GCM with different ice and liquid partitioning functions. This behavior still appears to be a robust feature of CMIP5 models (McCoy et al., 2015a). Ceppi et al. (2016a) further demonstrated that this linkage is causal and not coincidental by perturbing the mixed-phase microphysical parameterizations in GFDL-AM2.1 and CESM-CAM5 showing that decreased efficiency of liquid water sinks through mixed-phase processes played a critical role in the increase in midlatitude LWP with warming. It is interesting to note that there is not a consensus between GCMs participating in CMIP5 regarding whether the increase in LWP with warming is dominated by a simple repartitioning of condensate with warming, or if it is due to an increase in overall condensate mass in line with decreases in precipitation efficiency (Ceppi et al., 2016a; McCoy et al., 2015a). In some GCMs the increase in LWP with warming may be explained entirely by replacing ice with liquid in line with increasing atmospheric temperature and the curves shown in Fig. 2, while the increase in LWP in other GCMs is almost entirely due to increases in overall cloud condensate in line with suppression of frozen precipitation sinks as clouds move to a less glaciated state (McCoy et al., 2015a). Because of this model diversity, observations of changes in precipitation efficiency due to changes in prevalence of glaciated hydrometeors may be a useful constraint on the MPC feedback.

Viewing MPC properties in a zonal-mean sense is useful for discussing the large spread in cloud feedbacks in the Southern Ocean among GCMs. However, in order to provide more realistic model parameterization of MPC processes it is important to investigate how different cloud regimes contribute to the MPC feedback. Studies based on cyclone compositing in midlatitude regions reveal that the change in LWP with warming and changes in reflected SW are not tightly coupled (Bodas-Salcedo et al., 2016). The clouds in cyclone composites that are responsible for the bulk of the radiative response to warming are nonfrontal clouds, which are relatively thin and tend to be supercooled liquid, as opposed to the frontal clouds, which have significant amount of ice and liquid. Because the frontal clouds are already relatively opaque, increases in their optical depth are less important than increases in the optical depth of thin, nonfrontal clouds (Bodas-Salcedo et al., 2016).

Discussion of the MPC feedback tends to focus on mixed-, and ice-phase microphysics, but warm, liquid microphysics also have the potential to affect the MPC feedback. We have discussed the MPC feedback in the context of changes in LWP. However, cloud optical depth is controlled by both LWP and cloud droplet number concentration, which is, in turn, controlled by the availability of cloud condensation nuclei (CCN) (Bréon et al., 2002; Nakajima et al., 2001; Sekiguchi et al., 2003; Storelvmo et al., 2006; Twomey, 1977). It is interesting to speculate on how changes in the availability of cloud condensation nuclei (CCN) with warming will affect the MPC feedback. As noted before, liquid droplets are much smaller than ice-crystals, and thus a given mass of cloud liquid is brighter than the same mass of ice. However, the availability of CCN, and thus the number concentration in the deglaciated cloud, also significantly affects the strength of the MPC feedback by affecting how relatively bright the newly minted liquid is (McCoy et al., 2014).

The MPC feedback occurs in both the Northern and Southern midlatitudes. These are extremely different aerosol regimes. In the Northern Hemisphere anthropogenic CCN controls cloud microphysical properties (Carslaw et al., 2013). The Southern

Ocean is highly pristine and accurate representation of its aerosol sources is difficult (Hamilton et al., 2014). Sources of CCN in the Southern Ocean are primarily natural and composed of sea spray and the sulfate from biogenic dimethyl sulfide (DMS) (Ayers and Gras, 1991; Ayers and Cainey, 2007; Charlson et al., 1987; Kruger and Grassl, 2011; Lana et al., 2012; McCoy et al., 2015b; Meskhidze and Nenes, 2006, 2010; Vallina and Simó, 2007; Vallina et al., 2006). Because a complex web of organisms produces DMS it is difficult to precisely diagnose how changes in the ocean biome will affect its production. It seems likely that biogenic emissions of DMS will decrease with increasing ocean acidification in a warming world (Six et al., 2013), potentially blunting the negative midlatitude MPC feedback. The control of Southern Ocean CCN by sea-spray aerosol is particularly interesting because sea spray emissions are closely tied to wind speed (Grythe et al., 2014), and MPC parameterizations will affect wind speed through their control of the latitudinal gradient of absorbed SW radiation (Ceppi et al., 2014; McCoy et al., 2016), potentially yielding an interplay of these mechanisms.

Ultimately, the amount of liquid in a cloud plays a central role in determining its albedo. If the LWP in mixed-phase regions is so strongly controlled by the mixed-phase parameterization in a given GCM there must be another factor to counter-balance it and bring the planetary albedo into a reasonable agreement with observations. That is to say, the planetary albedo in a given GCM should be approximately consistent with observational estimates in the climate mean-state. If too little supercooled liquid is maintained in the clouds then this will lead to too low an albedo. Some other factor must increase the planetary albedo so that it is generally consistent with observations. It appears that, at least in the most recent generation of GCMs, this factor is the cloud fraction. It can be seen by regressing intermodel spread in cloud fraction on the mixed-phase characterization parameter, T5050, that models that glaciate clouds at warmer temperatures (higher T5050) both have lower LWP and a higher CF (McCoy et al., 2016). The correlation between T5050 and LWP is restricted to regions where a substantial amount of cloud exists above the melting level, but the intermodel correlation between T5050 and cloud area coverage appears to be a global phenomenon, which is clearly unphysical, especially since one would expect increased glaciation to decrease cloud cover (Heymsfield et al., 2009; McCoy et al., 2016). One possible explanation of this behavior is that the critical relative humidity (RH) that GCMs use to parameterize cloud cover (Bender, 2008; Mauritsen et al., 2012; Quaas, 2012) is adjusted to increase cloud cover and thus bring planetary albedo into a reasonable range. This is not an entirely unreasonable supposition and has been singled out as a common tuning parameter (Bender, 2008; Mauritsen et al., 2012). Anecdotally, it may be seen that in studies which have directly addressed the sensitivity of Southern Ocean cloud properties to mixed-phase parameterizations that the critical RH has been adjusted to yield a control climate that is in energy balance (Kay et al., 2016; Tan et al., 2016). Ultimately this tuning between MPCs and cloud fraction yields brighter subtropics and darker extratropics when model clouds glaciate at warmer temperatures (McCoy et al., 2016). It is interesting to note that this behavior is consistent with the emergent constraint on ECS offered by Volodin (2008) (see Klein and Hall, 2015 for a discussion of emergent constraints).

In mixed-phase regions this tuning between cloud cover and liquid content in MPCs also results in clouds that are both too few, covering too little area, as well as clouds that contain too much liquid and are too bright. In many GCMs this seesaw between cloud liquid and cloud area yields model cloud properties that agree poorly with observed cloud properties (McCoy et al., 2016).

The choices made regarding MPC parameterizations in GCMs have far ranging impacts on model behavior. Can we use observations of MPC temperature range to provide a so-called emergent constraint (Klein and Hall, 2015) on climate sensitivity in the current generation of GCMs? The T5050 that characterizes MPC parameterization does not correlate strongly across models with ECS (McCoy et al., 2016). This is because the subtropical cloud area feedback is more positive in models with a higher T5050, effectively counterbalancing the more negative cloud optical depth feedback in the midlatitudes (higher T5050 implies stronger increase in LWP with warming, see Fig. 4). It is not clear why models with a higher T5050 have a more positive subtropical cloud amount feedback. One potential mechanism may be the positive feedbacks between boundary-layer radiative cooling, relative humidity, and cloud cover—as described in Brient and Bony (2013)—thus linking climate mean-state cloud fraction to the response of cloud fraction to warming.

In summary, because of the wide variety of MPC behavior in the current GCMs cloud optical depth feedbacks are highly uncertain. However, GCMs must have a reasonable planetary albedo. Because of this necessity, uncertainty as to the amount of liquid in MPC cover results in a counterbalancing variability in cloud area. This seesaw between cloud area and MPC liquid results in cancelation between the negative optical depth feedback in the midlatitudes and the positive cloud area feedback in the subtropics. Investigation by Zelinka et al. (2016) in CMIP3 and CMIP5 GCMs that provided ISCCP simulator output showed a 17% decrease intermodel variance in net cloud feedback due to this anti-correlation between cloud amount and optical depth feedback. Given the robustness of the positive subtropical cloud area feedback (see Section 1) it is probable that this compensation between cloud amount and optical depth feedbacks leads to an underestimation of climate sensitivity in the current generation of GCMs. In the next section we will discuss observational constraints on the MPC feedback.

3. OBSERVATIONS OF THE MPC FEEDBACK

In the previous sections we have discussed the MPC feedback in the context of climate models. Can we observe the fingerprint of the MPC feedback in the observational record?

This task is somewhat hampered by the fact that the negative optical depth feedback should occur in the high- and midlatitudes. Passive remote sensing is subject to substantial errors at low sun angles in the high-latitude wintertime (Grosvenor and Wood, 2014). Further, the longer data records offered by ISCCP (Rossow and Schiffer, 1999) and PATMOS-x (Heidinger et al., 2014) are not stable in a climate sense and must be corrected for artifacts (Norris and Evan, 2015).

Even with these observational uncertainties, can we see optical depth increasing with increasing surface temperature in the satellite record? Several studies have shown a pronounced increase in optical depth with warming over land at low temperatures (Feigelson, 1978; Genio and Wolf, 2000; Tselioudis et al., 1992) while studies over ocean regions generally indicate no covariance between warming and optical depth, or a slight decrease (Norris and Iacobellis, 2005; Tselioudis et al., 1992). Gordon and Klein (2014) demonstrated that by comparing the optical depth feedback in GCMs with the optical depth-temperature relation detected by Tselioudis et al. (1992) that the strong negative cloud feedback diagnosed by GCMs was too negative.

Is there no evidence of a substantial increase in cloud optical depth with warming over oceans? The difficulty in robustly detecting an increase in optical depth with increasing surface temperature in the observational record may reflect observational limitations, but it may also be partially due to the fact that many different mechanisms affect boundary-layer maritime cloud cover in a warming world. As noted earlier, cloud amount, and to some extent LWP, should generally decrease with enhanced surface temperature in the absence of mixed-phase transitions (Bretherton and Blossey, 2014), and it should increase due to increased boundary layer stability, which increases with surface temperature (Klein and Hartmann, 1993; Myers and Norris, 2015; Qu et al., 2014, 2015a,b; Wood and Bretherton, 2006; McCoy et al., 2017). Given the limited resolution of remote-sensing instruments, observational artifacts engendered by attempting to disentangle changes in cloud area from cloud optical depth may potentially make detecting the sensitivity of cloud albedo to temperature difficult.

Despite these issues, recent investigation directed at exploring the possibility of a negative cloud feedback due to mixed phase transitions have diagnosed a near-zero to weak increase in cloud optical depth with temperature. While these studies disagree somewhat as to the strength of the midlatitude SW cloud feedback, they agree that the most negative SW cloud feedbacks in GCMs are not consistent with the current observational record (Ceppi et al., 2016b; Terai et al., 2016).

The observationally constrained range of the Southern Ocean SW cloud feedback (including both amount and optical depth components) inferred by Ceppi et al. (2016b) is more negative than the range inferred by Terai et al. (2016), even though these studies share observational data sets. It is probable that this difference is due to systematic differences in the approaches taken by these studies to diagnosing the sensitivity of cloud optical depth to temperature. Different predictor variables may partially explain the different results arrived at by these studies. Ceppi et al. (2016b) regressed upon low- to midtropospheric temperature alone, while Terai et al. (2016) regressed upon both estimated inversion strength (EIS; Wood and Bretherton, 2006) and temperature.

Strong and nonlinear covariation between EIS and tropospheric temperature (Myers and Norris, 2015) may lead to attributing variation in optical depth and cloud cover to tropospheric temperature that are due to variation in EIS if only temperature is used as a predictor. Another possible source of disagreement between these studies is that Terai et al. (2016) focused on the optical depth of low clouds, while Ceppi et al. (2016b) investigated changes in both cloud fraction and optical depth without restricting to low clouds. (Ceppi et al., 2016b) diagnosed increases in both cloud cover and optical depth with warming leading to a negative overall SW cloud feedback. For these studies to be compared they must both be cast in terms of the SW cloud feedback as a whole. When Terai et al. (2016) replaced the optical depth portion of the SW cloud feedback in GCMs with the optical depth sensitivities that they diagnosed from observations their results were in agreement with the overall SW cloud feedback range inferred by Ceppi et al. (2016b). This is summarized in Fig. 5 for the Southern Ocean in the latitude band 45–60°S.

Ultimately, it appears the observational record is in qualitative agreement that the most negative SW cloud feedbacks predicted by GCMs are too negative (Fig. 5).

This result is consistent with the results presented in the previous section: compared to observations, GCMs generally represent MPCs as too glaciated at warm temperatures and increase LWP with warming too strongly. This too-strong dependence of LWP on temperature is corroborated by investigation of the long data record of microwave-observed LWP (O'Dell et al., 2008) shown in Ceppi et al. (2016b). The dependence of LWP on temperature derived in this study is shown in Fig. 6. This provides a complimentary analysis to studies investigating the dependence of optical depth on temperature because optical depth is a function of both droplet number concentration and liquid water path. Showing that LWP is dependent on surface temperature disentangles possible trends in cloud microphysical properties.

4. CONSTRAINT OF MIXED-PHASE PROPERTIES IN GCMs

Evidently the mixed-phase optical depth feedback is consistent with the observational record. As discussed above, the decisions that GCMs make concerning the handling of MPC cover strongly affects the negative optical depth feedback. Because of the pronounced hemispheric contrast in IN and, subsequently cloud glaciation (Hu et al., 2010; Kanitz et al., 2011; Tan et al., 2014), GCMs should have a parameterization that responds to aerosol concentrations to properly represent MPC cover. One way to pursue this is to attempt to simply create the most advanced parameterization possible, but due to the complexity of MPC microphysics this has been exceedingly difficult to accomplish. Some processes that govern the mixed–phase system simply lack any strong observational

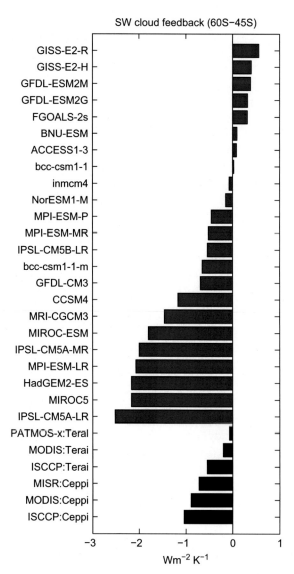

Fig. 5 SW cloud feedback from GCMs participating in CMIP5 (see Zelinka et al., 2013) compared to observationally constrained estimates. Estimates from Ceppi et al. (2016b) and Terai et al. (2016) note the instrument used. Averages are taken over the latitude band between 45°S and 60°S.

constraint and they may be thought of as a so-called tunable-parameter (Tan and Storelvmo, 2016).

Ultimately, the goal of adjusting the mixed-phase parameterization is to improve model biases in regional radiation budgets, and the global circulation (Grise et al., 2015; Kay et al., 2016; Trenberth and Fasullo, 2010). One approach that has been used

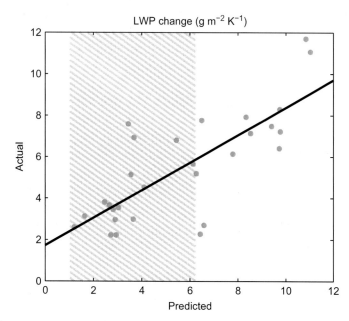

Fig. 6 LWP change in GCMs predicted by their temperature sensitivity versus their response to warming. The observational range inferred from long-term microwave measurements of LWP is indicated using cross-hatching. *(Adapted from Ceppi, P., McCoy, D.T., Hartmann, D.L., 2016b. Observational evidence for a negative shortwave cloud feedback in middle to high latitudes. Geophys. Res. Lett. 43(3), 1331–1339).*

to address uncertainty how to adjust the mixed-phase parameterization is to choose the parameters that govern MPCs in GCMs in such a way that the simulated CALIPSO supercooled liquid occurrence in the GCM matches observations (Tan and Storelvmo, 2016). In prognostic mixed-phase parameterizations there are many different factors that control the occurrence of supercooled liquid and there are many different combinations that may generate similar looking MPCs. To explore this Tan and Storelvmo (2016) utilized a quasi–Monte Carlo sampling approach to investigate how different combinations of mixed-phase parameters satisfied observational constraints on supercooled liquid occurrence. In the sensitivity analysis conducted by Tan and Storelvmo (2016) it was found that the vast majority of supercooled liquid occurrence in the CAM5.1 GCM was governed by the Wegener–Bergeron–Findeisen (WBF) process (Storelvmo and Tan, 2015). The importance of the WBF process inferred by Tan and Storelvmo (2016) is in agreement with the investigations of existing GCM parameterizations conducted by Cesana et al. (2015) and Komurcu et al. (2014), which also found that the WBF process exerted a significant control on the MPC behavior in an array of different GCMs. The version of CAM5.1 created by Tan and Storelvmo (2016) to agree best with CALIPSO was run with the fully coupled version of the model in Tan et al.

(2016) to investigate the response of the model to warming. It was found that this adjustment to bring the MPC parameterization into agreement with observed supercooled liquid occurrence raised the ECS substantially as it reduced the occurrence of glaciated cloud cover in the climate mean-state and reduced the negative midlatitude optical depth cloud feedback.

The creation of a MPC scheme that is tuned to agree with our best space-borne measures of mixed-phase behavior in a state-of-the-art GCM substantially increases the ECS within that model. What does this mean for the range on ECS offered by model intercomparison? It should be noted that many other factors determine the ECS of a given GCM. However, the increase in CESM's ECS in Tan et al. (2016)'s analysis indicates that misrepresentation of MPCs had led to an under-representation of ECS within that model. As noted earlier, in general, GCMs tend to glaciate MPCs at temperatures that are too warm in the global-mean relative to space-borne estimates (Cesana et al., 2015; McCoy et al., 2016). Tighter constraints on the mixed-phase parameterizations in these GCMs should lead to an increase in ECS in models with too little supercooled liquid as the magnitude of the MPC feedback is reduced.

5. SUMMARY

In this chapter we have discussed the negative cloud optical depth feedback that appears across GCMs in middle to high latitudes. This feedback is due to MPCs transitioning to a less glaciated state as the planet warms. The uncertainty in the MPC feedback results from the wide variety of mixed-phase parameterizations that exist in the current generation of GCMs. Models that glaciate at warmer temperatures have a larger reservoir of ice in their mixed phase cloud cover that is susceptible to warming, which transitions to liquid as the climate warms and produce a stronger negative optical depth feedback. Cloud fraction is higher in models that glaciate clouds at warmer temperatures. This appears to be a result of the fact that a good fit to the observed cloud reflectivity is a product of cloud fraction and cloud optical depth. If models have more ice and thus a lower cloud optical depth, then they must have a higher cloud fraction to produce a realistic planetary albedo. This indirect control of cloud cover by the mixed-phase parameterizations in GCMs also produce an artifact of cancelation between subtropical positive cloud feedback and midlatitude negative cloud feedback (Fig. 1).

We discuss several recent papers that use the satellite observational record to evaluate the strength of the MPC feedback. These studies agree in diagnosing a cloud feedback in the midlatitudes due to cloud optical depth changes that is either weakly negative or near zero. Overall, they agree in showing that many GCMs have SW cloud feedbacks that are too negative in the Southern Ocean (Fig. 5) (Ceppi et al., 2016b; Terai et al., 2016). We have also discussed studies that evaluate the mixed-phase temperature range in the current generation of GCMs. It was found that GCMs are generally unable to maintain supercooled liquid to low enough temperatures. Because of this GCMs generally over-represent the strength of the negative midlatitude cloud optical depth feedback (McCoy et al., 2016). This is also in agreement with evaluations made using a state-of-the-art GCM that has had its mixed-phase parameterization constrained to better agree with space-borne observations of supercooled liquid cloud occurrence (Tan et al., 2016).

The representation of MPCs in GCMs is important to the accurate prediction of 21st century climate change and to accurately represent the current climate. Overall, it is likely that this too-strong negative cloud optical depth feedback leads to an underestimation of climate sensitivity. Based on these different lines of investigation it seems clear that GCMs must carefully vet their mixed-phase parameterizations so that they agree, at least roughly, with observations of MPCs.

ACKNOWLEDGMENT

The authors wish to thank Ivy Tan for useful discussion during the preparation of this chapter.

REFERENCES

Atkinson, J.D., Murray, B.J., Woodhouse, M.T., Whale, T.F., Baustian, K.J., Carslaw, K.S., Dobbie, S., O'Sullivan, D., Malkin, T.L., 2013. The importance of feldspar for ice nucleation by mineral dust in mixed-phase clouds. Nature 498 (7454), 355–358.

Ayers, G.P., Cainey, J.M., 2007. The CLAW hypothesis: a review of the major developments. Environ. Chem. 4 (6), 366–374.

Ayers, G.P., Gras, J.L., 1991. Seasonal relationship between cloud condensation nuclei and aerosol methanesulphonate in marine air. Nature 353 (6347), 834–835.

Bender, F.A.M., 2008. A note on the effect of GCM tuning on climate sensitivity. Environ. Res. Lett. 3 (1), 014001.

Blossey, P.N., Bretherton, C.S., Zhang, M.H., Cheng, A.N., Endo, S., Heus, T., Liu, Y.G., Lock, A.P., de Roode, S.R., Xu, K.M., 2013. Marine low cloud sensitivity to an idealized climate change: the CGILS LES intercomparison. J. Adv. Model. Earth Syst. 5 (2), 234–258.

Bodas-Salcedo, A., Andrews, T., Karmalkar, A.V., Ringer, M.A., 2016. Cloud liquid water path and radiative feedbacks over the Southern Ocean. Geophys. Res. Lett. 43 (20), 10938–10946.

Bony, S., et al., 2006. How well do we understand and evaluate climate change feedback processes? J. Climate 19 (15), 3445–3482.

Bower, K.N., Moss, S.J., Johnson, D.W., Choularton, T.W., Latham, J., Brown, P.R.A., Blyth, A.M., Cardwell, J., 1996. A parametrization of the ice water content observed in frontal and convective clouds. Q. J. Roy. Meteorol. Soc. 122 (536), 1815–1844.

Bréon, F.-M., Tanré, D., Generoso, S., 2002. Aerosol effect on cloud droplet size monitored from satellite. Science 295 (5556), 834–838.

Bretherton, C.S., 2015. Insights into low-latitude cloud feedbacks from high-resolution models. Philos. Trans. R. Soc. Lond. A: Math. Phys. Eng. Sci. 373 (2054). https://doi.org/10.1098/rsta.2014.0415.

Bretherton, C.S., Blossey, P.N., 2014. Low cloud reduction in a greenhouse- warmed climate: results from Lagrangian LES of a subtropical marine cloudiness transition. J. Adv. Model. Earth Syst. 6 (1), 91–114.

Bretherton, C.S., Blossey, P.N., Jones, C.R., 2013. Mechanisms of marine low cloud sensitivity to idealized climate perturbations: a single-LES exploration extending the CGILS cases. J. Adv. Model. Earth Syst. 5 (2), 316–337.

Brient, F., Bony, S., 2013. Interpretation of the positive low-cloud feedback predicted by a climate model under global warming. Climate Dynam. 40 (9–10), 2415–2431.

Caldwell, P.M., Zhang, Y.Y., Klein, S.A., 2013. CMIP3 subtropical stratocumulus cloud feedback interpreted through a mixed-layer model. J. Climate 26 (5), 1607–1625.

Carro-Calvo, L., Hoose, C., Stengel, M., Salcedo-Sanz, S., 2016. Cloud glaciation temperature estimation from passive remote sensing data with evolutionary computing. J. Geophys. Res. Atmos. 121, 13,591–13,608. https://doi.org/10.1002/2016JD025552.

Carslaw, K.S., et al., 2013. Large contribution of natural aerosols to uncertainty in indirect forcing. Nature 503 (7474), 67–71.

Ceppi, P., Zelinka, M.D., Hartmann, D.L., 2014. The response of the Southern Hemispheric eddy-driven jet to future changes in shortwave radiation in CMIP5. Geophys. Res. Lett. 41 (9), 3244–3250.

Ceppi, P., Hartmann, D.L., Webb, M.J., 2016a. Mechanisms of the negative shortwave cloud feedback in middle to high latitudes. J. Climate 29 (1), 139–157.

Ceppi, P., McCoy, D.T., Hartmann, D.L., 2016b. Observational evidence for a negative shortwave cloud feedback in middle to high latitudes. Geophys. Res. Lett. 43 (3), 1331–1339.

Cesana, G., Waliser, D.E., Jiang, X., Li, J.L.F., 2015. Multi-model evaluation of cloud phase transition using satellite and reanalysis data. J. Geophys. Res. Atmos. 120, 7871–7892. https://doi.org/10.1002/2014JD022932.

Charlson, R.J., Lovelock, J.E., Andreae, M.O., Warren, S.G., 1987. Oceanic phytoplankton, atmospheric sulfur, cloud albedo and climate. Nature 326 (6114), 655–661.

Choi, Y.S., Ho, C.H., Park, C.E., Storelvmo, T., Tan, I., 2014. Influence of cloud phase composition on climate feedbacks. J. Geophys. Res.-Atmos. 119 (7), 3687–3700.

Clement, A.C., Burgman, R., Norris, J.R., 2009. Observational and model evidence for positive low-level cloud feedback. Science 325 (5939), 460–464.

Cober, S.G., Isaac, G.A., Korolev, A.V., Strapp, J.W., 2001. Assessing cloud-phase conditions. J. Appl. Meteorol. 40 (11), 1967–1983.

Eliasson, S., Buehler, S.A., Milz, M., Eriksson, P., John, V.O., 2011. Assessing observed and modelled spatial distributions of ice water path using satellite data. Atmos. Chem. Phys. 11 (1), 375–391.

Feigelson, E., 1978. Preliminary radiation model of a cloudy atmosphere. I—structure of clouds and solar radiation. Beitr. Phys. Atmos. 51 (3), 203–229.

Field, P.R., Heymsfield, A.J., 2015. Importance of snow to global precipitation. Geophys. Res. Lett. 42 (21), 9512–9520.

Furtado, K., 2017. Subgrid representation of mixed-phase clouds in a general circulation model. In: Andronache, C. (Ed.), Mixed-Phase Clouds: Observations and Modeling. Elsevier.

Genio, A.D.D., Wolf, A.B., 2000. The temperature dependence of the liquid water path of low clouds in the southern Great Plains. J. Climate 13 (19), 3465–3486.

Gettelman, A., Sherwood, S.C., 2016. Processes responsible for cloud feedback. Curr. Clim. Change Rep. 2 (4), 179–189.

Gordon, N.D., Klein, S.A., 2014. Low-cloud optical depth feedback in climate models. J. Geophys. Res. Atmos. 119 (10), 6052–6065.

Grise, K.M., Polvani, L.M., Fasullo, J.T., 2015. Reexamining the relationship between climate sensitivity and the Southern Hemisphere radiation budget in CMIP models. J. Climate 28 (23), 9298–9312.

Grosvenor, D.P., Wood, R., 2014. The effect of solar zenith angle on MODIS cloud optical and microphysical retrievals within marine liquid water clouds. Atmos. Chem. Phys. 14 (14), 7291–7321.

Grythe, H., Strom, J., Krejci, R., Quinn, P., Stohl, A., 2014. A review of sea-spray aerosol source functions using a large global set of sea salt aerosol concentration measurements. Atmos. Chem. Phys. 14 (3), 1277–1297.

Hamilton, D.S., Lee, L.A., Pringle, K.J., Reddington, C.L., Spracklen, D.V., Carslaw, K.S., 2014. Occurrence of pristine aerosol environments on a polluted planet. Proc. Natl. Acad. Sci. U. S. A. 111 (52), 18466–18471.

Hartmann, D.L., Short, D.A., 1980. On the use of earth radiation budget statistics for studies of clouds and climate. J. Atmos. Sci. 37 (6), 1233–1250.

Heidinger, A.K., Foster, M.J., Walther, A., Zhao, X., 2014. The pathfinder atmospheres-extended AVHRR climate dataset. Bull. Am. Meteorol. Soc. 95 (6), 909–922.

Heymsfield, A.J., Kennedy, P.C., Massie, S., Schmitt, C., Wang, Z., Haimov, S., Rangno, A., 2009. Aircraft-induced hole punch and canal clouds: inadvertent cloud seeding. Bull. Am. Meteorol. Soc. 91 (6), 753–766.

Hoose, C., Möhler, O., 2012. Heterogeneous ice nucleation on atmospheric aerosols: a review of results from laboratory experiments. Atmos. Chem. Phys. 12 (20), 9817–9854.

Hu, Y.X., Rodier, S., Xu, K.M., Sun, W.B., Huang, J.P., Lin, B., Zhai, P.W., Josset, D., 2010. Occurrence, liquid water content, and fraction of supercooled water clouds from combined CALIOP/IIR/MODIS measurements. J. Geophys. Res. Atmos. 115, 13.

Isaac, G.A., Schemenauer, R.S., 1979. Large particles in supercooled regions of northern Canadian cumulus clouds. J. Appl. Meteorol. 18 (8), 1056–1065.

Jiang, J.H., et al., 2012. Evaluation of cloud and water vapor simulations in CMIP5 climate models using NASA "A-Train" satellite observations. J. Geophys. Res. Atmos. 117 (D14), D14105.

Kanitz, T., Seifert, P., Ansmann, A., Engelmann, R., Althausen, D., Casiccia, C., Rohwer, E.G., 2011. Contrasting the impact of aerosols at northern and southern midlatitudes on heterogeneous ice formation. Geophys. Res. Lett. 38, 5.

Kay, J.E., Wall, C., Yettella, V., Medeiros, B., Hannay, C., Caldwell, P., Bitz, C., 2016. Global climate impacts of fixing the southern ocean shortwave radiation bias in the Community Earth System Model (CESM). J. Climate 29 (12), 4617–4636.

King, M.D., Menzel, W.P., Kaufman, Y.J., Tanre, D., Bo-Cai, G., Platnick, S., Ackerman, S.A., Remer, L.A., Pincus, R., Hubanks, P.A., 2003. Cloud and aerosol properties, precipitable water, and profiles of temperature and water vapor from MODIS. IEEE Trans. Geosci. Remote Sens. 41 (2), 442–458.

Klein, S.A., Hall, A., 2015. Emergent constraints for cloud feedbacks. Curr. Clim. Change Rep. 1 (4), 276–287.

Klein, S.A., Hartmann, D.L., 1993. The seasonal cycle of low stratiform clouds. J. Climate 6 (8), 1587–1606.

Klein, S.A., Hartmann, D.L., Norris, J.R., 1995. On the relationships among low-cloud structure, sea surface temperature, and atmospheric circulation in the summertime northeast Pacific. J. Climate 8 (5), 1140–1155.

Komurcu, M., Storelvmo, T., Tan, I., Lohmann, U., Yun, Y.X., Penner, J.E., Wang, Y., Liu, X.H., Takemura, T., 2014. Intercomparison of the cloud water phase among global climate models. J. Geophys. Res.-Atmos. 119 (6), 3372–3400.

Korolev, A., Isaac, G., 2003. Phase transformation of mixed-phase clouds. Q. J. Roy. Meteorol. Soc. 129 (587), 19–38.

Kruger, O., Grassl, H., 2011. Southern Ocean phytoplankton increases cloud albedo and reduces precipitation. Geophys. Res. Lett. 38L08809. https://doi.org/10.1029/2011GL047116.

Lana, A., Simo, R., Vallina, S.M., Dachs, J., 2012. Potential for a biogenic influence on cloud microphysics over the ocean: a correlation study with satellite-derived data. Atmos. Chem. Phys. 12 (17), 7977–7993.

Li, Z.-X., Le Treut, H., 1992. Cloud-radiation feedbacks in a general circulation model and their dependence on cloud modeling assumptions. Climate Dynam. 7 (3), 133–139.

Mauritsen, T., et al., 2012. Tuning the climate of a global model. J. Adv. Model. Earth Syst. 4, M00A01. https://doi.org/10.1029/2012MS000154.

McCoy, D.T., Hartmann, D.L., Grosvenor, D.P., 2014. Observed Southern Ocean cloud properties and shortwave reflection. Part II: phase changes and low cloud feedback. J. Climate 27 (23), 8858–8868.

McCoy, D.T., Hartmann, D.L., Zelinka, M.D., Ceppi, P., Grosvenor, D.P., 2015a. Mixed-phase cloud physics and Southern Ocean cloud feedback in climate models. J. Geophys. Res. Atmos. 120 (18), 9539–9554.

McCoy, D.T., Burrows, S.M., Wood, R., Grosvenor, D.P., Elliott, S.M., Ma, P.-L., Rasch, P.J., Hartmann, D.L., 2015b. Natural aerosols explain seasonal and spatial patterns of Southern Ocean cloud albedo. Sci. Adv. 1 (6), e1500157.

McCoy, D.T., Tan, I., Hartmann, D.L., Zelinka, M.D., Storelvmo, T., 2016. On the relationships among cloud cover, mixed-phase partitioning, and planetary albedo in GCMs. J. Adv. Model. Earth Syst. 8, 650–668. https://doi.org/10.1002/2015MS000589.

McCoy, D.T., Eastman, R., Hartmann, D.L., Wood, R., 2017. The change in low cloud cover in a warmed climate inferred from AIRS, MODIS, and ERA-Interim. J. Clim. 30 (10), 3609–3620.

Meskhidze, N., Nenes, A., 2006. Phytoplankton and cloudiness in the Southern Ocean. Science 314 (5804), 1419–1423.

Meskhidze, N., Nenes, A., 2010. Effects of ocean ecosystem on marine aerosol-cloud interaction. Adv. Meteorol. 2010, 13.

Mitchell, J.F.B., Senior, C.A., Ingram, W.J., 1989. CO_2 and climate – a missing feedback. Nature 341 (6238), 132–134.

Morrison, A.E., Siems, S.T., Manton, M.J., 2010. A three-year climatology of cloud-top phase over the Southern Ocean and North Pacific. J. Climate 24 (9), 2405–2418.

Morrison, H., de Boer, G., Feingold, G., Harrington, J., Shupe, M.D., Sulia, K., 2011. Resilience of persistent Arctic mixed-phase clouds. Nat. Geosci. 5 (1), 11–17.

Moss, S.J., Johnson, D.W., 1994. Aircraft measurements to validate and improve numerical model parametrisations of ice to water ratios in clouds. Atmos. Res. 34 (1–4), 1–25.

Mossop, S.C., Ono, A., Wishart, E.R., 1970. Ice particles in maritime clouds near Tasmania. Q. J. Roy. Meteorol. Soc. 96 (409), 487–508.

Murray, B.J., O'Sullivan, D., Atkinson, J.D., Webb, M.E., 2012. Ice nucleation by particles immersed in supercooled cloud droplets. Chem. Soc. Rev. 41 (19), 6519–6554.

Myers, T.A., Norris, J.R., 2013. Observational evidence that enhanced subsidence reduces subtropical marine boundary layer cloudiness. J. Climate 26 (19), 7507–7524.

Myers, T.A., Norris, J.R., 2015. On the relationships between subtropical clouds and meteorology in observations and CMIP3 and CMIP5 models. J. Climate 28 (8), 2945–2967.

Myers, T.A., Norris, J.R., 2016. Reducing the uncertainty in subtropical cloud feedback. Geophys. Res. Lett. 43, 2144–2148. https://doi.org/10.1002/2015GL067416.

Nakajima, T., Higurashi, A., Kawamoto, K., Penner, J.E., 2001. A possible correlation between satellite-derived cloud and aerosol microphysical parameters. Geophys. Res. Lett. 28 (7), 1171–1174.

Naud, C.M., Del Genio, A.D., Bauer, M., 2006. Observational constraints on the cloud thermodynamic phase in midlatitude storms. J. Climate 19 (20), 5273–5288.

Norris, J.R., Evan, A.T., 2015. Empirical removal of artifacts from the ISCCP and PATMOS x satellite cloud records. J. Atmos. Oceanic Tech. 32 (4), 691–702.

Norris, J.R., Iacobellis, S.F., 2005. North Pacific cloud feedbacks inferred from synoptic-scale dynamic and thermodynamic relationships. J. Climate 18 (22), 4862–4878.

Norris, J.R., Leovy, C.B., 1994. Interannual variability in stratiform cloudiness and sea surface temperature. J. Climate 7, 1915–1925.

Norris, J.R., Allen, R.J., Evan, A.T., Zelinka, M.D., O'Dell, C.W., Klein, S.A., 2016. Evidence for climate change in the satellite cloud record. Nature 536 (7614), 72–75.

O'Dell, C.W., Wentz, F.J., Bennartz, R., 2008. Cloud liquid water path from satellite-based passive microwave observations: a new climatology over the global oceans. J. Climate 21 (8), 1721–1739.

Qu, X., Hall, A., Klein, S., Caldwell, P., 2014. On the spread of changes in marine low cloud cover in climate model simulations of the 21st century. Climate Dynam. 42 (9–10), 2603–2626.

Qu, X., Hall, A., Klein, S., Caldwell, P., 2015a. The strength of the tropical inversion and its response to climate change in 18 CMIP5 models. Climate Dynam., 45 (1–2), 375–396.

Qu, X., Hall, A., Klein, S.A., DeAngelis, Anthony, M., 2015b. Positive tropical marine low-cloud cover feedback inferred from cloud-controlling factors. Geophys. Res. Lett. 42, 7767–7775. https://doi.org/10.1002/2015GL065627.

Quaas, J., 2012. Evaluating the "critical relative humidity" as a measure of subgrid-scale variability of humidity in general circulation model cloud cover parameterizations using satellite data. J. Geophys. Res. Atmos. 117 (D9), D09208. https://doi.org/10.1029/2012JD017495.

Rieck, M., Nuijens, L., Stevens, B., 2012. Marine boundary layer cloud feedbacks in a constant relative humidity atmosphere. J. Atmos. Sci. 69 (8), 2538–2550.

Rossow, W.B., Schiffer, R.A., 1999. Advances in understanding clouds from ISCCP. Bull. Am. Meteorol. Soc. 80 (11), 2261–2287.

Seethala, C., Norris, J.R., Myers, T.A., 2015. How has subtropical stratocumulus and associated meteorology changed since the 1980s? J. Climate 28 (21), 8396–8410.

Sekiguchi, M., Nakajima, T., Suzuki, K., Kawamoto, K., Higurashi, A., Rosenfeld, D., Sano, I., Mukai, S., 2003. A study of the direct and indirect effects of aerosols using global satellite data sets of aerosol and cloud parameters. J. Geophys. Res. 108, 4699. https://doi.org/10.1029/2002JD003359 D22.

Six, K.D., Kloster, S., Ilyina, T., Archer, S.D., Zhang, K., Maier-Reimer, E., 2013. Global warming amplified by reduced sulphur fluxes as a result of ocean acidification. Nat. Clim. Change 3 (11), 975–978.

Storelvmo, T., Tan, I., 2015. The Wegener-Bergeron-Findeisen process? Its discovery and vital importance for weather and climate. Meteorol. Z. 24 (4), 455–461.

Storelvmo, T., Kristjánsson, J.E., Myhre, G., Johnsrud, M., Stordal, F., 2006. Combined observational and modeling based study of the aerosol indirect effect. Atmos. Chem. Phys. 6 (11), 3583–3601.

Storelvmo, T., Tan, I., Korolev, A., 2015. Cloud phase changes induced by CO_2 warming—a powerful yet poorly constrained cloud-climate feedback. Curr. Clim. Change Rep. 1 (4), 1–9.

Tan, I., Storelvmo, T., 2016. Sensitivity study on the influence of cloud microphysical parameters on mixed-phase cloud thermodynamic phase partitioning in CAM5. J. Atmos. Sci. 73 (2), 709–728.

Tan, I., Storelvmo, T., Choi, Y.-S., 2014. Spaceborne lidar observations of the ice- nucleating potential of dust, polluted dust, and smoke aerosols in mixed-phase clouds. J. Geophys. Res. Atmos. 119, 6653–6665. https://doi.org/10.1002/2013JD021333.

Tan, I., Storelvmo, T., Zelinka, M.D., 2016. Observational constraints on mixed-phase clouds imply higher climate sensitivity. Science 352 (6282), 224–227.

Terai, C.R., Klein, S.A., Zelinka, M.D., 2016. Constraining the low-cloud optical depth feedback at middle and high latitudes using satellite observations. J. Geophys. Res. Atmos. 121, 9696–9716. https://doi.org/10.1002/2016JD025233.

Trenberth, K.E., Fasullo, J.T., 2010. Simulation of present-day and twenty-first-century energy budgets of the Southern Oceans. J. Climate 23 (2), 440–454.

Tselioudis, G., Rossow, W.B., Rind, D., 1992. Global patterns of cloud optical thickness variation with temperature. J. Climate 5 (12), 1484–1495.

Tsushima, Y., Emori, S., Ogura, T., Kimoto, M., Webb, M.J., Williams, K.D., Ringer, M.A., Soden, B.J., Li, B., Andronova, N., 2006. Importance of the mixed-phase cloud distribution in the control climate for assessing the response of clouds to carbon dioxide increase: a multi-model study. Climate Dynam. 27 (2–3), 113–126.

Twomey, S., 1977. Influence of pollution on shortwave albedo of clouds. J. Atmos. Sci. 34 (7), 1149–1152.

Vallina, S.M., Simó, R., 2007. Re-visiting the CLAW hypothesis. Environ. Chem. 4 (6), 384–387.

Vallina, S.M., Simo, R., Gasso, S., 2006. What controls CCN seasonality in the Southern Ocean? A statistical analysis based on satellite-derived chlorophyll and CCN and model-estimated OH radical and rainfall. Global Biogeochem. Cycles. 20(1). https://doi.org/10.1029/2005GB002597.

Vial, J., Dufresne, J.L., Bony, S., 2013. On the interpretation of inter-model spread in CMIP5 climate sensitivity estimates. Climate Dynam. 41 (11–12), 3339–3362.

Volodin, E.M., 2008. Relation between temperature sensitivity to doubled carbon dioxide and the distribution of clouds in current climate models. Izv. Atmos. Oceanic Phys. 44 (3), 288–299.

Webb, M.J., et al., 2006. On the contribution of local feedback mechanisms to the range of climate sensitivity in two GCM ensembles. Climate Dynam. 27 (1), 17–38.

Webb, M.J., Lambert, F., Gregory, J.M., 2013. Origins of differences in climate sensitivity, forcing and feedback in climate models. Climate Dynam. 40 (3–4), 677–707.

Winker, D.M., Vaughan, M.A., Omar, A., Hu, Y., Powell, K.A., Liu, Z., Hunt, W.H., Young, S.A., 2009. Overview of the CALIPSO mission and CALIOP data processing algorithms. J. Atmos. Oceanic Tech. 26 (11), 2310–2323.

Wood, R., Bretherton, C.S., 2006. On the relationship between stratiform low cloud cover and lower-tropospheric stability. J. Climate 19 (24), 6425–6432.

Zelinka, M.D., Klein, S.A., Hartmann, D.L., 2012. Computing and partitioning cloud feedbacks using cloud property histograms. Part II: attribution to changes in cloud amount, altitude, and optical depth. J. Climate 25 (11), 3736–3754.

Zelinka, M.D., Klein, S.A., Taylor, K.E., Andrews, T., Webb, M.J., Gregory, J.M., Forster, P.M., 2013. Contributions of different cloud types to feedbacks and rapid adjustments in CMIP5*. J. Climate 26 (14), 5007–5027.

Zelinka, M.D., Zhou, C., Klein, S.A., 2016. Insights from a refined decomposition of cloud feedbacks. Geophys. Res. Lett. 43 (17), 9259–9269.

CHAPTER 10

The Climatic Impact of Thermodynamic Phase Partitioning in Mixed-Phase Clouds

Ivy Tan*,[a], Trude Storelvmo*, Mark D. Zelinka[†]
*Yale University, New Haven, CT, United States
[†]Lawrence Livermore National Laboratory, Livermore, CA, United States

Contents

1. INTRODUCTION

As atmospheric CO_2 concentrations continue to rise, Earth's top of the atmosphere (TOA) radiation balance is perturbed in such a way as to ultimately result in an increase in the global mean surface temperature. The standard metric for quantifying the ultimate response of the climate system to doubled atmospheric CO_2 concentrations is the equilibrium climate sensitivity (ECS), defined as the change in the equilibrated surface temperature response to a doubling of atmospheric CO_2 concentrations. ECS estimates currently range from 2.0°C to 4.6°C and have a mean of 3.2°C across the CMIP5 models (Mauritsen and Stevens, 2015). This wide range in the ECS estimates has been attributed to disparate responses of clouds to global warming (Boucher et al., 2013).

The focus of this study is on the influence of mixed-phase clouds on ECS. Ubiquitous in Earth's atmosphere, mixed-phase clouds consist of a highly thermodynamically unstable mixture of supercooled liquid droplets and ice crystals at temperatures between 0°C

---

[a] Current affiliation: Climate and Radiation Branch, NASA Goddard Space Flight Center, Greenbelt, MD, USA.

Mixed-Phase Clouds
https://doi.org/10.1016/B978-0-12-810549-8.00010-6

and $\sim-35°C$, owing to the lower saturation vapor pressure over ice compared to that over liquid. Ice crystals will thus grow at the expense of the surrounding liquid droplets in these mixtures in a process known as the Wegener-Bergeron-Findeisen (WBF) process (Wegener, 1911; Bergeron, 1935; Findeisen, 1938). The majority of the ice crystals in mixed-phase clouds are formed with the aid of atmospheric ice nuclei (IN) that lower the temperature and humidity threshold for ice crystals to form (Pruppacher and Klett, 1997). Mineral dust is the most abundant aerosol type in Earth's atmosphere by mass (Ramanathan et al., 2001), and can nucleate ice via various heterogeneous nucleation pathways (condensation, deposition, contact, and immersion) (Vali, 1985). Secondary ice production mechanisms such as ice splintering (Hallett and Mossop, 1974) may also occur at warmer mixed-phase cloud temperatures. Mixed-phase clouds influence ECS through the fact that liquid droplets are much more abundant and smaller in size than ice crystals, which tend to be less abundant, larger in size, and precipitate faster (Sun and Shine, 1994). For a given total water path, clouds consisting of ice crystals allow more shortwave (SW) radiation to pass through and longwave (LW) radiation to escape to space than their liquid phase counterpart (Pruppacher and Klett, 1997). Thus, Earth's radiation balance will delicately depend on the thermodynamic phase partitioning of mixed-phase clouds.

One way that cloud thermodynamic phase partitioning in mixed-phase clouds impacts ECS is through the "cloud phase feedback" (Mitchell et al., 1989; Tsushima et al., 2006; Storelvmo et al., 2015; Tan et al., 2016). This negative feedback depends on the initial amount of ice present in clouds prior to CO_2 doubling—the smaller the amount of ice, the smaller the scope for cloud ice to be converted to liquid in response to warming. The cloud phase feedback occurs as a consequence of the deepening of the troposphere that shifts isotherms upward in altitude in response to global warming. This implies that there is an additional layer of liquid where ice was formerly present. The replacement of ice crystals with liquid droplets implies that more SW radiation is reflected back out to space, thereby counteracting the initial warming.

Mitchell et al. (1989) performed some of the first simulations to assess the impact of cloud thermodynamic phase partitioning in mixed-phase clouds on ECS using an atmospheric general circulation model (GCM) coupled to a slab ocean model that explicitly diagnosed the cloud phase. They found that their simulation that accounted for the thermodynamic phase of mixed-phase cloud water had an ECS of 2.7°C compared to the value of 5.2°C for their simulation that did not consider the phase of clouds and had simply diagnosed cloud cover based on a relative humidity scheme. They attributed the lower ECS estimate in the former simulation with increased reflection of SW radiation back to space owing to the cloud phase feedback. In much the same spirit, Li and LeTreut (1992) employed an atmospheric GCM that included a treatment of cloud phase at mixed-phase clouds temperatures to diagnose the impact of cloud phase partitioning on cloud feedbacks and ECS. Assuming the cloud freezing temperature to be 0°C in one case and $-15°C$ in another, they found a flip in the sign of the cloud feedback from positive to negative between the two experiments using the Cess and Potter (1988) method of evaluating cloud feedbacks. More recently, McCoy et al. (2014) estimated the impact of such thermodynamic phase transitions assuming fixed water content on upwelling shortwave radiation to be up to ~ 1 $Wm^{-2}K^{-1}$ over the Southern Ocean using a suite of satellite observations. These studies have highlighted the importance of cloud thermodynamic phase partitioning on Earth's radiation balance and ECS.

Traditionally, cloud feedbacks can be partitioned into contributions due to cloud optical depth, height, and amount (Schneider and Dickinson, 1974; Zelinka et al., 2012b). The cloud phase feedback, which is the focus of this study, is not equivalent to the cloud optical depth feedback, but is rather a subcategory within the cloud optical depth feedback that specifically pertains to how repartitioning of cloud liquid droplets and ice crystals affects mainly the reflectivity of mixed-phase clouds. Early in situ and satellite observations have noted increases in low cold cloud optical depth with temperature in the mid- and high-latitudes (Feigelson, 1978; Somerville and Remer, 1984; Tselioudis et al., 1992). These increases in cloud optical depth at the mid- and high-latitudes emerge as a robust feature among Cloud Feedback Model Intercomparison Project (CMIP) models (Zelinka et al., 2012b) and have been shown to arise from increases in low cloud liquid water content (Gordon and Klein, 2014; Ceppi and Hartmann, 2015). The role of cloud thermodynamic phase repartitioning in these increases in cloud optical depth and liquid water content is currently unclear (Gordon and Klein, 2014; Klein and Hall, 2015), although its importance has been demonstrated in a recent study that showed that the repartitioning between liquid and ice within mixed-phase clouds in the Southern Ocean in 19 models participating in Phase 5 of CMIP contributes 20%–80% of the increase in liquid water path (LWP) with global warming (McCoy et al., 2015).

In the current generation of GCMs, the partitioning of thermodynamic phase in mixed-phase clouds is uncertain and the role of mineral dust in ice nucleation through its chemical and physical properties is still unclear, primarily due to the lack of observations available. In particular, in situ observations of clouds and aerosols are naturally limited by their sparse spatial coverage. On the other hand, satellite observations offer high-resolution data on the global scale. Using satellite observations, recent studies have quantified the global distribution of cloud thermodynamic phase and have developed parameterizations of cloud thermodynamic phase based on temperature-ramp schemes (Hu et al., 2010). When implemented into an atmospheric model, the influence of the cloud thermodynamic phase parameterization was found to significantly affect cloud water contents (Cheng et al., 2012). More recently, several studies have shown that a multitude of GCMs underestimate the amount of supercooled liquid in mixed-phase clouds (Komurcu et al., 2014; Cesana et al., 2015; McCoy et al., 2016). A recent study (Tan et al., 2016) (henceforth TSZ) has taken advantage of global, vertically-resolved satellite observations of cloud thermodynamic phase available since 2006 to assess the impact of cloud thermodynamic phase partitioning on the climate and ultimately, the ECS. In Tan et al. (2016), five pairs of coupled climate simulations with full-depth oceans were run, two pairs of which were constrained by satellite observations of cloud thermodynamic phase, and all of which were run to statistical equilibrium. Here, we analyze the simulations presented in TSZ in greater depth, with a focus on the influence of LW radiation from low clouds on Arctic polar amplification and the strength of the cloud phase feedback in the Southern Ocean. A description of the simulations is provided in Section 2. This is followed by an analysis of the simulations in Section 3. Finally, a summary of this study is provided and conclusions are drawn in Section 4.

2. DESCRIPTION OF SIMULATIONS

This study employs the National Center for Atmospheric Research (NCAR)'s freely-available GCM, the Community Earth System Model (CESM1.0) (Hurrell et al., 2013) run with the 3-mode Modal Aerosol Module (MAM3) (Liu et al., 2011) in the fifth generation of the Community Atmosphere Model (CAM5.1) (Neale et al., 2012). This version of CESM is the GCM of choice for this study as the atmosphere component, CAM5, includes prognostic equations for cloud liquid and ice, a parameterization of the WBF process (Morrison and Gettelman, 2008) and a heterogeneous ice nucleation scheme. MAM3 classifies aerosols into three lognormal size distributions (Aitken, accumulation and coarse modes).

Five pairs of simulations with present-day (PD) and doubled CO_2 concentrations were run in this study, totaling 10 simulations altogether. These simulations were designed to range widely in the fraction of supercooled liquid to the total amount of liquid and ice present in their mixed-phase clouds, which will henceforth be referred to as the supercooled liquid fraction (SLF). The simulations include a control (default model) run (Control), two runs that were constrained by observations obtained by the Cloud Aerosol LIdar with Orthogonal Polarization (CALIOP) onboard the Cloud-Aerosol Lidar and Infrared Pathfinder Satellite Observations (CALIPSO) satellite ("CALIOP-SLF1" and "CALIOP-SLF2"), and two more simulations with unrealistically high (High-SLF) and low (Low-SLF) SLFs meant to serve as upper and lower bound simulations, respectively. In increasing order of global mean SLF, the simulations are: Low-SLF, Control, CALIOP-SLF1, CALIOP-SLF2, High-SLF. The five pairs of simulations are summarized in Table 1.

CALIOP is a dual-wavelength depolarization lidar that is capable of vertically measuring cloud and aerosol properties from space. It flies as part of A-train constellation in a 705-km sun-synchronous polar orbit, and thus has a 16-day repeat cycle (Winker et al., 2009). Its Nd:YAG laser produces coaligned pulses at 532 and 1064 nm, where the 532 nm pulse is linearly polarized with >99% polarization purity. Its horizontal (vertical)

Table 1 Summary of fully coupled CESM simulations listed in order of increasing global mean SLF

Simulation	General description
Low–SLF	IN concentrations increased by a factor of 75
Control	Default model
CALIOP–SLF1	Constrained by satellite observations
CALIOP–SLF2	Constrained by satellite observations
High–SLF	IN-free

resolution is 333 m (30 m) below 8.2 km, and the diameter its footprint on Earth's surface is 90 m. CALIOP contains a cloud–aerosol discrimination (CAD) algorithm (Liu et al., 2009) as well as a cloud phase discrimination algorithm (Hu et al., 2009) that relies heavily on the depolarization ratio. The two algorithms have been validated using various ground-based and aircraft observations (Liu et al., 2010; Tesche et al., 2013; Pappalardo et al., 2010; Kim et al., 2008; Hlavka et al., 2012). This study utilizes 79 months (Nov. 1, 2007 to Jun. 30, 2014) of nighttime data with medium and high confidence level CAD scores in the level 2 Vertical Feature Mask product to obtain information on cloud phase. CALIOP-observed SLFs are calculated as

$$\text{SLF} = \frac{f_{liquid}}{f_{liquid} + f_{ice}} \qquad (1)$$

where f is the number of detected layers (either liquid or ice as indicated by the subscript) identified by the Feature Classification Flag. Note that there is no mixed-phase cloud category. SLFs are calculated on isotherms using NCEP-DOE Reanalysis II data (Kanamitsu et al., 2002). The reader is referred to Tan et al. (2014) for a thorough description of the method of calculation of CALIOP-observed SLFs. To compute CAM5-modelled SLFs in a way that is consistent with the way they were computed using observations from CALIOP, SLFs are calculated by using cloud liquid and ice mixing ratios at cloud tops in place of frequency counts. SLFs were only calculated for cloud tops either without any overlying clouds or for those that only had optically thin (optical depth < 3) overlying clouds. Thus while the CALIOP observations are global in scope, they are mostly relevant to cloud tops. This limitation in lidar attenuation may be problematic for instances of where a supercooled liquid cloud with optical thickness greater than 3 overlies a precipitating ice layer as might be seen in the Arctic (see e.g., Morrison et al., 2012). In such cases, the mixed-phase clouds would be classified as liquid. Although not addressed in our study, it is possible to involve collocated radar observations from CloudSat (Marchand et al., 2008) that would likely be able to detect underlying ice clouds in such situations.

The simulations CALIOP-SLF1 and CALIOP-SLF2 were constrained by using the Quasi Monte Carlo (QMC) sampling method to select 256 combinations of six cloud microphysical parameters in CAM5.1 (Table 2), and selecting the two simulations that

Table 2 Description and values of the cloud microphysical parameters used to constrain CALIOP-SLF1 and CALIOP-SLF2

Process investigated	Relevant parameter	Default value	Investigated range	CALIOP-SLF1	CALIOP-SLF2
Fraction of dust aerosols active as IN	fin	1	[0, 0.5]	0.49	0.19
WBF process retardation factor for ice	epsi	1	$[10^{-6}, 1]$	0.024	0.80
WBF process retardation factor for snow	epss	1	$[10^{-6}, 1]$	0.024	0.80
Fraction of aerosols scavenged in stratiform clouds	sol_facti	1	[0.5, 1]	0.96	0.99
Fraction of aerosols scavenged in convective clouds	Sol_factic	0.4	[0.2, 0.8]	0.72	0.97
Related to the ice crystal fall speed	ai	700 s^{-1}	$[350, 1400] \text{ s}^{-1}$	354	371

produced SLFs that best match CALIOP-observed SLFs based on certain criteria. The primary criterion used to determine the best-matched simulations is the cumulative difference between modeled and observed SLF at the −10°C, −20°C, and −30°C isotherms. From the pool of simulations that satisfied this criterion, the next criterion was that the simulation should have substantially different values of the most influential parameters. Using the aforementioned criteria, CALIOP-SLF2 scored better than CALIOP-SLF1. A variance-based sensitivity analysis of the 256 simulations showed that out of the six parameters, the parameters representing the WBF process timescale for ice and snow (epsi and epss, respectively) and the fraction of large dust aerosols (diameter >0.5 μm) active as IN (fin) were identified as the most influential parameters for the variance in SLF. The reader is referred to Tan and Storelvmo (2016) for a thorough description of the methodology used to constrain the model with observations and the methodology of the variance-based sensitivity analysis.

In order to obtain low SLF values for the simulation Low–SLF, IN concentrations were increased by a factor of 75. High SLF values in the simulation High–SLF were obtained by eliminating all IN. In all but Control, which was run in the default configuration, the ice nucleation scheme was modified. The default parameterization of ice nucleation in the model follows Meyers et al. (1992), which parameterizes deposition and condensation freezing and implicitly assumes immersion freezing. It is, however, instead replaced by the DeMott et al. (2014) ice nucleation parameterization of immersion and deposition freezing in all simulations with the exclusion of Control. The latter scheme is preferred as it determines the concentration of large dust aerosols acting as IN, and is based on both surface-level and aircraft in situ observations obtained over a wide range of seasons and regions, as opposed to diagnosing IN based on temperature and supersaturation as in the Meyers et al. (1992) scheme. Thus, instead of parameterizing IN concentrations, N_{IN} (in units of L^{-1}), based on ice supersaturation (S_i) as

$$N_{IN} = e^{a + b(100(S_i - 1))} \qquad (2)$$

where $a = -0.639$, $b = 0.1296$, IN concentrations, N_{IN} (in units of L^{-1}) in this study is instead parameterized as

$$N_{IN} = Fn_{a>0.5\mu m}{}^{\alpha(273.16 - T_k) + \beta e^{\gamma(273.16 - T_k) + \delta}} \qquad (3)$$

where $F = 3$ is a calibration factor, $\alpha = 0.074$, $\beta = 3.8$, $\gamma = 0.414$, $\delta = -9.671$, T_k is the cloud temperature in degrees Kelvin, and $n_a > 0.5$ μm is the model-predicted number concentration of dust particles with diameters larger than 0.5 μm in units of cm^{-3}. Since the DeMott et al. (2014) scheme parameterizes immersion freezing the default parameterization for immersion freezing based on Bigg (1953) was deactivated in the simulations of this study. The default schemes for contact ice nucleation (Young, 1974) and the Hallet-Mossop process (Cotton et al., 1986) were both activated in this study.

In all but Control and Low-SLF, the detrainment scheme for shallow convection was also modified in order to bring the CAM5-modelled SLFs, which were severely underestimated relative to CALIOP observations, into reasonable agreement with these observations without interfering with any of the six cloud microphysical tuning parameters involved in the QMC sampling procedure. Modelled SLFs are compared with CALIOP SLF observations at the $-10°C$, $-20°C$, and $-30°C$ isotherms at three different latitude bands in Fig. 1 using a Taylor diagram (Taylor, 2001). The data were smoothed by averaging over 20° latitudinal bands for these calculations. Maps of the SLFs at the $-10°C$, $-20°C$, and $-30°C$ isotherms for each of the five simulations prior to CO_2 doubling are displayed in Fig. 2. Note that while CALIOP-SLF1 generally has a higher SLF compared to CALIOP-SLF2, CALIOP-SLF2 has a higher SLF than CALIOP-SLF1 at the $-10°C$.

It should be noted that any modification to the mean climate state will inevitably impact the TOA radiation balance, and possibly in such a way that a substantially different mean-state climate could result. As such, the model was retuned accordingly by changing the low cloud relative humidity threshold (RH_{crit}) in all pairs of simulations except Control in order to obtain PD climate states consistent with observations. The RH_{crit} values are provided in Table 3. A more thorough discussion of the impact of RH_{crit} on cloud feedbacks is found in Section 3.5. In addition to retuning the low cloud relative humidity threshold in CALIOP-SLF1, the ice cloud fraction calculation was also modified such that it is based on a fitted function of in-cloud ice water content obtained from in situ

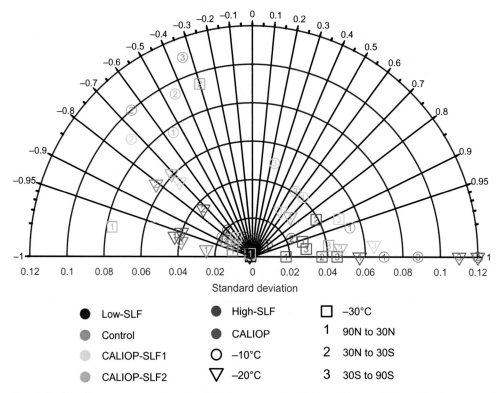

Fig. 1 Taylor diagram comparing the initial mean state SLFs at three different isotherms and three different latitude bands for the five simulations compare to CALIOP observations at the same locations.

observations of cirrus clouds (Schiller et al., 2008) instead of relative humidity (Slingo, 1987) as is done in the default model (Morrison and Gettelman, 2008).

The fully coupled simulations were run with active atmosphere (CAM5), land (CLM4), ocean (POP2) and sea ice (CICE4) components (Hurrell et al., 2013). CAM5 and CLM4 have a common horizontal resolution of 1.9° latitude by 2.5° longitude. POP2 and CICE4 are run with a Greenland displaced dipole grid at approximately one degree horizontal resolution. The number of vertical levels in CAM5 and POP2 are 30 and 60, respectively. CAM5 was run with the finite volume dynamical core. The criterion for statistical equilibrium used for each of the 10 simulations described in this study is a top-of-the-atmosphere balance of <0.3 Wm^{-2} averaged over the last 50 years of simulation.

3. RESULTS

Using the same five pairs of simulations described in Section 2, TSZ showed a progressive weakening of the cloud phase feedback with increasing SLF of the mixed-phase clouds

SLF

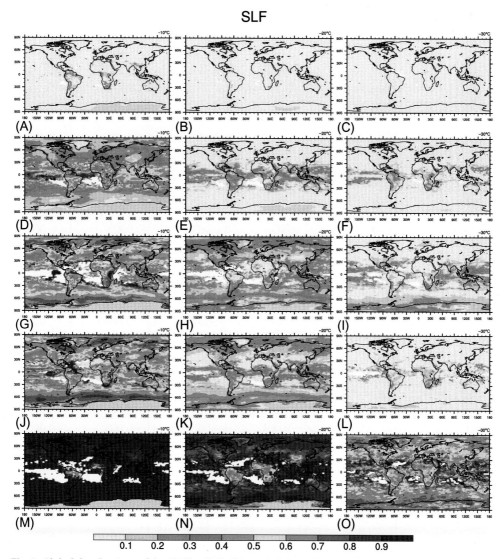

Fig. 2 Global distributions of the SLF for (A, B, C) Low-SLF, (D, E, F) Control, (G, H, I) CALIOP-SLF1, (J, K, L) CALIOP-SLF2, (M, N, O) High-SLF at the (left column) − 10°C isotherm, (middle column) − 20°C isotherm and (right column) − 30°C isotherm, with present-day CO_2 concentrations. Values were computed based on the last 50 years of each simulation after their TOA radiation budgets were balanced.

prior to CO_2 doubling (henceforth referred to as the "initial SLF"). Thus, Low-SLF exhibited the strongest cloud phase feedback, followed by Control, CALIOP-SLF1, CALIOP-SLF2 and finally High-SLF exhibited the weakest cloud phase feedback. It should be noted that the two observationally-constrained simulations, CALIOP-SLF1 and CALIOP-SLF2 have comparable SLFs overall, however, CALIOP-SLF2 has a

Table 3 Surface temperatures (T_s), liquid water paths (LWP), ice water paths (IWP), ISCCP simulator-predicted total cloud fraction (CF), SW cloud forcing, LW cloud forcing, and the relative humidity threshold for the formation of clouds used for all 10 simulations

Variable	Low-SLF	Control	CALIOP-SLF1	CALIOP-SLF2	High-SLF
T_s (°C)	16.5 (20.4)	16.8 (20.8)	13.6 (18.6)	14.7 (20.0)	15.0 (20.7)
Difference (ECS)	3.9	4.0	5.0	5.3	5.7
LWP (kg m^{-2})	0.047 (0.050)	0.045 (0.047)	0.056 (0.057)	0.056 (0.055)	0.050 (0.046)
Difference	0.003	0.002	−0.001	−0.001	−0.004
Extratropical LWP (kg m^{-2})	0.044 (0.049)	0.042 (0.045)	0.0639 (0.0636)	0.0562 (0.0564)	0.053 (0.046)
Difference	0.005	0.003	0.003	0.0002	−0.007
Tropical LWP (kg m^{-2})	0.043 (0.040)	0.042 (0.040)	0.049 (0.045)	0.052 (0.047)	0.043 (0.040)
Difference	−0.003	−0.002	−0.004	−0.005	−0.003
IWP (kg m^{-2})	0.072 (0.073)	0.016 (0.015)	0.027 (0.021)	0.025 (0.019)	0.0060 (0.0049)
Difference	0.001	−0.001	−0.0060	−0.0060	−0.0011
Extratropical IWP (kg m^{-2})	0.085 (0.083)	0.018 (0.016)	0.039 (0.029)	0.032 (0.023)	0.0057 (0.0047)
Difference	−0.002	−0.002	−0.01	−0.009	−0.001
Tropical IWP (kg m^{-2})	0.0454 (0.0449)	0.012 (0.011)	0.019 (0.016)	0.023 (0.018)	0.0054 (0.0047)
Difference	−0.0005	−0.001	−0.003	−0.005	−0.0007
CF (%)	59.5 (59.1)	52.7 (51.6)	57.5 (56.8)	62.3 (59.2)	62.0 (59.0)
Difference	−0.4	−1.1	−0.70	−3.1	−3.0
Extratropical CF	65.2 (65.0)	59.5 (58.5)	68.0 (65.6)	69.0 (65.7)	71.5 (66.9)
Difference	−0.2	−1.0	−2.4	−3.3	−4.6
Tropical CF	46.2 (44.5)	39.8 (38.2)	42.5 (41.5)	48.5 (44.4)	46.4 (44.7)
Difference	−1.7	−1.6	−1.0	−4.1	−1.7
SWCF (Wm^{-2})	−67.2 (−54.1)	−50.8 (−49.2)	−58.2 (−55.8)	−62.8 (−57.5)	−62.0 (−56.7)
Difference	0.2	1.6	2.4	5.3	5.3
Extratropical SWCF (Wm^{-2})	−60.0 (−60.1)	−46.2 (−44.7)	−60.0 (−56.2)	−58.3 (−53.1)	−61.2 (−54.2)
Difference	−0.1	1.5	3.8	5.2	7.0
Tropical SWCF (Wm^{-2})	−56.6 (−54.1)	−43.2 (−41.0)	−50.7 (−47.7)	−55.7 (−49.1)	−51.9 (−48.6)
Difference	2.5	2.2	3.0	6.6	3.3
LWCF (Wm^{-2})	38.6(39.1)	23.7 (22.6)	25.4 (25.2)	30.9 (28.2)	31.1 (28.9)
Difference	0.5	−1.1	−0.2	−2.7	−2.2

Table 3 Surface temperatures (T_s), liquid water paths (LWP), ice water paths (IWP), ISCCP simulator-predicted total cloud fraction (CF), SW cloud forcing, LW cloud forcing, and the relative humidity threshold for the formation of clouds used for all 10 simulations—cont'd

Variable	Low-SLF	Control	CALIOP-SLF1	CALIOP-SLF2	High-SLF
Extratropical LWCF (Wm^{-2})	35.6 (36.9)	21.4(21.0)	27.9 (26.6)	29.0 (27.3)	33.5 (31.1)
Difference	1.3	−0.4	−1.3	−1.7	−2.4
Tropical LWCF (Wm^{-2})	30.6 (29.9)	18.3 (17.1)	19.2 (19.0)	24.7 (21.4)	25.2 (23.1)
Difference	−0.7	−1.2	−0.2	−3.3	−2.1
RH_{crit}	0.8875	0.8875	0.8725	0.8475	0.86

Simulations with present-day CO_2 concentrations are on the left side and simulations with doubled CO_2 concentrations are in parentheses. Their differences (doubled CO_2 minus PD) are given in the line below. The values are globally averaged unless otherwise noted.

higher SLF at the −10°C isotherm especially over the Southern Ocean. This section presents a more detailed analysis of the simulations than was provided in TSZ.

3.1 Surface Temperature

Zonal averages of the differences in surface temperature between the doubled CO_2 and PD CO_2 states as a function of latitude are shown in Fig. 3 for the five pairs of simulations. All of the simulations warm after CO_2 doubling. The global averages are displayed in Table 3. In general, the zonally averaged surface temperature increases with increasing SLF in the Southern Hemisphere. This is particularly consistent for the regions in and around the Southern Ocean (40° to 75°S). In fact, a strong linear correlation

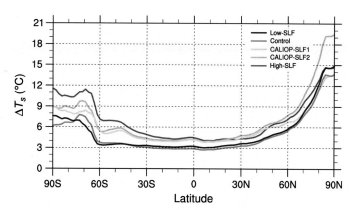

Fig. 3 Difference in zonal mean surface temperature between the doubled and the present-day CO_2 states as a function of latitude.

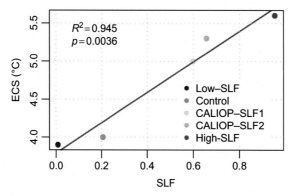

Fig. 4 Linear regression of ECS on extratropical SLF.

($R=0.98$, $p=0.0036$) exists between the PD mean (50-year average after equilibrium has been reached) extratropical (poleward of 60°) SLF and ECS (Fig. 4). Simulated climate states with lower initial mean extratropical SLFs also have lower ECS values. This behavior is consistent with a weakening of the negative cloud phase feedback in simulations with higher SLFs in their initial mean state. It is noteworthy that the satellite constrained simulations have ECS values that are closer to the extreme case, High-SLF, than to the control case. Global distributions of the difference in surface temperature between the doubled CO_2 and PD states reveal that the warming is stronger over the Northern Hemisphere continents, as expected (Fig. 5).

3.2 Hydrological Cycle

Zonal averages of the differences in precipitation minus evaporation (*P-E*) between the doubled CO_2 and PD CO_2 states as a function of latitude are shown in Fig. 6 (solid lines) for the five pairs of simulations along with the Clausius-Clapeyron-scaled *P-E* responses (dashed lines) (Held and Soden, 2006). The expected "wet gets wetter and dry gets drier" response to doubled CO_2 is evident from Fig. 6A. Also apparent for each simulation is the southward shift of the Intertropical Convergence Zonal (ITCZ) in all of the simulations relative to the thermodynamic response represented by Clausius-Clapeyron scaling. It is possible that the southward shift of the ITCZ is associated with the stronger positive cloud feedback in the southern hemisphere (Fig. 9). Kang et al. (2008) showed that ITCZ displacement towards the southern hemisphere can result from asymmetrically heating the extratropics of the southern hemisphere more relative to the northern hemisphere. The southward shift of the ITCZ was shown to be highly sensitive to a parameter in the convection scheme that was intimately linked to cloud feedbacks.

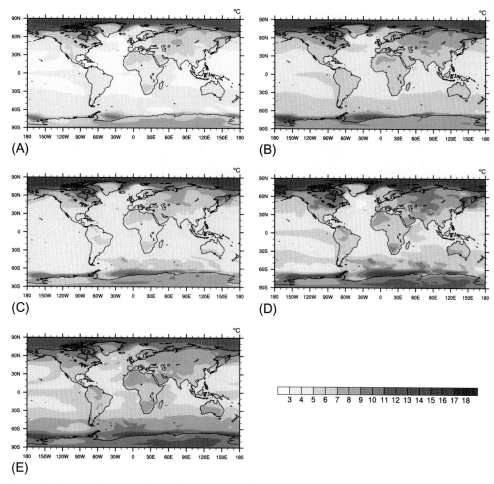

Fig. 5 Global distributions of the difference in surface temperature (in °C) between the doubled and the present-day CO_2 states. (A) Low-SLF. (B) Control. (C) CALIOP-SLF1. (D) CALIOP-SLF2. (E) High-SLF.

3.3 Climate Feedbacks

Surface temperature-mediated responses of the climate system play the important role of amplifying (positive feedback) or damping (negative feedback) the initial surface temperature response to CO_2 forcing. Through the use of radiative kernels (Soden et al., 2008; Shell et al., 2008; Zelinka et al., 2012a,b), these feedbacks can be separated into non–cloud (Planck, water vapor, lapse rate, and albedo) feedbacks and cloud (amount, optical depth, and height) feedbacks in terms of their contributions to both SW, LW, and net (SW plus LW) radiation. Thorough descriptions of the methodology and development of the radiative kernel technique are presented elsewhere (Soden et al., 2008; Shell et al., 2008;

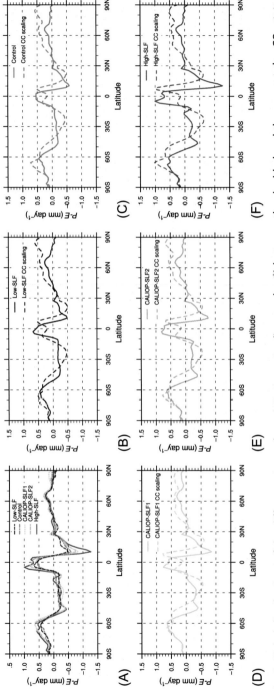

Fig. 6 Global distributions of the difference in total precipitation rates (in mm day *) between the doubled and the present-day CO$_2$ states.

Zelinka et al., 2012a,b). The basic idea of the radiative kernel technique is to decompose the change in the TOA radiation balance, R, with surface temperature, T_s, known as the climate feedback parameter, λ, into contributions from various climate variables. Since R is a function of T_s through the dependence of various other climate variables (labeled generically as x_i) of T_s, the chain rule thus allows λ to be written as

$$\lambda = \frac{dR}{d\overline{T}_s} = \sum_i \frac{\partial R}{\partial x_i} \frac{dx_i}{d\overline{T}_s}. \tag{4}$$

Here, $\overline{T}_s$ refers to the global mean surface temperature, the climate variables x_i refer to T_s, atmospheric temperature, T, specific humidity, q, surface albedo, α, and cloud fraction. The multiplicative term, $\dfrac{\partial R}{\partial x_i}$ is referred to as the "radiative kernel," which is in practice, computed separately using an offline radiative transfer model. Feedbacks are thus computed by multiplying the radiative kernels by changes in the climate variables of interest and then normalizing by the global mean surface temperature. Global mean feedbacks are computed by summing over all three spatial dimensions. In this way, the always-negative Planck feedback, λ_P, which damps the CO_2-induced surface warming by emitting LW radiation from the heated surface out to space, can be written as

$$\lambda_P = \frac{\partial R}{\partial T_s} \frac{dT_s}{d\overline{T}_s} + \frac{\partial R}{\partial T} \frac{dT_s}{d\overline{T}_s}. \tag{5}$$

The lapse rate feedback, λ_{LR}, which quantifies the vertical atmospheric temperature response to CO_2-induced surface warming is computed as

$$\lambda_{LR} = \frac{\partial R}{\partial T} \frac{dT}{d\overline{T}_s} - \frac{\partial R}{\partial T} \frac{dT_s}{d\overline{T}_s}. \tag{6}$$

The water vapor feedback, λ_{wv}, is computed as

$$\lambda_{WV} = \frac{\partial R}{\partial \ln q} \frac{d \ln q}{d\overline{T}_s}. \tag{7}$$

Note that the natural logarithm of q here is taken since the absorption of radiation by water vapor is proportional to the natural logarithm of specific humidity. Similarly, the surface albedo feedback is defined as

$$\lambda_\alpha = \frac{\partial R}{\partial \alpha} \frac{d\alpha}{d\overline{T}_s}. \tag{8}$$

Two different sets of radiative kernels were downloaded and employed. One set of radiative kernels was computed using the offline radiative transfer model version of version 3 of CAM (Shell et al., 2008) in order to compute non-cloud feedbacks. At each latitude, longitude and height gridpoint, and for each month, Shell et al. (2008) perturbed each variable of interest for the feedback one level at a time corresponding to a 1 K temperature change under PD conditions, saving the clear and all-sky radiative flux changes each time. The cloud feedback was calculated separately as the adjusted cloud radiative forcing, ΔCRF. Computing the cloud feedback as the ΔCRF involves adjusting the difference between all-sky and clear-sky radiative forcings by the radiative forcing masked by other non-cloud feedbacks. The cloud feedback computed in this manner will henceforth be referred to as the *adjusted cloud feedback*. Thus,

$$\Delta CRF = \Delta\left((R-R_c)_P + (R-R_c)_{LR} + (R-R_c)_{WV} + (R-R_c)_\alpha + (F-F_c)_{CO_2}\right), \quad (9)$$

where $(F-F_c)_{CO2}$ is the difference between all-sky and clear-sky CO_2 forcing.

The other set of radiative kernels used for analysis are cloud radiative kernels that were computed specifically for cloud feedbacks. This method essentially entails using the Fu-Liou radiative transfer model (Fu and Liou, 1992) to calculate the SW and LW TOA radiative forcing for a single cloud layer covering 100% of the column for 49 different cloud types classified according to seven different cloud optical depth and seven different cloud top pressure categories used by the International Satellite Cloud Climatology Project (ISCCP) simulator (Zelinka et al., 2012a). Changes in cloud fraction histograms after CO_2 doubling, output from the ISCCP simulator implemented in CAM5, were then manipulated to yield cloud fraction histograms changes by separately holding either cloud amount, cloud optical depth or cloud top pressure constant following the method of Zelinka et al. (2012b). These three types of cloud fraction histograms were then multiplied by the cloud radiative kernels and divided by the global mean surface temperature to obtain the total LW and SW cloud amount, optical depth, and height feedbacks.

Climate feedbacks calculated using the kernels by Shell et al. (2008) are displayed in Fig. 7. A noteworthy feature of the Planck feedback (Fig. 7A) is that it decreases monotonically with increasing SLF in the Southern Ocean region. The Planck feedback refers to the additional emission of LW radiation with increasing temperature. The LW radiation emitted per unit surface area follows the well-established Stefan-Boltzmann law, $F = \varepsilon \sigma T_s^4$, where ε is the emissivity of the surface, σ is the Stefan-Boltzmann constant, and T_s is the surface temperature.

The positive albedo feedback (Fig. 7B) is predominantly related to sea ice in the polar regions, with the feedback being much stronger in the northern hemisphere polar region than the southern hemisphere polar region. Polar projections of the Arctic show the corresponding losses in sea ice (Fig. 8). CALIOP-SLF1 stands out as it shows the smallest relative loss in sea ice of the five cases, and this is reflected in its much weaker albedo feedback in the Arctic. The strongly opposing negative Planck and positive albedo feedbacks largely cancel each other in the Arctic. Climate response in the Arctic to cloud thermodynamic phase changes will be topic of future research by the authors.

The positive water vapor feedback (Fig. 7C) arises from the greenhouse effect of water vapor. It is the strongest feedback in the tropics where the troposphere is close to saturation.

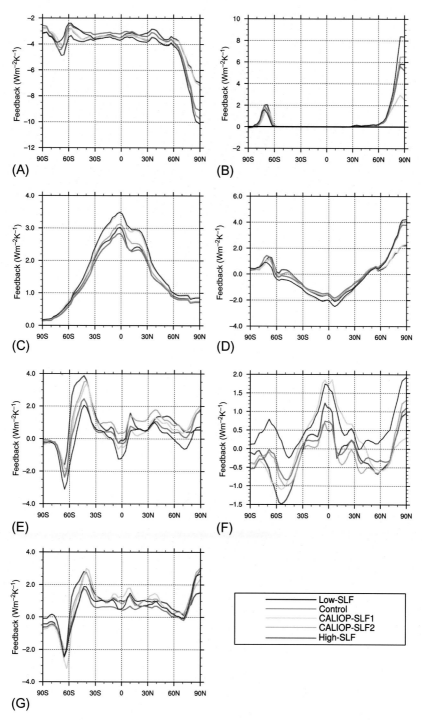

Fig. 7 Zonal mean (A) Planck, (B) albedo, (C) water vapor, (D) lapse rate, (E) SW cloud, (F) LW cloud, and (G) net cloud feedbacks computed using radiative kernels computed using CAM3 (Shell et al., 2008).

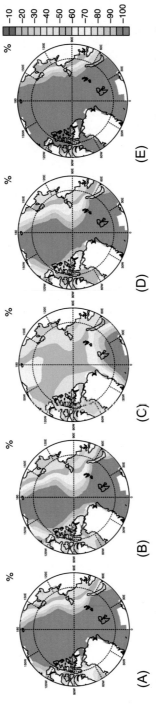

Fig. 8 Arctic projections of the relative loss in surface area covered by sea ice (expressed as a percentage) for the five pairs of simulations. (A) Low-SLF. (B) Control. (C) CALIOP-SLF1. (D) CALIOP-SLF2 (E) High-SLF.

The lapse rate feedback (Fig. 7D) is negative in the low latitudes and positive in the high latitudes. The release of latent heat via convection in the low latitudes causes more warming of the upper troposphere and thus eventually cools earth more through the emission of LW radiation back to space than a uniform heating profile would otherwise. The reduction of the moist adiabatic lapse rate in the low latitudes thus leads to a negative feedback. At high latitudes, stable stratification prevents the mixing of air near the surface with air aloft. This lack of coupling between the surface and upper troposphere implies that a larger increase in surface temperature in the Arctic is required to offset the TOA imbalance due to CO_2 forcing. Thus the lapse rate feedback is positive at high latitudes. The lapse rate feedback plays an important role in polar amplification. It has been previously shown to account for as much as 15% of Arctic amplification and 20% of Antarctic polar amplification (Graversen et al., 2014). Using radiative kernels computed from a surface perspective, it has also been shown that the lapse rate feedback is the largest contributor to Arctic amplification when one also considers the fact that the lapse rate feedback is locally negative in the tropics (PIthan and Mauritsen, 2014).

The adjusted SW cloud feedbacks (Fig. 7E) show a monotonic increase with increasing SLF in the extratropics (approximately 40°–70°) in both hemispheres. The adjusted SW cloud feedback is positive over almost all regions for all of the simulations, except approximately over the Southern Ocean (50°S–70°S), where it is negative. An increasingly strong cloud phase feedback with decreasing SLF implies that the total SW cloud feedback should become increasingly negative due to the replacement of portions of ice clouds with portions of optically thicker liquid clouds. In fact, the simulations with higher initial SLFs do appear to have increasingly positive (albeit still negative) SW cloud feedbacks, and this is consistent with the fact that higher initial SLFs result in weaker cloud phase feedbacks. Partitioning of the SW cloud feedback into the optical depth, amount, and altitude feedbacks using the cloud radiative kernel technique is presented below and further elucidate the role of the cloud phase feedback. The adjusted LW cloud feedback (Fig. 7F), on the other hand, would exhibit the opposite effect. Over the Southern Ocean, the adjusted LW cloud feedback would become increasingly positive because the optically thicker warm clouds also reduce the emission of LW radiation to space. It is interesting to note that the ordering of the LW feedback begins to switch moving toward the North Pole and eventually reverses at the high northern latitudes, excluding CALIOP-SLF2. This behavior is likely linked to cloud amount changes in the Arctic and is currently under investigation in a follow-up study. Overall, the mostly positive SW cloud feedback overweighs the mostly negative LW cloud feedback so that the net adjusted cloud feedback is mostly positive except over the Southern Ocean.

The SW cloud feedback computed using the cloud radiative kernel method (Fig. 9E) compares fairly well overall to the adjusted SW cloud feedback. The overall relative magnitudes and signs of the feedbacks for the simulations in relation to one another are comparable to the adjusted SW cloud feedback calculations. The same monotonic increase

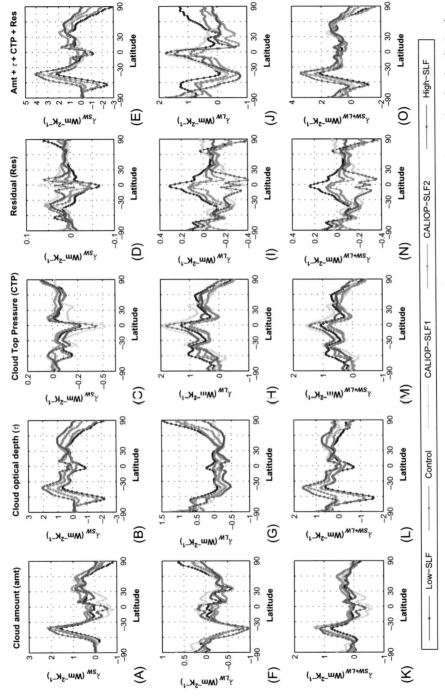

Fig. 9 Zonal mean cloud (A, F, K) amount, (B, G, L) optical depth, and (C, H, M) height feedbacks computed using the cloud radiative kernel method (Zelinka et al., 2012a,b). The (D, I, N) residual term and (E, J, O) total cloud feedbacks are displayed in the second to rightmost and rightmost columns, respectively. The top, middle, and bottom rows display the contributions of the SW, LW, and net (SW+LW) radiation, respectively, of each term (cloud amount, optical depth, height, residual, and total).

with increasing SLF over the extratropics (approximately 40°–70°) is seen when using both methods, however, there is one striking difference resulting from the two methods. While the adjusted SW cloud feedback is mostly positive and increasing in the Arctic, it decreases to more negative values when computed using the cloud radiative kernel method. This difference can be attributed to the fact that while the cloud radiative kernel method computes cloud feedbacks directly from satellite simulator outputs of cloud fraction, the adjusted cloud feedback requires adjustment due to the effect due to the cloud masking effect of non-cloud feedbacks. Adjustment due to non-cloud feedbacks may be problematic over regions with large surface albedo changes, such as the Arctic (Zelinka et al., 2012a). Thus the SW cloud feedback computed using the radiative kernel method is likely to be more accurate. Even outside the extratropics, the SW cloud feedbacks computed using the two methods closely resemble one another in terms of both sign and magnitude.

In addition to the total SW cloud feedback, the contributions from the horizontal cloud amount (Fig. 9A), optical depth (Fig. 9B), cloud top pressure (Fig. 9C), and residual (Fig. 9D), computed using the cloud radiative kernel method are also displayed. The net SW cloud feedback closely resembles the optical depth feedback. The SW cloud amount feedback also appears to contribute substantially to the net SW cloud feedback. Taken together, these two facts, in consideration with the fact that both feedbacks are positive over most latitudes excluding the Arctic, suggest that clouds may be becoming both thinner and sparser in spatial coverage with global warming, which therefore amplifies global warming. However, the magnitude of the positive overall cloud feedback appears to be extremely sensitive to cloud phase prior to CO_2 doubling. It is noteworthy that the cloud radiative kernel method predicts a more positive SW cloud feedback approximately over the Southern Ocean (50°S–70°S) than the adjusted SW cloud feedback.

The LW cloud feedback is essentially the reverse of the SW cloud feedback. The LW cloud feedback is negative over regions where the SW cloud feedbacks are positive and vice versa. A positive (negative) SW cloud amount feedback results from a reduction (increase) in the horizontal coverage of clouds after global warming, allowing for more sunlight to penetrate to earth's surface. This leads to a negative (positive) LW feedback at the same time because it reduces (increases) the downward LW radiation emitted back to the surface. A positive (negative) SW optical depth feedback results from a reduction (increase) in the optical depth of clouds, which also allows more sunlight to penetrate to the earth's surface. This leads to a negative (positive) LW cloud optical depth feedback at the same time because it reduces (increases) the greenhouse effect of the clouds. A positive (negative) LW cloud top pressure (CTP) or equivalently, altitude feedback results from an increase in the altitude of clouds, which acts to increase (decrease) the greenhouse effect of the clouds by decreasing (increasing) the amount of LW radiation emitted back to space. The SW altitude feedback is much smaller in magnitude compared to the LW altitude feedback, which is very sensitive to changes in cloud top temperatures.

Negative (positive) values of the SW altitude feedback occur when clouds increase (decrease) in altitude because higher clouds absorb less SW radiation than lower clouds due to their lower water content.

3.4 Water Paths, Cloud Fraction, and Cloud Forcings

The liquid and ice water paths, cloud fractions, and SW and LW cloud forcings for the five pairs of simulations are displayed in Table 3. The changes in the LWPs of Low-SLF and Control after global warming are positive. These increases in LWPs are dominated by contributions from the extratropics (poleward of 30°) that are likely responsible for the negative extratropical optical depth feedbacks (Fig. 9). Coupled with the fact that the IWPs for these (and all other) simulations are mostly negative, a strong cloud phase feedback is occurring in Low-SLF and Control. The LWPs in the tropics for all simulations are consistently negative across all simulations.

In contrast with Low-SLF and Control, the simulations that have shown a much higher degree of warming, CALIOP-SLF1, CALIOP-SLF2, and High-SLF, all show LWP decreases in the tropics and in the global average with increasing SLFs. These three simulations also have positive extratropical optical depth feedbacks (Fig. 9) that are caused by a weakened cloud phase feedback.

Cloud fraction decreases both in the extratropics and the tropics for all of the simulations. These decreases in cloud cover contribute to a positive cloud amount feedback. Decreases in cloud fraction roughly between 55°S and 60°N are a robust feature seen in many climate models (Zelinka et al., 2012b). Consistent with this finding, the decreases in cloud fraction in the extratropics shown in Table 3 are dominated by decreases in cloud fraction equatorward of 70° in both hemispheres (not shown).

The increase in shortwave cloud forcings (SWCFs) in the extratropics are consistent with what one would expect for simulations exhibiting a monotonic weakening with increasing SLFs prior to global warming. The replacement of liquid droplets with ice crystals in mixed-phase clouds that normally occurs in the cloud phase feedback would result in a more negative SW cloud forcing after global warming. However, when the cloud phase feedback weakens, the opposite effect is expected. Indeed, the SWCFs in the extratropics, where the cloud phase feedback is more pronounced, gradually decrease with increasing initial SLFs in the simulations (Table 3). The longwave cloud forcing (LWCF) shows the opposite effect, i.e. a steady increase in a cooling effect with increasing initial SLF in the extratropics. There appears to be no clear pattern in the SWCF and LWCF in the tropics.

3.5 Robustness of Results to Retuning

When improvements to certain aspects of GCMs are implemented, usually related to parameterizations, there is often a need to retune the model if the resulting TOA radiation imbalance becomes too large, in order to avoid climate drift (Mauritsen et al., 2012). Although there are different methods of retuning the TOA radiation imbalance, the typical method of retuning the TOA radiation imbalance is to retune cloud-related parameters (Mauritsen et al., 2012; McCoy et al., 2016). As previously mentioned in Section 2, in perturbing the six microphysical parameters in the two CALIOP-constrained simulations to obtain SLFs that better agree with CALIOP observations (Table 2), the TOA radiation imbalance was reduced by tuning RH_{crit}, the relative humidity threshold for the formation of low clouds (Table 3). This was achieved by adjusting RH_{crit} such that the globally averaged outgoing LW radiation and absorbed SW radiation of the simulations are in better agreement with the present-day observed values of 235 Wm^{-2} (Trenberth et al., 2009). After retuning RH_{crit}, the simulations were found to be off balance by <0.2 Wm^{-2} for both simulations.

To determine to what extent retuning RH_{crit} influences the simulations, the five pairs of simulations were run once more in the Atmosphere Model Intercomparison Project (AMIP) configuration with prescribed SSTs and sea ice. AMIP4K simulations (Gates, 1992), where SSTs were increased uniformly by 4 K were also run as the analogue for the doubled CO_2 simulations. Cloud feedbacks were then calculated once again using the average over the years 1980–2006 (Fig. 10). The AMIP/AMIP4K simulations are used as a method to gauge the influence of RH_{crit} on the cloud feedbacks since no retuning is required to rebalance the TOA energy budget.

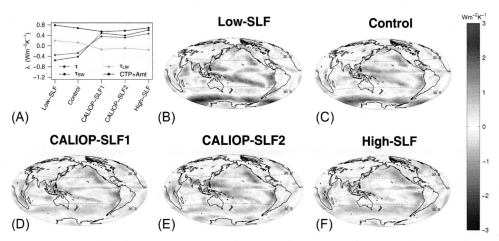

Fig. 10 Cloud feedbacks computed using AMIP4k simulations. (A) Mean net; extratropical optical depth feedback (red), LW optical depth feedback (green), SW optical depth feedback (blue), and sum of the altitude and amount feedbacks (black) for each of the five simulations, shown in order of increasing SLF. (B) Global distribution of the net optical depth feedback for Low-SLF. (C) Global distribution of the net optical depth feedback for Control. (D) Global distribution of the net optical depth feedback for CALIOP-SLF1. (E) Global distribution of the net optical depth feedback for CALIOP-SLF2. (F) Global distribution of the net optical depth feedback for High-SLF.

The results of the AMIP simulations shown in Fig. 10 should be compared with Fig. 3 in TSZ. Many qualitative similarities between the two sets of simulations exist. The same increase in the net extratropical SW optical depth feedback is observed to dominate the changes in the net extratropical optical depth feedback, while the net extratropical LW optical depth feedback and the sum of the net extratropical amount and altitude feedbacks stay relatively constant for all of the simulations. One can thus conclude that retuning using RH_{crit} in the coupled simulations to avoid climate drifts likely does not impact the final conclusions of TSZ that are centrally based on cloud feedback calculations.

4. SUMMARY AND CONCLUSIONS

Additional analysis and new simulations following up on TSZ were presented. Of particular importance are the additional climate feedback analyses and AMIP simulations presented herein. Although there are regional differences, the adjusted cloud feedback generally compares favorably with the cloud feedbacks computed using cloud radiative kernel method in terms of both the magnitude and sign of the feedbacks. The only striking difference between the two methods occurs in the Arctic, where the large surface albedo due to sea ice must be taken into account and adjusted for when applying the adjusted cloud feedback method.

The same simulations run in the AMIP/AMIP4K configuration were used to gauge the influence of the parameter, RH_{crit} that was tuned to avoid climate drifts associated with TOA imbalances arising from adjusting the six cloud microphysical tuning parameters in order to obtain SLFs more comparable to that of CALIOP. Cloud feedback calculations using the radiative kernel method for the AMIP/AMIP4K simulations were qualitatively similar to those of TSZ. Thus although RH_{crit} is likely to have influenced the results to some extent, the main conclusions presented in TSZ are robust to the tuning of RH_{crit}.

It is worth noting that the extratropical net and SW optical depth feedbacks of the two observationally constrained simulations presented herein and in TSZ are distinctly positive (Fig. 9A). A comparison with recent studies reveals that there is still a high degree of uncertainty regarding the sign of the extratropical optical depth feedback. The positive extratropical SW optical depth feedback suggested by TSZ and the AMIP4K simulations presented herein are in agreement with the results of Gordon and Klein (2014) who demonstrated that the extratropical SW optical depth feedback for low clouds in models is likely too negative. Terai et al. (2016), who extended the analysis of Gordon and Klein (2014), used a combination of satellite retrievals to constrain an ensemble of AMIP models using the timescale invariant property related to the change in optical depth with temperature change described in Gordon and Klein (2014). In agreement with the positive extratropical SW optical depth feedback found in this study and TSZ, Terai et al. (2016) found that when replacing the timescale invariant parameter related to optical depth sensitivity in the models with that of the satellites, the negative SW optical depth feedback is reduced by at least 50% in part of the extratropics. However, when analyzing Phase 5 of the Cloud Model Intercomparison Project (CMIP5) models and satellite observations separately, Ceppi et al. (2016) found that both models and satellite observations suggest a negative extratropical SW optical depth feedback.

Given the uncertainty in the SW optical depth feedback in the extratropics, future efforts should focus on better understanding the processes involved in the SW feedback. For a comprehensive review of the current state of knowledge of the extratropical SW cloud optical depth feedback, the reader is referred to McCoy et al. (2017). The results of this study and TSZ suggest that cloud thermodynamic phase plays an important role in the SW optical depth feedback in the extratropics, and ultimately influences climate change. Continuing efforts to improve the representation of cloud phase in climate models with the aid of global observations will be important moving forward.

ACKNOWLEDGMENTS

This work was supported by NASA Headquarters under the NASA Earth and Space Science Fellowship Program—Grant NNX14AL07H. We would also like to acknowledge high-performance computing support from Yellowstone provided by NCAR's Computational and Information Systems Laboratory, sponsored by the National Science Foundation under Grant 1352417. The effort of M.D.Z. was supported by the Regional and Global Climate Modeling Program of the Office of Science at the US Department of Energy (DOE) under Grant DE-SC0012580 and was performed under the auspices of DOE by Lawrence Livermore National Laboratory under Contract DE-AC52-07NA27344. IM release #LLNL-JRNL-727698. CALIOP data are available online at https://eosweb.larc.nasa.gov/order-data. NCEP-DOE Reanalysis 2 data are also available online at http://www.ersl.noaa.gov/psd/data/gridded/data.ncep.reanalysis2.pressure.html. The authors would like to thank Stephen A. Klein at Lawrence Livermore National Laboratory for helpful suggestions.

REFERENCES

Bergeron, T., 1935. Proces Verbaux de l'Association de Meteorologie. International Union of Geodesy and Geophysics, Paris, France.

Bigg, E.K., 1953. The supercooling of water. Proc. Phys. Soc. 66B (8), 688–694.

Boucher, O., et al., 2013. Clouds and aerosols. In: Climate Change 2013: The Physical Science Basis. Contribution of Working Group I to the Fifth Assessment Report of the Intergovernmental Panel on Climate Change.

Ceppi, P., Hartmann, D.L., 2015. Connections between clouds, radiation, and midlatitude dynamics: a review. Curr. Clim. Change Rep. 1 (2), 94–102.

Ceppi, P., McCoy, D.T., Hartmann, D.L., 2016. Observational evidence for a negative shortwave cloud feedback in middle to high latitudes. Geophys. Res. Lett. 43 (3), 1331–1339.

Cesana, G., Waliser, D.E., Jiang, X., Li, J.L., 2015. Multimodel evaluation of cloud phase transition using satellite and reanalysis data. J. Geophys. Res. Atmos. 120 (15), 7871–7892.

Cess, R.D., Potter, G.L., 1988. A methodology for understanding and intercomparing atmospheric climate feedback processes in general circulation models. J. Geophys. Res. 93 (D), 8305–8314.

Cheng, A., Xu, K., Hu, Y., Kato, S., 2012. Impact of a cloud thermodynamic phase parameterization based on calipso observations on climate simulation. J. Geophys. Res. Atmos. 117 (D09103).

Cotton, W.R., Tripoli, G.J., Rauber, R.M., Mulvihill, E.A., 1986. Numerical simulation of the effects of varying ice crystal nucleation rates and aggregation processes on orographic snowfall. J. Appl. Meteorol. 25 (11), 1658–1680.

DeMott, P.J., et al., 2014. Integrating laboratory and field data to quantify the immersion freezing ice nucleation activity of mineral dust particles. Atmos. Chem. Phys. 15 (1), 393–409.

Feigelson, E.M., 1978. Preliminary radiation model of a cloudy atmosphere. Part I: structure of clouds and solar radiation. Beitr. Phys. Atmos. 51, 203–229.

Findeisen, W., 1938. Kolloid-meteorologische vorgange bei niederschlags-bildung. Meteor. Z. 55, 121–133.

Fu, Q., Liou, K.N., 1992. On the correlated k-distribution method for radiative transfer in nonhomogeneous atmospheres. J. Atmos. Sci. 49, 2139–2156.

Gates, W.L., 1992. The atmospheric model intercomparison project. Bull. Am. Meteorol. Soc. 73 (12), 1962–1970.

Gordon, N.D., Klein, S.A., 2014. Low-cloud optical depth feedback in climate models. J. Geophys. Res. Atmos. 119 (10), 6052–6065.

Graversen, R.G., Langen, P.L., Mauritsen, T., 2014. Polar amplification in ccsm4: contributions from the lapse rate and surface albedo feedbacks. J. Clim. 27 (12), 4433–4450.

Hallett, J., Mossop, S.C., 1974. Production of secondary ice particles during the riming process. Nature 249, 26–28.

Held, I.M., Soden, B.J., 2006. Robust responses of the hydrological cycle to global warming. J. Clim. 19 (21), 5686–5699.

Hlavka, D.L., Yorks, J.E., Young, S.A., Vaughan, M.A., Kuehn, R.E., McGill, M.J., Rodier, S.D., 2012. Airborne validation of cirrus cloud properties derived from calipso lidar measurements: optical properties. J. Geophys. Res. Atmos. 116 (D19207).

Hu, Y., et al., 2009. CALIPSO/CALIOP cloud phase discrimination algorithm. J. Atmos. Ocean. Technol. 26 (11), 2293–2309.

Hu, Y., Rodier, S., Xu, K., Sun, W., Huang, J., Lin, B., Zhai, P., Josset, D., 2010. Occurrence, liquid water content, and fraction of supercooled water clouds from combined caliop/iir/modis measurements. J. Geophys. Res. Atmos. 115 (D00H34), 2293–2309.

Hurrell, J.W., et al., 2013. The community earth system model: a framework for collaborative research. Bull. Am. Meteorol. Soc. 94 (9), 1339–1360.

Kanamitsu, M., Ebisuzaki, W., Woollen, J., Yang, S.-K., Hnilo, J., Florino, M., Potter, G.L., 2002. NCEP-DOE AMIP-II reanalysis (R-2). Bull. Am. Meteorol. Soc. 83 (11), 1631–1643.

Kang, S.M., Held, I.M., Frierson, D.M., Zhao, M., 2008. The response of the ITCZ to extratropical thermal forcing: idealized slab-ocean experiments with a gcm. J. Clim. 21 (14), 3521–3532.

Kim, S.-W., Berthier, S., Raut, J.-C., Chazette, P., Dulac, F., Yoon, S.-C., 2008. Validation of aerosol and cloud layer structures from the space-borne lidar caliop using a ground-based lidar in Seoul, Korea. Atmos. Chem. Phys. 8 (13), 3705–3720.

Klein, S.A., Hall, A., 2015. Emergent constraints for cloud feedbacks. Curr. Clim. Change Rep. 1 (4), 276–287.

Komurcu, M., et al., 2014. Intercomparison of the cloud water phase among global climate models. J. Geophys. Res. Atmos. 119 (6), 3372–3400.

Li, Z.-X., LeTreut, H., 1992. Cloud-radiation feedbacks in a general circulation model and their dependence on cloud modelling assumptions. Clim. Dyn. 7 (3), 133–139.

Liu, Z., et al., 2009. The calipso lidar cloud and aerosol discrimination: Version 2 algorithm and initial assessment of performance. J. Atmos. Ocean. Technol. 26 (7), 1198–1213.

Liu, Z., et al., 2010. The CALIPSO cloud and aerosol discrimination: Version 3 algorithm and test results. In: 25th International Laser Radar Conference, St. Petersburg, Russia.

Liu, X., et al., 2011. Toward a minimal representation of aerosols in climate models: description and evaluation in the community atmosphere model cam5. Geosci. Model Dev. 5 (3), 709–739.

Marchand, R., Mace, G.G., Ackerman, T., Stephens, G., 2008. Hydrometeor detection using cloudsat—an earth-orbiting 94-ghz cloud radar. J. Atmos. Ocean. Technol. 25 (4), 519–533.

Mauritsen, T., Stevens, B., 2015. Missing iris effect as a possible cause of muted hydrological change and high climate sensitivity in models. Nat. Geosci. 8 (5), 346–351.

Mauritsen, T., et al., 2012. Tuning the climate of a global model tuning the climate of a global model. J. Adv. Model. Earth Sy. 4, M00A01.

McCoy, D.T., Hartmann, D.L., Grosvenor, D.P., 2014. Observed southern ocean cloud properties and shortwave reflection. Part II: phase changes and low cloud feedback. J. Clim. 27 (23), 8858–8868.

McCoy, D.T., Hartmann, D.L., Zelinka, M.D., Ceppi, P., Grosvenor, D.P., 2015. Mixed-phase cloud physics and southern ocean cloud feedback in climate models. J. Geophys. Res. Atmos. 120 (18), 9539–9554.

McCoy, D.T., Tan, I., Hartmann, D.L., Zelinka, M.D., Storelvmo, T., 2016. On the relationships among cloud cover, mixed-phase partitioning, and planetary albedo in GCMs. J. Adv. Model. Earth Sy. 8, 650–668.

McCoy, D.T., Hartmann, D.L., Zelinka, M.D., 2017. Mixed-Phase Cloud Feedbacks. Elsevier.

Meyers, M.P., DeMott, P.J., Cotton, W.R., 1992. New primary ice-nucleation parameterizations in an explicit cloud model. J. Appl. Meteorol. 31 (7), 708–721.

Mitchell, J.F.B., Senior, C.A., Ingram, W.J., 1989. CO_2 and climate: a missing feedback? Nature 341, 132–134.

Morrison, H., Gettelman, A., 2008. A new two-moment bulk stratiform cloud microphysics scheme in the community atmosphere model, version 3 (cam3). Part I: Description and numerical tests. J. Clim. 21 (15), 3642–3659.

Morrison, H., de Boer, G., Feingold, G., Harrington, J., Shupe, M.D., Sulia, K., 2012. Resilience of persistent arctic mixed-phase clouds. Nat. Geosci. 5 (1), 11–17.

Neale, R. B., et al. 2012: Description of the near community atmosphere model (cam5.0). NCAR Tech. Note NCAR/TN-486+STR.

Pappalardo, G., Wandinger, U., Mona, L., Hiebsch, A., Mattis, I., Amodeo, A., Ansmann, A., Seifert, Patric, et al., 2010. EARLINET correlative measurements for CALIPSO: first intercomparison results. J. Geophys. Res. Atmos. 115 (D00H19).

PIthan, F., Mauritsen, T., 2014. Arctic amplification dominated by temperature feedbacks in contemporary climate models. Nat. Geosci. 4 (3), 181–184.

Pruppacher, H.R., Klett, J.D., 1997. Microphysics of Clouds and Precipitation, 2nd ed. Kluwer Academic Publishers, Boston, MA. ISBN: 0-7923-4211-9.

Ramanathan, V.C., Crutzen, P.J., Kiehl, K.T., Rosenfeld, D., 2001. Aerosols, climate, and the hydrological cycle. Science. 294 (5549), 2119–2124.

Schiller, C., Kramer, M., Afchine, A., Spelten, N., Sitnikov, N., 2008. Ice water content of arctic, midlatitude, and tropical cirrus. J. Geophys. Res. Atmos. 113 (D24208).

Schneider, S.H., Dickinson, R.E., 1974. Climate modeling. Rev. Geophys. 12 (3), 447–493.

Shell, K.M., Kiehl, J.T., Shields, C.A., 2008. Using the radiative kernel technique to calculate climate feedbacks in ncar's community atmospheric model. J. Clim. 21 (10), 2269–2282.

Slingo, J.M., 1987. The development and verification of a cloud prediction scheme for the ecmwf model. Q. J. R. Meteorol. Soc. 113 (477), 899–927.

Soden, B.J., Held, I.M., Colman, R., Shell, K.M., Kiehl, J.T., Shields, C.A., 2008. Quantifying climate feedbacks using radiative kernels. J. Clim. 21, 3504–3520.

Somerville, R.C.J., Remer, L.A., 1984. Cloud optical thickness feedbacks in the CO_2 climate problem. J. Geophys. Res. Atmos. 89 (D6), 9668–9672.

Storelvmo, T., Tan, I., Korolev, A.V., 2015. Cloud phase changes induced by co2 warming—a powerful yet poorly constrained cloud-climate feedback. Curr. Clim. Change Rep. 1 (4), 288–296.

Sun, Z., Shine, K.P., 1994. Studies of the radiative properties of ice and mixed-phase clouds. Q. J. R. Meteorol. Soc. 120 (515), 111–137.

Tan, I., Storelvmo, T., 2016. Sensitivity study on the influence of cloud microphysical parameters on mixed-phase cloud thermodynamic phase partitioning in cam5. J. Atmos. Sci. 73 (2), 709–728.

Tan, I., Storelvmo, T., Choi, Y.-S., 2014. Spaceborne lidar observations of the ice-nucleating potential of dust, polluted dust, and smoke aerosols in mixed-phase clouds. J. Geophys. Res. Atmos. 119 (11), 6653–6665.

Tan, I., Storelvmo, T., Zelinka, M.D., 2016. Observational constraints on mixed-phase clouds imply higher climate sensitivity. Science 352 (6282), 224–227.

Taylor, K.E., 2001. Summarizing multiple aspects of model performance in a single diagram. J. Geophys. Res. 106 (D7), 7183–7192.

Terai, C.R., Klein, S.A., Zelinka, M.D., 2016. Constraining the low-cloud optical depth feedback at middle and high latitudes using satellite observations. J. Geophys. Res. Atmos. 121 (16), 9696–9716.

Tesche, M., Wandinger, U., Ansmann, A., Althausen, D., Muller, D., Omar, A.H., 2013. Ground-based validation of calipso observations of dust and smoke in the cape verde region. J. Geophys. Res. Atmos. 118 (7), 2889–2902.

Trenberth, K.E., Fasullo, J.T., Kiehl, J., 2009. Earth's global energy budget. Bull. Am. Meteorol. Soc. 90 (3), 311.

Tselioudis, G., Rossow, W.B., Rind, D., 1992. Global patterns of cloud optical thickness variation with temperature. J. Clim. 5 (12), 1484–1495.

Tsushima, Y., et al., 2006. Importance of the mixed-phase cloud distribution in the control climate for assessing the response of clouds to carbon dioxide increase: a multi-model study. Clim. Dyn. 27 (2–3), 113–126.

Vali, G., 1985. Nucleation terminology. Bull. Am. Meteorol. Soc. 66 (11), 1426–1427.

Wegener, A., 1911. Thermodynamik der Atmosphare. Verlag Von Johann Ambrosius Barth, Leipzig.

Winker, D.M., Vaughan, M.A., Omar, A., Hu, Y., Powell, K.A., Liu, Z., Hung, W.H., Young, S.A., 2009. Overview of the calipso mission and caliop data processing algorithms. J. Atmos. Ocean. Technol. 26 (11), 2310–2323.

Young, K.C., 1974. The role of contact nucleation in ice phase initiation of clouds. J. Atmos. Sci. 31 (3), 768–776.

Zelinka, M.D., Klein, S.A., Hartmann, D.L., 2012a. Computing and partitioning cloud feedbacks using cloud property histograms. Part I: Cloud radiative kernels. J. Clim. 25 (11), 3715–3735.

Zelinka, M.D., Klein, S.A., Hartmann, D.L., 2012b. Computing and partitioning cloud feedbacks using cloud property histograms. Part II: attribution to changes in cloud amount, altitude, and optical depth. J. Clim. 25 (11), 3736–3754.

AUTHOR INDEX

Note: Page numbers followed by *f* indicate figures, *t* indicate tables, and *np* indicate footnotes.

SUBJECT INDEX

Note: Page numbers followed by *f* indicate figures, *t* indicate tables, and *np* indicate footnotes.

during field campaigns, 102–110
properties
  from previous experiments, 131–136, 132*f*
  in situ characterization, 131
theoretical models of, 194–201
types, 99–100
Mixed-phase clouds feedbacks
  in GCMs, 217–225, 219*f*
  observations, 225–227, 228–229*f*
Mixed-phase microphysical process, 153–154
MMCR. *See* Millimeter-wavelength Cloud Radar
  (MMCR)
Moderate Resolution Imaging Spectroradiometer
  (MODIS), 3–4, 46–47, 47*f*, 220–221
Molecular dynamics (MD), 29
MOUDI. *See* Micro-orifice uniform-deposit
  impactor (MOUDI)
M-PACE. *See* Mixed-Phase Arctic Cloud
  Experiment (M-PACE)
MPCs. *See* Mixed-phase clouds (MPCs)
MPL. *See* Micropulse Lidar (MPL)
Multiple stochastic models (MCSMs), ice
  nucleation, 22–23
MWR. *See* Microwave Radiometer (MWR)

N

Nucleates ice, in mixed-phase clouds, 30–34
Nucleation theory
  heterogeneous ice nucleation, 21
    application of CNT to, 22
    comparison and summary of models, 25–26
    FROST, 24–25
    multiple component stochastic models, 22–23
    single component stochastic models, 22
    singular models, 23–24
  homogeneous ice nucleation, 18–21, 19–20*f*
Nucleators, biological ice, 31–33
Numerical weather prediction (NWP), 2, 108, 116

O

OAPs. *See* Optical array probes (OAPs)
Oceanic planetary boundary layer (PBL), 215
Operational weather forecast models, 101
Optical array probes (OAPs), 75, 79–80
Ornstein-Uhlenbeck process, 195, 198–199

P

Particle size distribution (PSD), 3–4, 157
  in MPCs, 134–135, 134*f*
Passive radiometry, 2–3

POLARCAT. *See* Polar Study Using Aircraft,
  Remote Sensing Surface Measurements and
  Models of Climate, Chemistry, Aerosols and
  Transport (POLARCAT)
Polar nephelometer (PN) scattering, 137
Polar Study Using Aircraft, Remote Sensing Surface
  Measurements and Models of Climate,
  Chemistry, Aerosols and Transport
  (POLARCAT), 107–108
Precipitation forecast, 101
Principal component analysis (PCA), 137
Probability density function (PDF), 186
Prognostic cloud scheme (PC2), 206–207
PSD. *See* Particle size distribution (PSD)
Pseudomonas syringae, 31–32

Q

Quasi Monte Carlo (QMC) sampling method,
  228–230, 241–242
Quasisteady supersaturation, 190*np*

R

RACEPAC campaign, 137
Radar, 49, 107–108
RALI, airborne radar-lidar instrument, 107
Relative humidity (RH), 224–225
Remote sensing, 45–46, 98, 105
  evaluation, 137–142
  instrument, 113*t*
  techniques, 2–3
    ground-based, capabilities, 4
RID. *See* Rosemount icing detector (RID)
Riming, 155–156, 167, 172, 177–178
Rosemount icing detector (RID), 74

S

Santa Barbara DISORT Atmospheric Radiative
  Transfer (SBDART) model, 52–53
Satellite radiometer, cloud phase determination
  from passive, 46–48
Satellite remote sensing. *See* Remote sensing
SCMs. *See* Single-column models (SCMs)
SDE. *See* Stochastic differential equation (SDE)
SHEBA Program. *See* Surface Heat Budget of the
  Arctic (SHEBA) Program
Shortwave (SW) cloud feedback, 215–216, 216*f*,
  226–227, 228*f*, 255, 257
Shortwave cloud forcing (SWCF), 258
Shortwave infrared (SIR) band, 48, 51–52
Shortwave (SW) radiation, 6, 215